STEPHEN I. WARSHAW
435 BEACH 137th STREET
ROCKAWAY BEACH 94, N. Y.

*Principles
of
Modern Physics*

NEW YORK · JOHN WILEY & SONS, INC.
London · Chapman & Hall, Limited

Principles
of
Modern Physics

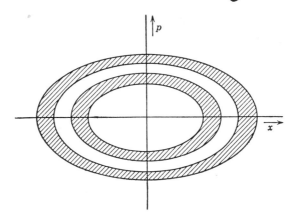

A. P. FRENCH

Professor of Physics, University of South Carolina

Formerly Fellow of Pembroke College, Cambridge
and Lecturer at the Cavendish Laboratory

SECOND PRINTING, NOVEMBER, 1959

Copyright © 1958 by John Wiley & Sons, Inc.

All Rights Reserved. This book or any part thereof must not be reproduced in any form without the written permission of the publisher.

Library of Congress Catalog Card Number: 58-7898

Printed in the United States of America

There are more things in man's philosophy

Than are dreamt of in heaven and earth!

Shakespeare

somewhat modified by H. F. Ellis

It is the customary fate of new truths

to begin as heresies and end as superstitions.

T. H. Huxley

Preface

It has been my aim in this book to give some feeling for the way in which our picture of the physical world has been constructed. The astonishing development of physics during the last fifty years has tended to obscure the magnitude of our debt to the more distant past. It has also tended to distort the perspective of the subject as a whole, and my choice of topics has been a deliberate attempt to counteract this. Thus, rather than present a wealth of facts at the expense of discussion, I have preferred in the space available to trace the progress of some of our more important physical concepts with the help of a minimum of detailed information. I have also tried, wherever possible, to indicate the transfer of concepts and methods from one field into another. In the chapter on the nucleus, for example, I have endeavored to suggest, perhaps more strongly than is customary, an isomorphism between the study of nuclear physics and the study of atomic physics that (for the most part) preceded it. I hope that in this way the reader will obtain a better feeling for the logical coherence—"unity" would be too optimistic a word at our present stage of understanding—that makes the study of physics so fascinating.

The seasoned reader will recognize my debt to a large number of different sources and writers in this field. It would be impossible to do justice to all of them, but I hope that the bibliography is fairly representative and will supplement the text itself in a significant way. I have appended a few problems to each chapter; there is no attempt to secure uniformity of standards, and I hope that there will be something for all tastes. Some of the problems are fairly lengthy and difficult.

This is essentially a book about the individual atom and its structure. It does not attempt to discuss the properties of matter in bulk, except for a brief mention of the diffraction processes that played an important role in the study of X rays and particle waves, and for some discussion of the quantum statistics of solids, with its place in the development of quantum theory. Needless to say, in a restricted choice of topics I have indulged my own whim to a certain extent, but I hope that the over-all result is to give a reasonably balanced account of the more important ideas that form the basis of what is called (very ineptly) "modern physics."

I might add that the standard of presentation is intended to be suitable for advanced undergraduate students or beginning graduate students. The book had its origin in a one-semester course given at the University of South Carolina, but in its final form, and with the collateral reading that should accompany it, there is material enough for two semesters of work.

It is a pleasure to acknowledge the excellent work of Mrs. Kay Rast and Mrs. R. L. Turbeville in typing the manuscript.

A. P. FRENCH

March 1958

Contents

Chapter	1	The Atomic Theory of Matter	1
	2	Light and the Electromagnetic Field	28
	3	The Atomicity of Electric Charge	44
	4	Thermal Radiation and the Quantum Theory	69
	5	Quanta and Atoms	99
	6	Relativity	137
	7	Wave Mechanics	174
	8	Some Applications of Quantum Mechanics	206
	9	The Nucleus	254

Appendices   318

Bibliography 343

1 | The Atomic Theory of Matter

1.1 THE EARLY HISTORY OF ATOMIC THEORY

Atomic theory began with the purely hypothetical approach of the ancient Greeks, and the famous names in this connection are those of Leucippus (440 B.C.) and Democritus (420 B.C.). These were the first atomists; they believed everything to be composed of indivisible entities, called atoms, and the void, or empty space. The atoms were imagined as being constantly in motion. According to the Roman poet Lucretius (100 B.C.), who wrote a lengthy poetical treatise on the nature of the universe (*De Rerum Natura*), the philosopher Epicurus (300 B.C.) revived the atomistic idea, but of course it was a purely conceptual matter, beyond the reach of any experimental test at that time.

Atomic theory with a scientific basis did not emerge until chemistry had established itself as a science about the end of the eighteenth century, and then some important general laws appeared. The first of these was the law of conservation of mass in chemical reactions (Lavoisier), and at nearly the same time Gay-Lussac put forward the law of combining volumes, stating that gases combined in simple ratios by volume. These were simply the statements of experience, but over the period 1803–1808 the chemist Dalton sought an explanation of what was observed. He proposed the two essentials of a sound atomic theory, namely, that (*a*) the basic units of a substance are identical with each other, and (*b*) the basic units of a substance are indivisible.

The explanation of much that was known concerning macroscopic chemical reactions would then follow, but some difficulties still remained. These were resolved when, in 1811, Avogadro drew a distinction between *atoms* and *molecules*. The atom was thought of as the smallest unit of a substance that could take part in a chemical reaction, whereas the molecule was defined as the smallest unit of an element or compound that could exist stably by itself. Avogadro then put forward the hypothesis that equal volumes of gases under the same conditions of temperature and pressure contain equal numbers of molecules.

Nothing was known at this time of the size or mass of the atoms, although in 1805 Thomas Young gave a theory of cohesion in fluids, made more precise by Laplace (1806), which made possible some kind of estimate of molecular size.

1.2 SIMPLE KINETIC THEORY OF GASES

The kinetic theory of gases seems to have become a live scientific theory in 1739, when Daniel Bernoulli explained Boyle's law by a simple picture. He considered a gas to be composed of atoms (molecules) all moving at the same speed. Let us suppose that there are n molecules per cubic centimeter; then of these, on the average, $n/3$ are moving parallel to each of three mutually perpendicular axes (x, y, z). Thus $n/6$ molecules per cm^3 are moving parallel to the positive x direction. If each molecule has speed c, all such molecules in a cylinder of length c and cross section 1 cm^2 (Fig. 1.1) will strike 1 cm^2 of any containing wall in 1 sec. Thus the number of molecules making an impact on the containing wall is equal to $nc/6$ per cm^2 per sec.* Each molecule has an initial momentum $+mc$, which becomes $-mc$ as the result of an elastic rebound. Hence we have

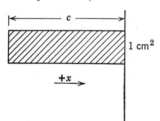

Figure 1.1. The simplest picture of the impact of molecules on a containing wall.

$$\text{Momentum change per sec per } cm^2 \text{ of wall} = \frac{nc}{6} \cdot 2mc$$

But this represents the force per square centimeter due to the molecular

---

* But see Problem 6 at the end of this chapter.

bombardment, and so is to be identified with the pressure p.

$$\therefore \quad p = \tfrac{1}{3}nmc^2 = \tfrac{1}{3}\rho c^2$$

where we introduce the density ρ of the gas. The molecular velocity is thus given by

$$c = \left(\frac{3p}{\rho}\right)^{1/2}$$

Now $\rho = M/V$ for a mass M of gas occupying a volume V. We can therefore rewrite our result in the form

$$pV = \tfrac{1}{3}Mc^2$$

which is Boyle's law for the expansion of a gas at a fixed temperature.

Bernoulli did not have access to reliable values of gas densities; so he did not estimate the magnitude of the velocity c, but we can do this. Taking air as an example, we have

$$\rho \approx 1.3 \text{ mg/cm}^3 \text{ at } p = 1 \text{ atm} \approx 10^6 \text{ dynes/cm}^2$$

$$\therefore \quad c \approx 5 \cdot 10^4 \text{ cm/sec} \approx 2000 \text{ ft/sec}$$

i.e., about the speed of a rifle bullet.

This was a simple and attractive description of a gas, but it remained a conjectural matter until the botanist Robert Brown (1827) discovered the constant agitation of small inanimate particles when suspended in a liquid, and so for the first time put in evidence the reality of a kinetic picture of matter (although Brown himself does not appear to have recognized the true implications of his discovery).

1.3 MOLECULAR SIZES

We have referred to the work of Young and Laplace in estimating the sizes of molecules, and we shall now consider this in some detail. It is well known that liquids exhibit characteristic features of (a) latent heat of vaporization, (b) surface tension. Latent heat can be described in terms of the amount of thermal energy L required to vaporize unit mass of the liquid. Surface tension takes the form of a force exerted at the boundary of a liquid surface, and work must be done against this force to increase the surface area; thus we can define the surface tension S as the mechanical work (i.e., energy) needed to create 1 cm^2 of new surface area. We shall consider both these effects as arising from the attractions of individual molecules. In order to do this we shall postulate (i) a range of interaction d between any two molecules, (ii) a force of attraction between them, which assumes a constant value f when the distance between the centers of the molecules is less than or equal to d, and which is zero for all separations greater than d. These are gross oversimplifications, of course, but they are nevertheless of value in obtaining an insight into the problem.

(1) Latent Heat

We consider the transfer of a molecule from within the body of the liquid to a point a distance d above the free surface (Fig. 1.2). The

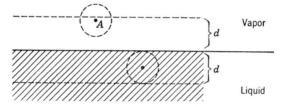

Figure 1.2. To illustrate the latent heat of vaporization of a liquid in terms of short-range molecular forces. Molecule A is about to escape from all attractions due to molecules within the liquid.

molecule, in being evaporated, has to move a distance $2d$ against the mean molecular attraction f. Thus the work per molecule = $2fd$.

$$\therefore \quad \text{Work per gram} = \frac{2fd}{m}$$

where m is the mass of one molecule.

Hence
$$L = \frac{2fd}{m}$$

(2) Surface Tension

In this case we shall find the work done in bringing up molecules from within the body of a liquid to create 1 cm^2 of new free surface. So long as a molecule is more than a distance d from the surface, it is not, on the average, subject to a resultant force in any particular direction. But, as it enters the surface layer, it begins to experience a net force pulling it back into the main body of liquid (Fig. 1.3). In creating 1 cm^2 of boundary layer we have to bring up $\rho d/m$ molecules, and the average distance moved by each molecule is $\tfrac{1}{2}d$ against a force f. Thus the work per molecule to bring it into the surface layer is $\tfrac{1}{2}fd$, and so we have

$$\text{Work to create 1 cm}^2 \text{ of surface} = S = \frac{\rho f d^2}{2m}$$

From these two results we can find d. For we have

$$d = \frac{4S}{\rho L}$$

For liquid water at room temperature,

$$S = 75 \text{ dynes/cm} = 75 \text{ ergs/cm}^2$$
$$L \approx 500 \text{ cal/g} \approx 2 \cdot 10^{10} \text{ ergs/g}$$
$$\rho = 1 \text{ g/cm}^3$$
$$\therefore \ d \approx 1.5 \cdot 10^{-8} \text{ cm}$$

If we assume that the molecules in water are closely packed (and the highly incompressible nature of water would suggest this), then we

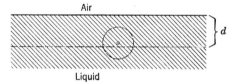

Figure 1.3. A molecule entering the surface layer of thickness d, and so beginning to acquire surface energy.

may take the figure $1.5 \cdot 10^{-8}$ cm as a rough value of the molecular diameter.

1.4 PRESSURE OF A PERFECT GAS

During the years 1857–1859 the physicists Clausius and Clerk Maxwell put the kinetic theory of gases on an exact footing. In particular they dropped Bernoulli's artificial assumption that the molecules of a gas at a given temperature all have the same speed. To illustrate this new approach to the problem we shall rederive the expression for the pressure of a gas, and this time we shall suppose that the molecules have a certain velocity spectrum described by $f(v)$, such that, if there is a total of n molecules per cm^3, the number of molecules with speeds between v and $v + dv$ is given by

$$dn(v) = n\,f(v)\,dv$$

with
$$\int_{v=0}^{\infty} f(v)\,dv = 1$$

The problem can be discussed with the help of Fig. 1.4. We consider the impacts on an area dA of the wall of the containing vessel during time dt. Let us first restrict our attention to molecules with speeds in the range v to $v + dv$. Then all such molecules within a hemisphere of radius $v\,dt$ have a chance to reach dA during dt. Within this hemisphere, consider the molecules that are initially in a small volume element in the form of a ring contained within a range of distance dr at r from dA, and between the directions θ and $\theta + d\theta$. The volume of this ring is thus $2\pi r \sin\theta \cdot r\,d\theta \cdot dr$.

If the directions of the molecules are entirely random, the chance that any one molecule will reach dA from the volume element is given by the ratio $dA\cos\theta/4\pi r^2$. Thus the number of molecules reaching dA from this ring is given by

$$n\,f(v)\,dv \cdot 2\pi r^2 \sin\theta\,d\theta\,dr\,\frac{dA\cos\theta}{4\pi r^2}$$

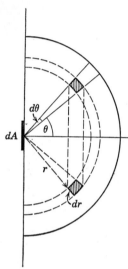

Figure 1.4. The impact of molecules from random directions on an element dA of a containing wall. The radius of the outer hemisphere is $v\,dt$.

The momentum transfer normal to $dA = 2mv\cos\theta$ per molecule. The net momentum transfer to dA in time dt is given by

$$\int_{v=0}^{\infty}\int_{r=0}^{vdt}\int_{\theta=0}^{\pi/2} 2mv\cos\theta \cdot n\,f(v)\,dv \cdot 2\pi r^2 \sin\theta\,d\theta\,dr\,\frac{dA\cos\theta}{4\pi r^2}$$

Thus, if the pressure exerted on dA by the gas is p, we have

$$p\,dt = \int_{v=0}^{\infty}\int_{r=0}^{vdt}\int_{\theta=0}^{\pi/2} n\,f(v) \cdot mv\,dv \cdot dr \cdot \cos^2\theta \sin\theta\,d\theta$$

$$= \int_{v=0}^{\infty}\int_{r=0}^{vdt} n\,f(v)\,mv\,dv \cdot dr \left[\frac{\cos^3\theta}{3}\right]_{\pi/2}^{0}$$

whence
$$p = \tfrac{1}{3} nm \int_{v=0}^{\infty} v^2 f(v)\, dv$$

The integral in this equation is mathematically a definition of the mean squared velocity of the molecules, and we therefore put

$$p = \tfrac{1}{3} nm\overline{v^2} = \tfrac{1}{3} nmc^2$$

For the evaluation of $\overline{v^2}$ once $f(v)$ is known, see Appendix I.

1.5 VELOCITY DISTRIBUTION IN A GAS

Our calculation is able to say nothing about the function $f(v)$, but in 1859 Maxwell solved this problem by considering in detail the process of collision between molecules. He found

$$f(v) = \text{const} \times v^2 \exp\left(-\frac{mv^2}{2kT}\right)$$

where
$$k = \text{Boltzmann's constant}$$
$$= R/N = 1.38 \cdot 10^{-16} \text{ erg/}°\text{K}$$

(R = gas constant per mole; N = Avogadro's number). The shape of this velocity distribution is shown in Fig. 1.5.

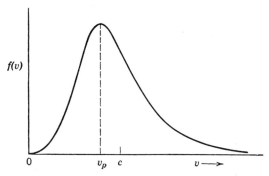

Figure 1.5. The form of the Maxwell velocity distribution. v_p is the most probable velocity, defined by the maximum of the curve. The relative value of the root-mean-square velocity c is shown for comparison.

Sixty years passed before an experimental test of this result became technically feasible, and measurements of real precision were not achieved until Zartman (1931) and Ko (1934) worked on the problem, using essentially similar methods (Fig. 1.6). A hollow cylinder C, with a slit S parallel to its axis, was rotated at high speed in vacuum. An "oven" O containing some suitable material (Bi was found convenient) was used to supply a stream of evaporated atoms. These were collimated by a slit system S', so that once in each revolution of the cylinder a pulse of atoms was admitted through S and traversed the

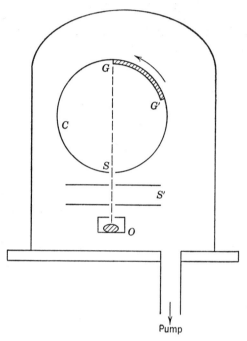

Figure 1.6. Determination of the molecular velocity distribution. The rotor C is shown admitting a pulse of atoms through the slit S.

cylinder. The atoms were received on a glass plate GG' and built up a deposit which corresponded to the velocity spectrum, since fast atoms would arrive near G and slow ones near G'. The density of the deposit was measured photometrically, and agreed well with Maxwell's formula. A very precise experiment of this type was done much later by Miller and Kusch (1955).

But perhaps the most beautiful demonstration of Maxwell's distribution was obtained by Rainwater and Havens (1946), who measured the velocity distribution of neutrons that had been brought into approximate thermal equilibrium with matter. The essentials of their arrangement are shown in Fig. 1.7. Fast neutrons were produced by bombarding a beryllium target with deuterons (nuclei of heavy hydrogen) in a cyclotron:

$$Be^9 + D \rightarrow B^{10} + n$$

With the help of electronic controls, the neutrons were produced in

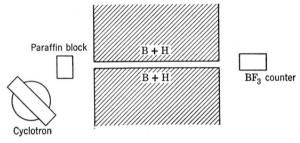

Figure 1.7. Schematic of arrangement for neutron time of flight experiment.

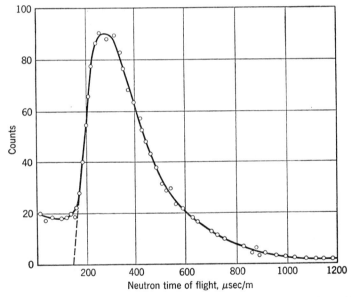

Figure 1.8. Number of neutrons versus time of flight ($\propto 1/v$) from data of Rainwater and Havens [*Phys. Rev.* **70**, 136 (1946)].

short bursts of only a few microseconds' duration (1 μsec = 10^{-6} sec). The neutrons were slowed down in a block of paraffin wax, behaving in it like a gas whose atoms were able to wander through this solid material and, as a result of numerous collisions, to assume its temperature. These "thermal neutrons" then diffused out of the block of wax. With the help of a collimator, composed effectively of a mixture of hydrogen (which slows neutrons down) and boron (which captures them), a beam of thermal neutrons was selected and fell upon a BF_3 counter. This counter, like the cyclotron, was made active for only a few microseconds at a time, but at some variable interval t after the burst of neutrons had been initially produced. If the distance from paraffin block to BF_3 counter is l, this system therefore responds only to neutrons of a speed equal to l/t. By plotting the number of pulses recorded in the counter as a function of the time of flight t (Fig. 1.8), the velocity spectrum of the neutrons was in effect traced out. This has been one of the most accurate tests yet made of the correctness of Maxwell's formula.

1.6   THE ISOTHERMAL ATMOSPHERE

We shall not follow through the details of Maxwell's derivation of the velocity distribution law, but will present here a treatment, due to

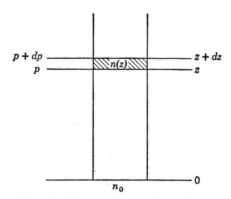

Figure 1.9.   Equilibrium in an isothermal atmosphere.

Boltzmann (1876) based on a consideration of the dynamic equilibrium in a gas, at the same temperature throughout, exposed to gravitational forces.

First, we find the variation of pressure with height, and we shall assume the problem to be one-dimensional (Fig. 1.9). If the density of the gas at height z is ρ, we have, for equilibrium of a slice of thickness dz,

$$dp + \rho g \, dz = 0$$

[dp is of course negative, but it is mathematically desirable to put

The Atomic Theory of Matter

$p(z + dz) = p(z) + dp$.] But by the gas laws

$$p = \frac{\rho RT}{M}$$

$$\therefore \frac{dp}{p} = -\frac{gM}{RT} dz$$

and so
$$p(z) = p_0 \exp\left(-\frac{gM}{RT} z\right)$$

Now the density ρ is strictly proportional to p if T is constant (Boyle's law), and $\rho = nm$, where n is the number of molecules per cubic centimeter of mass m. Thus we can just as well put

$$n(z) = n_0 \exp\left(-\frac{gM}{RT} z\right)$$

$n(z)$ is thus a smooth function of z. Now this would not happen if all the molecules had the same speed, for then we should have (by conservation of energy) $\frac{1}{2}mv^2 + mgz = \frac{1}{2}mv_0^2$, and no molecule could rise higher than $z_{\max} = v_0^2/2g$. Furthermore, the drop of speed with increase of z would imply a fall of temperature with height, which is contrary to our assumptions. We therefore introduce a velocity distribution function $f(v)$, such that $n f(v) dv$ is the number of molecules per cubic centimeter at height z whose velocity, assumed to be purely vertical, is between v and $v + dv$.

Now the density of molecules changes by an amount dn between z and $z + dz$, where

$$dn(z) = -\frac{gM}{RT} n_0 \exp\left(-\frac{gM}{RT} z\right) dz$$

This deficit at $z + dz$ compared with z must arise from molecules stopping between these levels, and the number so stopping per square centimeter per second is given by

$$\bar{v}\, dn(z) = \frac{gM}{RT} n_0 \bar{v} \exp\left(-\frac{gM}{RT} z\right) dz$$

where $\bar{v}$ is the mean molecular velocity and (by the isothermal condition) is independent of z.† Corresponding to this rate of arrival

---

† It may appear paradoxical that $\bar{v}$ can remain constant despite the fact that the individual velocities become smaller with increasing z. The explanation is, of course, that the slowest molecules are lost on the way up, and $\bar{v}$ is the average taken over those that survive.

of molecules in the layer dz at z there must be a certain number of molecules leaving the level $z = 0$ with vertical velocities just sufficient to bring them to rest between z and $z + dz$. This condition implies

$$v_0{}^2 = 2gz$$

$$v_0 \, dv_0 = g \, dz$$

The number per square centimeter per second of such molecules starting from $z = 0$ is given by $v_0 \times n_0 f(v_0) \, dv_0 = n_0 f(v_0) g \, dz$. Hence in a

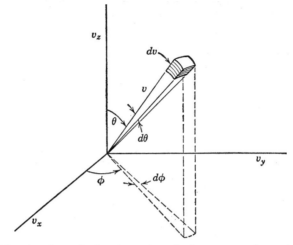

Figure 1.10. An element of volume in velocity space, using spherical polar co-ordinates.

state of dynamic equilibrium,

$$n_0 f(v_0) g \, dz = \bar{v} \, dn(z) = \frac{gM}{RT} n_0 \bar{v} \exp\left(-\frac{gM}{RT} z\right) dz$$

$$\therefore \quad f(v_0) = \frac{M\bar{v}}{RT} \exp\left(-\frac{Mgz}{RT}\right) = \text{const} \times \exp\left(-\frac{mv_0{}^2}{2kT}\right)$$

We see that g disappears from the result, and so plays no *essential* part in the problem.

So far our calculation deals with vertical velocities only; but, if now we consider the three-dimensional case, in which the molecules have *total* velocities v with components v_x, v_y, v_z, then‡

‡ The immediate extension from one dimension to three dimensions assumes that the velocity components are truly independent. This was accepted by Maxwell but is open to criticism. (See, for example, Jeans, *The Dynamical Theory of Gases*, New York: Dover Publications, 1954.)

The Atomic Theory of Matter

$$dn(v) \sim \exp\left(-\frac{mv_x^2}{2kT}\right) dv_x \exp\left(-\frac{mv_y^2}{2kT}\right) dv_y \exp\left(-\frac{mv_z^2}{2kT}\right) dv_z$$

$$= \exp\left[-\frac{m}{2kT}(v_x^2 + v_y^2 + v_z^2)\right] dv_x\, dv_y\, dv_z$$

$$= \exp\left(-\frac{mv^2}{2kT}\right) dv_x\, dv_y\, dv_z$$

If we are interested only in the magnitude of v and not in its separate components, we replace $dv_x\, dv_y\, dv_z$ by the corresponding volume element (in velocity space) that is obtained (Fig. 1.10) by taking as co-ordinates the magnitude v and its direction defined by two angles (θ, ϕ):

$$dv_x\, dv_y\, dv_z \equiv v\, d\theta \cdot v \sin\theta\, d\phi\, dv$$
$$= v^2\, dv \cdot \sin\theta\, d\theta\, d\phi$$

If, further, we are not interested in the direction of v, we integrate over all θ and ϕ, thus getting

$$f(v)\, dv \sim \exp\left(-\frac{mv^2}{2kT}\right) v^2\, dv \int_{\theta=0}^{\pi} \sin\theta\, d\theta \int_{\phi=0}^{2\pi} d\phi$$

$$= 4\pi \exp\left(-\frac{mv^2}{2kT}\right) v^2\, dv$$

Thus $\quad f(v) \sim v^2 \exp\left(-\frac{mv^2}{2kT}\right)$

which is Maxwell's distribution in the form in which we have already quoted it.

1.7 MEAN FREE PATHS

Air molecules have speeds of the order of 400 m/sec; so one might expect them to disappear rather quickly from any open-ended vessel in which they are contained. Their failure to do so (as demonstrated, for example, by the extreme sluggishness of diffusive processes) was offered as an objection to the acceptance of a kinetic theory of gases. It was soon realized, however, that the objection loses its force if molecules are endowed with a finite size instead of being regarded as geometrical points.

Let us suppose that two molecules can be said to collide if their centers come within a distance d (cf. the discussion of vaporization and surface tension, Section 3 of this chapter). Then, if we fix attention on one particular molecule, we can think of it as sweeping out a cylinder of *radius d* as it goes along (Fig. 1.11), and any other molecule whose center lies within the cylinder will be struck. A moving molecule of speed v thus sweeps out a volume $\pi d^2 \, v \, dt$ in time dt, and, if there are n molecules per cubic centimeter, the number of collisions is

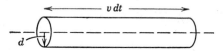

Figure 1.11. To illustrate the mean free path concept.

given by $n\pi d^2 \, v \, dt$. To describe this process we can introduce a *mean free path* λ, which is the mean distance between collisions under these conditions. Then another expression for the number of collisions during dt is simply $v \, dt/\lambda$. Equating these two expressions, we have

$$\lambda = \frac{1}{n\pi d^2}$$

In this treatment we have ignored the motions of all molecules except one; a more correct treatment, taking account of relative motions, introduces a further factor $\sqrt{2}$ in the denominator of the formula for λ if the molecules have a Maxwellian distribution of velocities.

Clearly for any given molecule there is a statistical distribution, given by the laws of chance, of the distances traveled between successive collisions. Let us suppose that the probability of a molecule going a distance x *without* collision is $p(x)$. In a further short distance dx, the probability of a collision is proportional simply to dx, and equal, say, to $\alpha \, dx$, where α is some constant. Then the probability of *no* collision in $dx = (1 - \alpha \, dx)$.

Now the probability of (*a*) going a distance x without collision, and (*b*) going a further distance dx without collision, is equal to the probability of going a total distance $(x + dx)$ without collision.

$$\therefore \quad p(x + dx) = p(x) \times (1 - \alpha \, dx)$$

Using Taylor's expansion to two terms only (dx being small), we have

$$p(x) + \frac{dp}{dx} \, dx = p(x) - \alpha \, p(x) \, dx$$

whence
$$p(x) = \text{const} \times e^{-\alpha x}$$

The Atomic Theory of Matter

Since $p(0) = 1$, by the definition of absolute certainty in probability theory, we have

$$p(x) = e^{-\alpha x}$$

We must now determine α and, in particular, relate it to the mean free path λ as already defined. To do this we consider the following argument:

The probability for a molecule to go through x without collision $= p(x)$

The probability of a collision in $dx = \alpha\, dx$

$\therefore$ Probability of collision in dx at $x = p(x)\alpha\, dx$

$$\therefore \quad \text{Mean free path } \lambda = \frac{\int_{x=0}^{\infty} x\, p(x)\alpha\, dx}{\int_{x=0}^{\infty} p(x)\alpha\, dx}$$

$$= \frac{\int_{x=0}^{\infty} x e^{-\alpha x}\, dx}{\int_{x=0}^{\infty} e^{-\alpha x}\, dx}$$

$$\therefore \quad \lambda = \frac{-\frac{1}{\alpha}[xe^{-\alpha x}]_0^{\infty} + \frac{1}{\alpha}\int_0^{\infty} e^{-\alpha x}\, dx}{\int_0^{\infty} e^{-\alpha x}\, dx}$$

The definite integral $[xe^{-\alpha x}]_0^{\infty}$ vanishes at both limits; so we see at once that
$$\lambda = 1/\alpha$$

Hence finally
$$p(x) = \exp(-x/\lambda)$$

The chance of a really long path between successive collisions is small; for example, in only $\frac{2}{3}\%$ of all cases will a molecule travel a distance 5λ without collision.

1.8 MAGNITUDE OF MEAN FREE PATH

To estimate the value of λ, we need to know the values of n and d for the problem. To get an order of magnitude answer, let us use our knowledge that 1 g of H_2O ($= \frac{1}{18}$ g molecule) occupies 1 cm^3 in the condensed state. We have already used this to deduce that $d \approx 2 \cdot 10^{-8}$ cm.

∴ Volume occupied by 1 molecule of $H_2O \approx d^3 \approx 10^{-23}$ cm^3

∴ No. of molecules in 1 g ($\equiv$ 1 cm^3) of liquid $H_2O \approx 10^{23}$

∴ No. of molecules in 1 g molecule $\approx 2 \cdot 10^{24} = N$

Now 1 mole (i.e. 1 g molecule) of any gas or vapor occupies 22.4 l at NTP, and by Avogadro's hypothesis contains this same number N of molecules. Hence for 1 cm^3 of gas at NTP we find

$$n \approx \frac{2 \cdot 10^{24}}{2 \cdot 10^4} = 10^{20} \text{ per cm}^3$$

(This is too high by a factor of about 3, because our crudely deduced value of N is too large by this amount.) Thus very roughly for $p = 1$ atm $= 760$ mm Hg,

$$\lambda_{NTP} \approx \frac{1}{10^{20} \times \pi \times 4 \cdot 10^{-16}} \approx 10^{-5} \text{ cm}$$

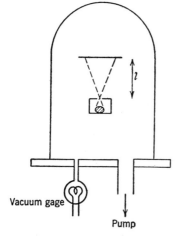

Figure 1.12. Direct determination of mean free path of evaporated atoms through a gas.

We see, then, that the mean free path in gases at atmospheric pressure is comparable with the wavelength of visible light, and to obtain a directly measurable path we must reduce the pressure ($\lambda \propto 1/n \propto 1/p$). For example, with $p = 10^{-3}$ mm Hg, a pressure that is readily attained by a mechanical vacuum pump, we have $\lambda \approx 10^6 \times \lambda_{NTP}$; i.e. about 10 cm.

Direct measurements of mean free path are rare, but Bielz (1925) found λ for silver atoms in air at low pressure, by measuring the mass of silver that was deposited on a plate exposed to silver atoms evaporated from an oven (Fig. 1.12.) He varied the pressure and compared the masses deposited in equal times; the experiment was a direct application of the formula

$$p(l) = \exp(-l/\lambda)$$

The results are summarized in Table 1.1, and the approximate constancy of the product $P\lambda$ may be seen.

The Atomic Theory of Matter

TABLE 1.1

Pressure, P, mm Hg	λ, cm	$P\lambda \times 10^3$
$1.4 \cdot 10^{-3}$	8.4	11.8
$3.2 \cdot 10^{-3}$	3.22	10.3
$4.0 \cdot 10^{-3}$	2.34	9.4
$7.0 \cdot 10^{-3}$	1.35	9.45

1.9 TRANSPORT PHENOMENA

Under this heading we have the behavior of a gas subjected to various nonequilibrium conditions. We shall consider two of these:

(a) The existence of a velocity gradient; i.e. a streaming motion of parallel planes of molecules relative to each other, which leads to viscous effects.

(b) The existence of a density gradient, which will tend to destroy itself by the process of diffusion.

(1) Viscosity of a Gas

Consider molecules crossing 1 cm² of a plane A (Fig. 1.13) that lies in the plane of the streaming motion. Let the streaming velocity at A be u (supposed much less than the random thermal velocities).

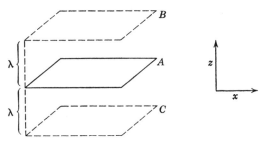

Figure 1.13. Diagram for the consideration of transport phenomena.

Now molecules arriving at A and making a collision there (thus becoming assimilated into the motion at this level) will on the average have had their last previous collision a distance λ away, and will bring with them the streaming velocity characteristic of planes B or C, distant λ above or below A. Suppose the mean molecular velocity is $\bar{v}$ at all levels ($\bar{v} \gg u$); its value for a Maxwell velocity distribution is derived in Appendix I. Then $\tfrac{1}{6}n\bar{v}$ molecules per cm² per sec come down from B, or up from C.§ The stream velocity at $B = u + \lambda(du/dz)$, and, at C, $= u - \lambda(du/dz)$.

∴ Momentum transferred *downward* through A per cm² per sec

$$= \tfrac{1}{6}n\bar{v} \cdot m \left(u + \lambda \frac{du}{dz}\right) - \tfrac{1}{6}n\bar{v} \cdot m \left(u - \lambda \frac{du}{dz}\right)$$

$$= \tfrac{1}{3}n\bar{v}m\lambda \frac{du}{dz}$$

But this represents the viscous shearing stress, which is also defined by $\eta \, (du/dz)$, where η is the coefficient of viscosity.

Hence

$$\eta = \tfrac{1}{3}nm\bar{v}\lambda = \tfrac{1}{3}\rho\bar{v}\lambda = \tfrac{1}{3}m\bar{v}(n\lambda)$$

This result led to one of the early triumphs of kinetic theory. We notice that the last form of the expression for η contains the product $n\lambda$, which as we have seen should be a constant. Thus we have the surprising result that the viscosity of a gas should be independent of its pressure or density. Maxwell verified this by studying the viscous damping of the oscillations of a set of plates, hung from a torsion fiber and interleaved with fixed plates as shown in Fig. 1.14. More extensive work has shown that, when the pressure becomes very small,

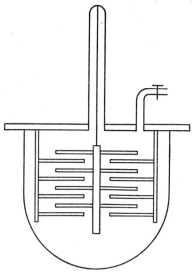

Figure 1.14. Maxwell's apparatus to demonstrate that viscosity is independent of pressure for a gas.

the value of η does begin to fall off (see Fig. 1.15). This departure from constancy happens when the mean free path becomes long enough to be comparable with spacings between plates, and represents a situation to which our theory of viscous effects no longer applies.

(2) **Diffusion**

This time we consider a *density* gradient parallel to the z axis, but the treatment of the problem is much the same as for viscosity, and we can refer to the same diagram (Fig. 1.13). Let the density of

§ This is a naive statement of the case. See Problem 7 at the end of this chapter.

molecules at A be n per cm^3. Then

$$\text{Density} = n + \lambda \frac{dn}{dz} \text{ at } B, \quad n - \lambda \frac{dn}{dz} \text{ at } C$$

Therefore the number of molecules transferred downward from B per square centimeter per second

$$= \tfrac{1}{6}\bar{v}\left(n + \lambda \frac{dn}{dz}\right)$$

and the number transferred upward from C per square centimeter per second

$$= \tfrac{1}{6}\bar{v}\left(n - \lambda \frac{dn}{dz}\right)$$

Therefore the number of molecules transferred per square centimeter per second in the direction of *increasing* z is given by $-\tfrac{1}{3}\bar{v}\lambda(dn/dz)$.

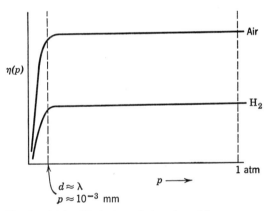

Figure 1.15. Variation of viscosity with pressure.

The negative sign indicates that the flow is away from the regions of higher concentration. But the rate of transfer is also described by $-D\,(dn/dz)$, where D is the coefficient of diffusion

$$\therefore \quad D = \tfrac{1}{3}\bar{v}\lambda$$

It may be noted that D is expressed in square centimeters per second, suggesting very directly the "spreading out" that is characteristic of diffusion processes. By comparing our formulas for η and D, we see that

$$D = \frac{\eta}{\rho}$$

Thus another test of the correctness of a kinetic picture of matter lies in this relation between quantities that are not very obviously connected otherwise. To take one example, for oxygen at 20° C, we have

$$\eta = 1.9 \cdot 10^{-4} \text{ cgs}, \qquad \rho = 1.4 \cdot 10^{-3} \text{ g/cm}^3$$

$$\therefore \frac{\eta}{\rho} = 0.14 \text{ cm}^2/\text{sec}$$

The observed value of D is $0.18 \text{ cm}^2/\text{sec}$.

1.10 BROWNIAN MOTION. DETERMINATION OF N

The Brownian motion was first recognized as a symptom of molecular agitation by Delsaux (1877) and Gouy (1888). The effects of the Brownian motion are: (1) When large numbers of small particles are suspended in a fluid they do not all sink to the bottom, but distribute themselves in "sedimentation equilibrium." (2) Any individual particle undergoes random successive displacements ("the drunkard's walk"). Each of these can be used to determine Avogadro's number from direct observation.

(1) Sedimentation Equilibrium

The essential feature is that a collection of small identical particles suspended in a fluid behaves like a gas, and comes into thermal equilibrium with the fluid. Now Maxwell showed that a mixture of gases is in equilibrium when the mean translational energy of each type of molecule is the same:

$$\tfrac{1}{2}\mu_1 c_1^2 = \tfrac{1}{2}\mu_2 c_2^2 = \tfrac{3}{2}kT$$

We shall assume that this equation holds good for the molecules of a liquid or a gas (denoted by 1) in equilibrium with a suspension of particles that are equivalent to very large molecules (denoted by 2). Suppose that, at height z, measured from some arbitrary zero, the pressure is p, and the numbers of particles per cm^3 of the two types are $n_1(z)$ and $n_2(z)$. Then, for equilibrium of a small slice of thickness dz and cross section 1 cm^2 (Fig. 1.16), we have

$$dp + n_1 \mu_1 g \, dz + n_2 \mu_2 g \, dz = 0$$

Also, by the law of partial pressures,

$$p = \tfrac{1}{3} n_1 \mu_1 c_1^2 + \tfrac{1}{3} n_2 \mu_2 c_2^2 = (n_1 + n_2)kT$$

Eliminating p between these two equations, and rearranging, we have

$$\frac{dn_1}{dz} + \frac{n_1\mu_1 g}{kT} = -\left(\frac{dn_2}{dz} + \frac{n_2\mu_2 g}{kT}\right)$$

From the complete symmetry of the problem as between the two types of particle, we deduce that

$$\frac{dn}{dz} + \frac{n\mu g}{kT} = 0$$

(Otherwise we should arrive at different solutions for n_1 and n_2.)

Hence $\quad n(z) = n(0) \exp\left(-\frac{\mu g}{kT}z\right) = n(0) \exp\left(-\frac{\mu N g}{RT}z\right)$

We see then that each type of molecule behaves as though it alone were present (as in our atmosphere, composed of different types of

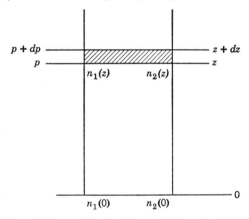

Figure 1.16. Sedimentation equilibrium.

gases). We must, however, qualify this statement slightly for the suspended particles. Being so large compared to the molecules of the medium in which they are immersed, they experience a normal buoyancy effect in addition to the random molecular bombardments. For these particles the quantity μ in the equation represents the effective mass, as given by the volume times the density difference between the particles and the suspending medium. Thus, if these densities are ρ, ρ', and if the particles have radius a, we can put $\mu_{\text{eff}} = \frac{4}{3}\pi a^3(\rho - \rho')$, and our equation can be written

$$\log_e n(z) = \log_e n(0) - N\frac{4}{3}\frac{\pi a^3(\rho - \rho')g}{RT}z$$

The experiment is conducted by focusing a microscope on various planes within a suspension and counting the numbers of particles in a certain fixed portion of the field of view. A plot of log $n(z)$ against z (Fig. 1.17) then gives a straight line from whose slope N can be found.

The most extensive measurements were made by Perrin (1912), who made a series of studies with particles about 10^{-4} to 10^{-5} cm in diameter. It is perhaps worth noting that in a typical experiment the

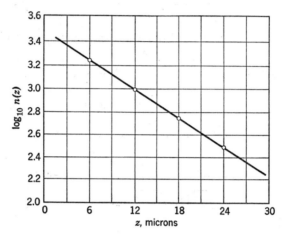

Figure 1.17. Determination of Avogadro's number from sedimentation equilibrium (from one of Perrin's experiments).

total height of the liquid column was only about 1 mm, and $n(z)$ was measured for values of z differing successively by about $3 \cdot 10^{-3}$ cm. Of the auxiliary quantities needed, g may be taken as known, and ρ' and T are easily found. The real difficulties lie in finding the density and the radius of the suspended particles, and several different methods were used for each of these quantities. ρ, for example, could be found by suspending the particles in a salt solution, whose concentration was adjusted until centrifuging failed to separate out the particles; and one method of finding a was to observe the rate of fall of the particles in a liquid (they descended at the rate of a few millimeters per day), and then to apply Stokes's law for motion of a sphere in a liquid:

$$F = 6\pi a \eta v = \tfrac{4}{3}\pi(\rho - \rho')a^3 g$$

The final answer obtained by Perrin was

$$N = 6.8 \cdot 10^{23}$$

We know today that this is too high by rather more than 10%, but

The Atomic Theory of Matter

Perrin's measurements have the special interest that they are so nearly a direct study of molecules in dynamic equilibrium.

(2) The Random Walk Problem (Einstein—Smoluchowski Equation)

The difficulty of the sedimentation experiment is to obtain vast numbers of very small particles which are of exactly the same size and are nearly enough spherical to conform to Stokes's law. Such problems do not arise if we study the motion of a single particle immersed in a medium. We can suppose that the forces acting on it are of two kinds:

(a) Forces due to lack of balance in the molecular bombardment upon the particle.

(b) A viscous force proportional to the velocity of the particle (Stokes's law).

Let us simplify the question by considering motion along the x axis only. Let X = resultant at any instant of forces due to bombardment. Let $F = -6\pi a\eta (dx/dt) = -\mu(dx/dt)$ (say) be the viscous retarding force. Then the equation of motion is

$$m\frac{d^2x}{dt^2} = -\mu\frac{dx}{dt} + X$$

Multiply both sides of the equation by $2x$:

$$m \cdot 2x \frac{d^2x}{dt^2} = -\mu \cdot 2x \frac{dx}{dt} + 2Xx \quad (1.1)$$

Now

$$2x\frac{dx}{dt} = \frac{d}{dt}(x^2)$$

and, differentiating this,

$$2\left(\frac{dx}{dt}\right)^2 + 2x\left(\frac{d^2x}{dt^2}\right) = \frac{d^2}{dt^2}(x^2)$$

$$\therefore \quad 2x\frac{d^2x}{dt^2} = \frac{d^2}{dt^2}(x^2) - 2\left(\frac{dx}{dt}\right)^2$$

Substituting in equation (1.1) for both $2x(dx/dt)$ and $2x(d^2x/dt^2)$, we find

$$m\frac{d^2}{dt^2}(x^2) - 2m\left(\frac{dx}{dt}\right)^2 = -\mu\frac{d}{dt}(x^2) + 2Xx \quad (1.2)$$

We shall now average the quantities in equation (1.2) over a long period of time. By "long" in this context we mean long compared with the interval between successive molecular impacts on the particle. Now a particle of radius 10^{-5} cm in air at NTP will be struck by about 10^{12} molecules per sec (evaluated roughly from $\tfrac{1}{6}nv \times \pi a^2$), so that the characteristic time interval here is about 10^{-12} sec, which is therefore very short indeed compared with any time of practical interest. Over an extended period the size and direction of the resultant force on the particle will change at random; we therefore put

$$\overline{Xx} = 0$$

Also
$$\overline{(dx/dt)^2} = \overline{v_x^2}$$

will represent the mean square of the x component of the velocity of the particle. But in thermal equilibrium we have

$$\overline{v_x^2} = \overline{v_y^2} = \overline{v_z^2} = \tfrac{1}{3}\overline{v^2}$$

and
$$\tfrac{1}{2}m\overline{v^2} = \tfrac{3}{2}kT = \frac{3}{2}\frac{R}{N}T$$

Thus
$$m\overline{\left(\frac{dx}{dt}\right)^2} = \frac{RT}{N} \quad \text{simply}$$

The equation of practical interest to us is therefore a simplification of equation (1.2): viz.,

$$m\frac{d^2}{dt^2}(\overline{x^2}) - \frac{2RT}{N} = -\mu \frac{d}{dt}(\overline{x^2})$$

In this we put

$$\frac{d}{dt}(\overline{x^2}) = w$$

so that the equation becomes

$$m\frac{dw}{dt} = \frac{2RT}{N} - \mu w$$

i.e.,
$$\frac{dw}{w - 2RT/N\mu} = -\frac{\mu}{m}dt$$

whence
$$w(t) = \frac{2RT}{N\mu} + A \exp(-t/\tau) \qquad (1.3)$$

where A is a constant of integration, and $\tau = m/\mu$ is what is called a "relaxation time." Evidently the exponential term becomes unimportant in a time t comparable with τ ($t = 5\tau$, say). Now

$$\tau = \frac{m}{\mu} = \frac{\frac{4}{3}\pi a^3 \rho}{6\pi a \eta} = \frac{2}{9}\frac{a^2 \rho}{\eta}$$

In a typical case, $a \approx 10^{-4}$ cm, $\rho \approx 1$, $\eta \approx 10^{-2}$ cgs (liquid medium) or $\eta \approx 10^{-4}$ cgs (gaseous medium). Thus τ (liquid) $\approx 2 \cdot 10^{-7}$ sec, τ (gas) $\approx 2 \cdot 10^{-5}$ sec. In either case the exponential term will have died out during any reasonable time of observation.

Equation (1.3) can, therefore, for practical purposes be rewritten as

$$w = \frac{d}{dt}(\overline{x^2}) = \frac{2RT}{N\mu}$$

$$\therefore \quad \overline{x^2} = \frac{2RT}{N\mu}t = \frac{RT}{N \cdot 3\pi a \eta}t \tag{1.4}$$

Equation (1.4) is the Einstein–Smoluchowski equation (1905). We see that the mean-squared displacement increases proportionally to the time. To see whether the displacements are large enough to be measurable, let us put $t = 1$ min for a particle of radius 10^{-4} cm in water ($\eta \approx 10^{-2}$). This gives

$$\sqrt{\overline{x^2}} = \left(\frac{8 \cdot 10^7 \times 300 \times 60}{6 \cdot 10^{23} \times 10 \times 10^{-4} \times 10^{-2}}\right)^{\frac{1}{2}} \approx 5 \cdot 10^{-4} \text{ cm}$$

Thus the rms displacement in this time is about 5 microns (1 $\mu = 10^{-3}$ mm). This can be measured to perhaps 0.2 μ with a good microscope. Experiments of this type, also, were carried out by Perrin, and again led to about the correct value for N.

1.11 SUMMARY: THE ATOMIC THEORY

We have seen that the hypothesis that matter is built up of discrete atoms led to a satisfactory kinetic theory of gases, and that, when direct investigation of the structure of matter became possible, the sizes of molecules, and the number of atoms or molecules in a given volume of matter, could be determined. The development of kinetic theory made it appear certain that heat, and the propagation of sound, were manifestations of molecular motion, heat in particular being identified with the kinetic energy of the molecules. Two things remain outside this picture: viz., (a) the nature of electricity and magnetism, and (b) the nature of light. Perhaps the greatest triumph of nineteenth century physics was Maxwell's work in demonstrating a connection between these two, and our next step is to see how this was done.

Problems

1.1. A certain gas at 27° C has a density of 0.08 g/cm³ at a pressure of 100 atm. Calculate the root-mean-square velocity, and identify the gas by finding its molecular weight.

1.2. Consider a gas at 300° K under a pressure of 10^6 dynes/cm² (i.e., roughly standard conditions), and find the length of edge of a cubical volume that contains a number of molecules equal to the human population of the world.

1.3. Find the temperature at which the mean translational energy of a molecule is equal to the kinetic energy of an electron accelerated through 0.01 volt. (300 volts = 1 esu; electronic charge = $4.8 \cdot 10^{-10}$ esu.)

1.4. Find the variation of pressure with height (a) in an atmosphere composed of a mixture of two sorts of molecules, the temperature being independent of height, and (b) in an atmosphere composed of only one sort of molecules, the temperature at height z being given by $T(z) = T_0(1 - \alpha z)$, where α is a constant and z is always less than $1/\alpha$.

1.5. Material is evaporated from an "oven" and collected on a plate a distance x away, the intervening space being filled with a gas at a variable pressure. It is found that 1.359 mg of material is deposited on the plate in 1 hr at a pressure such that the mean free path is 7 mm, and that 0.500 mg is deposited in 1 hr at a pressure such that the mean free path is 5 mm. Find x.

1.6. Suppose that the area dA of Fig. 1.4 represents a small aperture in the wall of a vessel. By carrying out a calculation along the lines of Section 4 of this chapter, show that the rate of escape of molecules through the aperture is given by $\frac{1}{4}n\bar{v}\,dA$, where $\bar{v}$ is defined by

$$\bar{v} = \int_{v=0}^{\infty} v f(v)\, dv$$

and represents the mean speed of the molecules. (Note the factor $\frac{1}{4}$, as against the factor $\frac{1}{6}$ that would be suggested by the simple approach of Section 1.2.)

1.7. The calculation of transport properties in Section 1.9 assumes that molecules are transferred between parallel planes a distance λ apart. A somewhat better treatment is to assume that molecules entering an element dA of a given layer during time dt have come from a spherical shell of radius λ and thickness $\bar{v}\,dt$. The molecules bring with them the value of the relevant property at a vertical distance $\lambda \cos\theta$, and an integral is to be taken from $\theta = 0$ (straight up) to $\theta = \pi$ (straight down). By following out an analysis in this way (guided by Section 4) show that $\eta = \frac{1}{3}\rho\bar{v}\lambda$ and that $D = \frac{1}{3}\bar{v}\lambda$. (This helps to confirm that, for transport problems, the molecules behave *as if* one sixth of them were moving in a specified direction at any instant.)

1.8. Show that the ratio of pressure to viscosity coefficient gives an approximately correct figure for the number of collisions per unit time experienced by a given molecule. Evaluate the collision rate for air under standard conditions. ($\eta = 1.8 \cdot 10^{-4}$ cgs.)

1.9. Two vessels of equal volume V are connected by a tube of length l and cross section A. Both vessels are at the same total pressure P. One vessel con-

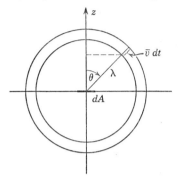

Problem 1.7

tains initially a small partial pressure p_0 of a gas X, the rest of its contents being a gas Y of the same molecular weight as X (e.g., CO and N_2); the other vessel is entirely filled with gas Y. Each gas therefore has the partial pressure difference p_0, and diffusion can take place without any change of total pressure or mass in either vessel. If the coefficient of diffusion of X into Y (and of Y into X) is D, show that the partial pressure of X in the first vessel at time t is given by

$$p(t) = \tfrac{1}{2}p_0\left[1 + \exp\left(-\frac{2DA}{Vl}t\right)\right]$$

1.10. (a) Evaluate the characteristic time $Vl/2DA$ of Problem 1.9 under the following conditions: The gases are CO and N_2; $D = 0.192$ cm^2/sec; $V = 1$ liter; the connecting tube is of length 10 cm and diameter 0.2 mm. (b) To recognize how slow diffusion is compared to viscous flow under most conditions, compare the characteristic time in (a) above with the corresponding relaxation time for annihilation of a pressure difference p set up between the two vessels after the mixing by diffusion has been completed. Assume that the mean pressure P is 1 atm. The coefficients of viscosity for both CO and N_2 are $1.75 \cdot 10^{-4}$ cgs.

1.11. An experiment on sedimentation equilibrium is carried out with particles of mean radius 0.24 μ and density 1.2 g/cm^3 suspended in water at 27° C. The following values were observed for the numbers of particles per field of view of a microscope focused at various depths in the suspension:

Depth, microns	0	18	33	46	57	71
Number of particles	131	214	312	437	625	877

Make an appropriate graph of these data, and from its slope deduce the value of Avogadro's number.

2 | Light and the Electromagnetic Field

2.1 THEORIES OF LIGHT: THE SEVENTEENTH CENTURY

Any satisfactory theory of light as envisaged at the time of Newton had to account for a variety of facts, the chief of which were the following:

1. Light is a form of energy.
2. It can travel through empty space.
3. It travels in straight lines.
4. It obeys Snell's laws of reflection and refraction.
5. It may undergo simultaneous reflection and refraction at a boundary between two media.
6. It travels through space with a characteristic speed ($3 \cdot 10^{10}$ cm/sec).*
7. It can be broken by a prism into different colors.

Only two descriptions of the nature of light have ever been put forward, and in classical physics they represent mutually exclusive points of view. These are the corpuscular and the wave theories, respectively, and between them they exhaust the possibilities presented by a mechanical picture of the universe.

The corpuscular theory was favored by Newton (1642–1727). It

---

* A fact that was recognized by Römer in 1676 as a result of his observations on the eclipses of Jupiter's moons.

Light and the Electromagnetic Field

accounts readily for properties 1, 2, and 3, although the propagation of light through solids might seem to present some difficulty. Newton imagined the corpuscles of light to be perfectly elastic, in which case the law of reflection follows at once. The known law of refraction demands that the corpuscles should have a higher speed in the optically denser medium, a supposition that was not open to experimental test in Newton's time. Property 5 is hard to explain on a corpuscular theory, and Newton had to postulate that the corpuscles on striking the boundary between two media might be in different states which he called "fits" of easy reflection or transmission. Property 6 is perhaps a little surprising, in that the speed of light might be expected to depend on the speed of the source, by analogy, say, with bullets fired from a moving vehicle. One might also think that a hot source would emit faster corpuscles than a relatively cool one. Finally, the analysis of white light into a spectrum was explained by Newton in terms of corpuscles of different sizes.

The wave theory was supported by Newton's Dutch contemporary, Huygens (1629–1695). This theory deals readily with properties 1, 4, 5, and 6, since these have their analogs in the familiar behavior of sound waves in ordinary material media. In contrast to the corpuscular theory, however, the law of refraction requires that light should have a *smaller* velocity in the optically denser medium. The propagation of light in vacuum demands the idea that what we call empty space is really a medium of some kind, endowed with special properties. But the real objection to a wave theory, at least in Newton's eyes, was the fact of rectilinear propagation, since all known waves showed the property of spreading rather than traveling in straight lines. The production of a spectrum from white light could be explained by assuming waves with a variety of wavelengths.

It may be seen, then, that each theory can claim a partial success in accounting for the facts. But the great authority of Newton, and the apparent impossibility of explaining rectilinear propagation by waves, led to an acceptance of the corpuscular theory, and it was not seriously questioned for about a hundred years.

2.2 RESURGENCE OF THE WAVE THEORY

It is perhaps surprising that Newton decided so firmly against a wave theory, for he knew of an experimental result that ought to have spoken strongly in its favor. It was found by Grimaldi (1618–1663) that the shadow of a small object (e.g. a needle) illuminated by parallel light is slightly larger than the object itself and shows other details of structure—bright and dark bands. Newton verified some of these observations, but thought he could account for them by corpuscular theory. He also discovered the colored circular fringes (Newton's rings) formed between a lens and a flat surface, which today are recognized as characteristic of interference in waves.

Diffraction phenomena, of the type discovered by Grimaldi, were studied extensively over the period 1700–1800, but without leading to any new interpretations. Then Thomas Young, in the years 1801–1804, attacked the problem by comparing acoustic phenomena with light. Interference of sound waves was a familiar fact. Young tried to explain Newton's rings in the same terms, regarding these rings as equivalent to a standing wave pattern. From Newton's measurements he could deduce the wavelength of the light. Young made the position clear by deliberately causing interference between two beams, obtaining dark bands that vanished when one source was cut off. This work, and that of Fresnel (1788–1827), completely established the picture of light as a transverse wave motion, although it was still required to show that light did travel slower, rather than faster, when it entered a refracting medium from air or vacuum. In 1850 Foucault succeeded in verifying this, and the corpuscular theory was finally abandoned.

2.3 THE ETHER

It was still necessary to define the properties of the medium that carried light across space which was apparently devoid of ordinary matter. The far-reaching success of Newtonian mechanics made physicists think that every phenomenon could be explained by means of these concepts alone, and the ether was pictured as an elastic solid, endowed with highly unlikely properties so as to make the quantity (elasticity/density)$^{1/2}$ equal to $3 \cdot 10^{10}$ cm per sec, by analogy with the formula for the speed of sound waves. By the middle of the nineteenth century the attempts to describe the ether and the propagation of light had gone to fantastic lengths. Electricity, too, could not be readily fitted into the mechanical scheme of things. But during the period 1865–1870 Clerk Maxwell succeeded in relating optical radiation to electricity, and apparently resolved these outstanding difficulties of physics. We shall sketch the main features of his electromagnetic theory.

Light and the Electromagnetic Field

2.4 BASIC ELECTRICITY AND MAGNETISM

It is well known that electrostatics and magnetostatics, which began and developed as separate subjects, were brought together when it was recognized that electrical forces could be brought about by purely magnetic means, and vice versa. Granted this interrelationship, the whole of electricity and magnetism could be stated in terms of four laws, describing (1) a steady electric field, (2) a steady magnetic field, (3) the magnetic effect of a varying electric field, and (4) the electric effect of a varying magnetic field.

Let us amplify these statements somewhat. The basis of electrostatics is Coulomb's law of force between charges:

$$F = \frac{q_1 q_2}{\epsilon r^2}$$

From this is derived Gauss's theorem, which relates the charge to the electric induction $\mathbf{D}(= \epsilon \mathbf{E})$ produced by it

$$\int \mathbf{D} \cdot d\mathbf{S} = \int D_n \, dS = 4\pi Q$$

An electric current represents a flow of electric charge past a given point, and so is related to the change with time of E. We thus have access to a description of charges, and of currents, in terms of electric fields and so, through Coulomb's law, in terms of mechanical forces. In magnetostatics one possible starting point is the law of force between magnetic poles:

$$F = \frac{m_1 m_2}{\mu r^2}$$

This law allows us to define a unit pole, and hence the magnetic field H as the field producing unit force on unit pole. We next bring in the experimental fact that a current-bearing wire experiences a force in a magnetic field, and arrive at an electromagnetic unit of current (and therefore charge) defined in terms of the mechanical force acting on it in a known magnetic field. Thus a given current can be expressed as the rate of transport of charge measured in either electrostatic units Q or electromagnetic units Q'. The ratio of these measures is a large number c, very nearly equal to $3 \cdot 10^{10}$ if the cgs system is used.

2.5 MAXWELL'S EQUATIONS

(1) Ampère's Circuital Theorem

We begin with the Biot (Ampère) law for the magnetic field due to a current element (Fig. 2.1):

$$dH = \frac{i'\, ds\, \sin\theta}{r^2}$$

(i' = current in emu.) By considering an infinitely long straight wire

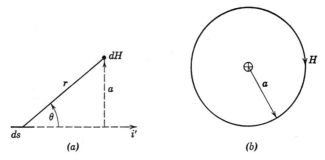

Figure 2.1. (a) Field due to an element ds of a straight wire. (b) Circular path around current, following magnetic field.

we obtain from this formula the result that the magnetic field due to such a wire is given by

$$H = \frac{2i'}{a}$$

We then consider carrying a magnetic pole of strength m once round the wire, and have, for the work done,

$$W = 2\pi a F = 2\pi a \frac{2mi'}{a} = m \cdot 4\pi i'$$

From the disappearance of a from the result we infer that the work done is independent of the precise path. This is the circuital theorem.

Let us now generalize this result by considering the situation anywhere within an arbitrary distribution of currents and magnetic fields. We apply the circuital theorem to an elementary closed path $OABC$ (Fig. 2.2) lying in the xy plane, and we introduce the concept of the current *density* j' (emu) such that the current enclosed by the path is

$j'\,dx\,dy$ (normal to $OABC$). The magnetic field H is a function of position, and the average values of the components of H along the various elements of the path are:

Along OA, $\qquad\qquad H = H_x$

Along AB, $\qquad\qquad H = H_y + \dfrac{\partial H_y}{\partial x} dx$

Along CB, $\qquad\qquad H = H_x + \dfrac{\partial H_x}{\partial y} dy$

Along OC, $\qquad\qquad H = H_y$

Thus in carrying unit magnetic pole once round the loop $OABC$ the

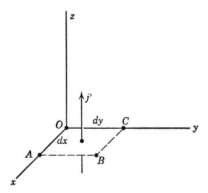

Figure 2.2. Rectangular closed path in the xy plane, enclosing current $j'dx\,dy$.

work done is given by

$$dW = dx \cdot H_x + dy\left(H_y + \frac{\partial H_y}{\partial x} dx\right) - dx\left(H_x + \frac{\partial H_x}{\partial y} dy\right) - dy \cdot H_y$$

$$= dx\,dy\left(\frac{\partial H_y}{\partial x} - \frac{\partial H_x}{\partial y}\right)$$

By equating this to 4π times the enclosed current we have

$$\frac{\partial H_y}{\partial x} - \frac{\partial H_x}{\partial y} = 4\pi j_z'$$

Similarly, by considering paths lying entirely in the yz and zx planes, we find

$$\frac{\partial H_z}{\partial y} - \frac{\partial H_y}{\partial z} = 4\pi j_x'$$

$$\frac{\partial H_x}{\partial z} - \frac{\partial H_z}{\partial x} = 4\pi j_y'$$

These three equations are all included in the vector equation

$$\text{curl } \mathbf{H} = 4\pi \mathbf{j}' \tag{2.1}$$

(2) Displacement Currents

The concept of displacement current was introduced by Maxwell to dispose of an unnecessary distinction between real and apparent

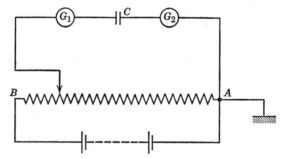

Figure 2.3. To illustrate the concept of a displacement current providing continuity in an electric circuit.

currents. Let us consider the charging of a capacitor C (Fig. 2.3) by the movement of a slider along a potentiometer wire AB. While the slider is in motion, the galvanometers G_1, G_2 on opposite sides of the capacitor register identical currents which are to be regarded as real currents in metallic conductors. The capacitor, regarded as a perfect insulator, carries no current as such, but the field strength E between the plates increases steadily, and this corresponds to some current i, providing continuity with the conduction currents in the connecting wires. This current i is Maxwell's displacement current. We have

$$E = \frac{4\pi Q}{\epsilon A}$$

where Q = charge, A = area of capacitor plate, ϵ = dielectric constant.

Light and the Electromagnetic Field

$$\therefore \frac{dE}{dt} = \frac{4\pi}{\epsilon A}\frac{dQ}{dt} \equiv \frac{4\pi}{\epsilon} j \qquad (2.2)$$

where j = equivalent current density in esu

Thus
$$j = \frac{\epsilon}{4\pi}\frac{dE}{dt}$$

and this description of current in terms of a changing electric field can be extended. Consider, for example, the charging of a sphere in a medium of dielectric constant ϵ. At any point, distant r from the center of the sphere,

$$E = \frac{q}{\epsilon r^2} \quad \text{instantaneously}$$

$$\therefore \frac{dE}{dt} = \frac{1}{\epsilon r^2}\frac{dq}{dt} = \frac{4\pi}{\epsilon} \cdot \frac{1}{4\pi r^2}\frac{dq}{dt}$$

Fixing attention on the surface of an imaginary sphere of radius r, we see, therefore (by comparison with equation 2.2), that the rate of change of field corresponds to an effective current density given by

$$j = \frac{\epsilon}{4\pi}\frac{dE}{dt} = \frac{1}{4\pi r^2}\frac{dq}{dt}$$

In general, we may put

$$\frac{d}{dt}(\epsilon E) = \frac{dD}{dt} = 4\pi j \qquad (2.3)$$

(3) Maxwell's Field Equations

We now combine equations (2.1) and (2.3). Restricting ourselves to the z components of all our vectors, we have

$$\frac{\partial H_y}{\partial x} - \frac{\partial H_x}{\partial y} = 4\pi j_z' = \frac{4\pi}{c} j_z \quad \text{with} \quad j_z = \frac{\epsilon}{4\pi}\frac{\partial E_z}{\partial t}$$

Hence
$$\frac{\epsilon}{c}\frac{\partial E_z}{\partial t} = \frac{\partial H_y}{\partial x} - \frac{\partial H_x}{\partial y}$$

If we consider the whole vectors, we have

$$\frac{\epsilon}{c}\frac{\partial \mathbf{E}}{\partial t} = \text{curl } \mathbf{H} \qquad (2.4)$$

Thus we have here a description of electromagnetic effects solely in terms of the fields generated by charges and currents. In arriving at this result, we have made use of the laws of electrostatics and magnetostatics, together with the laws describing the magnetic field of a current (regarded as the source of a varying electric field). But there is also

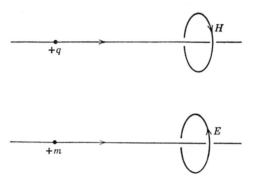

Figure 2.4. Illustrating the directions of induced magnetic and electric fields from moving electric and magnetic monopoles.

the law describing the electric effect of a varying magnetic field—Faraday's law of induction. By applying this law, similarly, to an elementary loop, we find

$$\frac{\mu}{c}\frac{\partial H_z}{\partial t} = -\left(\frac{\partial E_y}{\partial x} - \frac{\partial E_x}{\partial y}\right)$$

or

$$\frac{\mu}{c}\frac{\partial \mathbf{H}}{\partial t} = -\operatorname{curl} \mathbf{E} \tag{2.5}$$

The appearance of a negative sign in this result (which is another of Maxwell's field equations) represents the experimental facts about the signs of induced fields caused by charges or magnetic poles in motion (Fig. 2.4).

2.6 ELECTROMAGNETIC WAVES

Once we have Maxwell's field equations (2.4) and (2.5), it is easy to demonstrate the possibility of progressive waves consisting of oscillating electric and magnetic fields in space. Let us look for a very simple solution, in which the electric field component E_x varies with time.

Light and the Electromagnetic Field 37

Since we have

$$\frac{\epsilon}{c}\frac{\partial E_x}{\partial t} = \frac{\partial H_z}{\partial y} - \frac{\partial H_y}{\partial z}$$

it is necessary for either or both of H_y, H_z to be nonzero. Suppose that H_y exists, $H_z = 0$. We can also put $H_x = 0$ since its value does

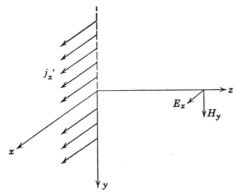

Figure 2.5. Generation of a plane electromagnetic wave along z by an oscillating current sheet in the xy plane.

not affect E_x. Let us further demand that $E_y = 0$, $E_z = 0$. Then our Maxwell equations reduce to the following:

$$\frac{\epsilon}{c}\frac{\partial E_x}{\partial t} = -\frac{\partial H_y}{\partial z} \tag{2.6}$$

$$\frac{\mu}{c}\frac{\partial H_y}{\partial t} = -\frac{\partial E_x}{\partial z} \tag{2.7}$$

$$\frac{\partial H_y}{\partial x} = 0$$

$$\frac{\partial E_x}{\partial y} = 0$$

We may seem to have imposed an excessive number of conditions on the field, so it is perhaps worth noting that the supposed form of the field could be brought about by a flow of current, entirely parallel to the x axis, in an infinite sheet lying in the xy plane (see Fig. 2.5). The current must of course vary with time.

We now differentiate equation (2.6) with respect to t, and equation

(2.7) with respect to z. Then

$$-\frac{\partial^2 H_y}{\partial z \, \partial t} = \frac{\epsilon}{c}\frac{\partial^2 E_x}{\partial t^2} = \frac{c}{\mu}\frac{\partial^2 E_x}{\partial z^2}$$

$$\therefore \quad \frac{\partial^2 E_x}{\partial z^2} = \frac{\epsilon\mu}{c^2}\frac{\partial^2 E_x}{\partial t^2} \tag{2.8}$$

This is the characteristic equation for a disturbance traveling along the direction of the z axis. Let us suppose

$$E_x = f(z \pm vt) = f(w) \quad \text{say}$$

which describes such a disturbance traveling with velocity v. Then

$$\frac{\partial^2 E_x}{\partial z^2} = \frac{d^2 f}{dw^2}$$

$$\frac{\partial^2 E_x}{\partial t^2} = v^2 \frac{d^2 f}{dw^2}$$

(Note the distinction between partial and total derivatives.)

i.e., $$\frac{\partial^2 E_x}{\partial z^2} = \frac{1}{v^2}\frac{\partial^2 E_x}{\partial t^2}$$

We therefore have the result

$$v = \frac{c}{\sqrt{\epsilon\mu}}$$

which describes the velocity of an electromagnetic wave in any given medium. For free space, $\epsilon = 1$, $\mu = 1$, and so $v = c$. We must remember that c has been defined purely as the ratio of electromagnetic and electrostatic units of charge or current. The fact that the observed velocity of light in vacuum is numerically identical with this figure is therefore a convincing demonstration that light is an electromagnetic disturbance.

We see that an essential feature of the theory is that the vibrations are transverse, and so may exhibit characteristic polarizations in a way that is impossible for longitudinal vibrations of the sort found in sound waves. This removes one of the objections that Newton had to a wave theory of light; in his corpuscular theory he was able to postulate nonspherical corpuscles to account for the facts of double refraction (discovered by Bartholinus in 1669), whereas he did not think that any theory of waves in an isotropic medium could encompass this phenomenon.

Maxwell's theory was put forward in 1865, but it was not until 1888 that H. Hertz demonstrated the production of radiation by recognizably electrical means. He produced a current distribution (of the type we have discussed theoretically) by inducing a discharge between two spheres to which were attached large metal sheets (Fig. 2.6). This generated a plane-polarized wave with its electric vector

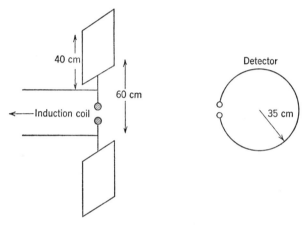

Figure 2.6. Hertz's method for generating and detecting electromagnetic waves.

parallel to the gap. Hertz detected the radiation with a broken loop of wire which acted as a resonator. The electric field of the waves generated a sufficiently strong potential gradient across the detector gap to break down its insulation and cause a spark. Hertz found further that the detector failed to respond if its gap was perpendicular to the source gap, and that its sensitivity was greatest when the two gaps were parallel. The state of polarization of the radiation was thus clearly shown. With Hertz's arrangement the wavelength of the radiation was about 5 m, i.e. 10^7 times longer than ordinary light, and it was a clear tribute to the success of Maxwell's theory that it could account for both of them in the same terms.

2.7 ENERGY AND MOMENTUM IN WAVES

Let us take

$$E_x = f(z - vt) = f(w)$$

as before, and let us suppose that it is possible to put

$$H_y = \alpha f(z - vt)$$

Then, by equation (2.6) of the previous section,

$$-\frac{\epsilon}{c} v \frac{df}{dw} = -\alpha \frac{df}{dw}$$

$$\therefore \alpha = \frac{\epsilon}{c} \cdot \frac{c}{\sqrt{\epsilon\mu}} = \left(\frac{\epsilon}{\mu}\right)^{1/2}$$

i.e.,
$$H_y = \left(\frac{\epsilon}{\mu}\right)^{1/2} E_x$$

Now Electrostatic energy per cm^3 = $\dfrac{\epsilon E_x^2}{8\pi}$

and Magnetic energy per cm^3 = $\dfrac{\mu H_y^2}{8\pi} = \dfrac{\epsilon E_x^2}{8\pi}$

Thus Total energy per cm^3 = $\dfrac{\epsilon E_x^2}{4\pi} = \dfrac{\mu H_y^2}{4\pi}$

$$= \frac{\sqrt{\epsilon\mu}}{4\pi} E_x H_y = W \quad \text{say}$$

The velocity of transport of this energy is v. Thus the rate of energy flow across 1 cm^2 normal to the direction of propagation is given by a quantity S, where

$$S = vW = \frac{c}{\sqrt{\epsilon\mu}} \cdot \frac{\sqrt{\epsilon\mu}}{4\pi} E_x H_y$$

$$S = \frac{c}{4\pi}(E_x H_y)$$

More generally, we put

$$\mathbf{S} = \frac{c}{4\pi} \mathbf{E} \times \mathbf{H}$$

This is Poynting's vector, and it describes the instantaneous flux density of electromagnetic radiation.

Evidently a unidirectional flow of energy implies a momentum, and the further working out of electromagnetic theory shows that the momentum density G (i.e. momentum per cubic centimeter) is given by

$$G = \frac{\epsilon\mu}{c^2} S = \frac{\epsilon\mu}{4\pi c}(E_x H_y) = \frac{W}{v}$$

2.8 THE PRESSURE OF LIGHT

Since light carries energy and momentum, it must be expected to exert a pressure on anything in its path. It was suggested, quite correctly, by Kepler in about 1600 that this explained why the tails of comets appeared to be blown away from the sun. But the matter was not tested quantitatively until Lebedew (1900) and Nichols and Hull (1903) carried out laboratory experiments on the effect.

The theory is very simple. If light falls on a surface and is, for argument's sake, totally absorbed, the pressure is equal to the momentum destroyed per square centimeter per second. Hence

$$p = vG = W$$

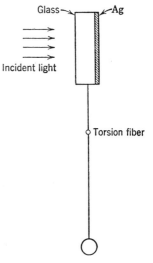

Figure 2.7. Basic apparatus for measuring the pressure of radiant energy.

We can easily assess the order of magnitude of the effect. Let us suppose that 1 watt per cm² of radiant energy falls on an absorbing surface in vacuo. Then

$$S = 1 \text{ watt/cm}^2 = 10^7 \text{ erg/cm}^2/\text{sec}$$

$$\therefore \quad p = W = \frac{S}{c} = \frac{10^7}{3 \cdot 10^{10}} \approx 3 \cdot 10^{-4} \text{ dynes/cm}^2$$

Nichols and Hull actually worked with intensities several times smaller than this. The experimental arrangement was very ingenious (see Fig. 2.7). A transparent glass plate was silvered on one side and hung on a torsion balance system. Light was brought in first from one side of the plate and then from the other. The purpose of this was to allow for spurious "radiometer" effects, due to the heating of the silver (and hence of gas molecules adjacent to it) as a result of absorption of the light. This radiometer pressure occurred in the same direction for both directions of incidence of the light, whereas the direction of the true light pressure was reversed. In order to compare the observed magnitude of the pressure with what was expected, the energy flow S was measured by absorbing the incident light on a blackened disk and measuring the rate of temperature rise. As an example of the results we quote the following:

$$\text{Observed pressure} = (7.01 \pm 0.02) \cdot 10^{-5} \text{ dyne/cm}^2$$
$$\text{Pressure calculated from } S/c = (7.05 \pm 0.03) \cdot 10^{-5} \text{ dyne/cm}^2$$

Problems

2.1. A current i' flows in a long straight wire perpendicular to the plane of the diagram, at the point marked with a cross. Verify that the work done in following the path $ABCD$ is zero. The use of θ as variable will be found convenient. (We see here a case of the circuital theorem when no current is enclosed by the path considered.)

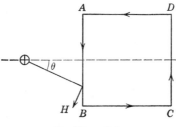

Problem 2.1

2.2. By the application of the induction law to an elementary rectangular loop, obtain the result

$$\frac{\mu}{c}\frac{\partial H_z}{\partial t} = -\left(\frac{\partial E_y}{\partial x} - \frac{\partial E_x}{\partial y}\right)$$

quoted in equation (2.5).

2.3. A luminous intensity of 10 watts falls on a glass plate. One tenth of the light is reflected at the front surface of the plate, one ninth of what is left is absorbed within the plate, and the rest of the radiant energy escapes from the back of the plate. What is the total force on the plate? (1 watt = 10^7 erg/sec.)

2.4. If the luminous intensity in Problem 3 is uniformly spread over 2 cm², what is the strength of the electric field amplitude in volts per centimeter in the incident beam?

Light and the Electromagnetic Field

2.5. A beam of light of intensity 0.1 watt falls upon a small perfectly reflecting mirror, which is mounted on a horizontal arm suspended on a thin vertical torsion wire. The horizontal distance from mirror to torsion wire is 2 cm. The torsion wire is of length 10 cm and diameter 0.01 mm, and its rigidity modulus is $6 \cdot 10^{11}$ dynes/cm². Through what angle is the horizontal arm turned by the light pressure? If the moment of inertia of the system is $1 \text{ g} \cdot \text{cm}^2$, what is the period of free torsional oscillation of the system? (This would represent a possible experimental arrangement for measuring the pressure of radiant energy.)

2.6. The sun has a mass of about $2 \cdot 10^{33}$ g, and radiates energy from its surface at the rate of about $6 \cdot 10^{10}$ erg/cm²/sec. Find the radius of a particle of density 2.5 g/cm³ that would be repelled as strongly by the sun's radiation pressure as it is attracted by the sun's gravitational force. ($G = 6.66 \cdot 10^{-8}$ dyne cm² g^{-2}; sun's diameter $= 1.39 \cdot 10^6$ km.)

$a = 2.7 \times 10^{-5}$ cm

3 | The Atomicity of Electric Charge

3.1 ELECTROLYSIS

After the invention of the voltaic cell as a source of current, it was found that chemical changes accompanied the passage of a current through a solution. (This did not happen, of course, in metallic conductors.) This process was called electrolysis, and was thoroughly investigated by Faraday during the years 1831–1834. He found that equal amounts of electricity liberated masses of elements proportional to their chemical equivalents. The amount of charge that must be passed across a solution to deposit one gram-equivalent of a substance has come to be called the Faraday.

$F = 1$ faraday $= 96{,}500$ coulombs. (1 coulomb = 1 ampere for 1 sec.) Faraday said, " $\cdots$ if we adopt the atomic theory or phraseology, then the atoms of bodies which are equivalent to each other in their ordinary chemical action have equal quantities of electricity naturally associated with them." In 1873 Maxwell wrote* "It is extremely natural to suppose that the currents of the ions are convection currents of electricity, and, in particular, that every molecule of the cation is charged with a certain fixed quantity of positive electricity, which is the same for the molecules of all cations $\cdots$." He went on to deduce that the charge per ion is $1/N$, where the unit is the Faraday (and N, of course, is Avogadro's number). He thought

---

*Quoted, by permission, from J. Clerk Maxwell, *Treatise on Electricity and Magnetism*, Vol. I, Oxford: The Clarendon Press, 1892.

The Atomicity of Electric Charge   45

this was a convenient idea, but then says,* "It is extremely improbable however that when we come to understand the true nature of electrolysis we shall retain in any form the theory of molecular charges." This was one of the few lapses in Maxwell's scientific intuition.

In 1874 Johnstone Stoney took the positive step of assuming a characteristic elementary charge whose size was given by dividing the faraday by Avogadro's number. Using the best figures available to him he deduced $e \approx 3 \cdot 10^{-11}$ esu. We now know that this figure is too small by a factor of about 15.

3.2 THE PHOTOELECTRIC EFFECT

In 1887 Hertz, during his studies of electromagnetic radiation (Chapter 2, Section 6), noticed that the critical gap length for sparking in his detector loop was diminished if a glass plate came between source and detector. He thought this might indicate an influence due to ultraviolet light emitted by the source. He then tried deliberately illuminating the detector gap from an independent source of ultraviolet light. He found that this restored the sensitivity of the detector gap.

In 1888 Hallwachs found that a freshly polished zinc plate, illuminated by ultraviolet light, lost its charge if it was originally negative, but retained it if it was positive. He found further that a neutral plate acquired a positive charge by irradiation. Thus light caused expulsion of negative electricity from a neutral plate.

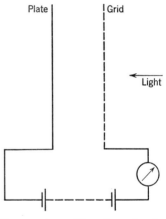

Figure 3.1. The photoelectric effect.

In 1890 Stoletow demonstrated the continuous passage of a current between a plate and a grid if the plate were made negative (Fig. 3.1). The photoelectric current was observed to flow in the highest attainable vacuum. It was concluded that the current flow was due to the motion of negative carriers.

Were these charges to be identified with the negative ions of electrolysis? In 1900 Lenard made an experiment to decide this question. He caused a photoelectric current to flow between sodium amalgam

* See footnote on page 44.

(Na/Hg) and a platinum (Pt) wire, the Na being the cathode. The total transfer of charge was about $3 \cdot 10^{-6}$ coulombs. Now the electrochemical equivalent of sodium is 23 g.

$$\therefore \text{ Expected mass transfer} = 23 \times \frac{3 \cdot 10^{-6}}{F}$$

$$= \frac{23 \times 3 \cdot 10^{-6}}{96{,}500} \text{ g} = 0.7 \cdot 10^{-6} \text{ mg}$$

This could have been noticed in a flame test on the Pt wire but was not found, and so the nature of the carriers remained a mystery.

3.3 THE ZEEMAN EFFECT

In 1862 Faraday had looked for the possible influence of a magnetic field on a light source, by placing a sodium flame between the poles of

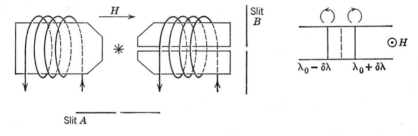

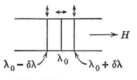

Figure 3.2. The classical Zeeman effect (otherwise known as the "normal" Zeeman effect), showing the lines and polarizations as viewed along and perpendicular to H.

a magnet and studying the D lines with a spectroscope. He failed to observe anything, but in 1896 Zeeman repeated the experiment under improved conditions, and found a splitting of a single spectral line into several components (see Fig. 3.2). He further found that the components had characteristic polarizations. The effect was explained (including the polarizations) by H. A. Lorentz. If we suppose the line spectrum to be produced by a charged particle oscillating in an orbit of some kind, we can investigate the effect of a magnetic field on this motion. Supposing the motion to be in general an ellipse, we resolve this into components parallel and perpendicular to H, i.e. into three linear oscillations, as indicated in Fig. 3.3. The z oscillation (parallel to H) we leave as a linear motion. But we regard the x and y linear motions as being each compounded of two opposite circular motions, initially (i.e. in the absence of H) having the same frequency.

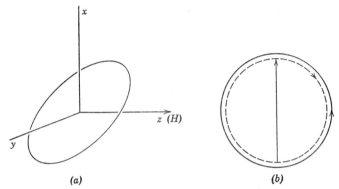

Figure 3.3. (a) A general elliptic motion referred to xyz axes. (b) A plane-polarized vibration resolved into two circular motions.

This is not simply an artifice; it was demonstrated as physically real by Fresnel, when he split a plane-polarized beam into two oppositely circularly polarized beams in a quartz prism—quartz is either right- or left-handed, and transmits these two beams at different speeds and so with different indices of refraction.

[Fresnel's Research on Rotary Polarization

Fresnel cemented together two prisms cut from right-handed quartz and one prism cut from left-handed quartz, in the way shown in Fig. 3.4. A plane-polarized beam becomes two circularly polarized beams traveling at different speeds in the right-handed prism R_1. With respect to the right-handed circularly polarized beam (r), R_1 is the

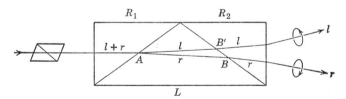

Figure 3.4. Fresnel's analysis of plane-polarized light into two circularly polarized beams.

fast medium, L the slow. Therefore r is bent down (and l is bent up) at A. The effect is accentuated at B, B'. If the emergent light is examined with a Nicol prism, the two images remain of constant brightness as it is rotated. If a $\frac{1}{4}$-wave plate is inserted, the images extinguish for different positions of the Nicol, 90° apart, thus proving the opposite senses of rotation.]

Let us now revert to the problem of the Zeeman effect. We need not consider the x and y vibrations separately; it is enough to know that each of them is composed of a right-handed and a left-handed circular vibration.

Now we consider the effect of applying a field. A moving charge looks like a current and so experiences a force given by the left-hand rule:

$$\mathbf{F} = \mathbf{i} \times \mathbf{H} = \frac{q}{c} \mathbf{v} \times \mathbf{H}$$

where q = charge in esu (See "Point Charges in Motion," Section 8 of this chapter). For the component vibration parallel to H, $F = 0$;

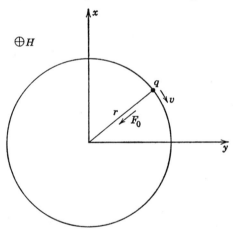

Figure 3.5. The motion of a point charge in a circle under the action of a central force F_0 and a magnetic field H.

this motion is therefore unaffected. Such a vibration leads to radiation perpendicular to H, but not parallel to it, as the latter would correspond to a longitudinal vibration in the wave, whereas light is essentially transverse. We can therefore understand the line of unmodified frequency appearing at A but not at B. (See Fig. 3.2.)

Now consider motion in the xy plane. Suppose that, before application of the magnetic field, the angular velocity is ω_0 under a central force F_0, and consider the right-handed rotation (Fig. 3.5). Then $F_0 = m\omega_0^2 r$. If the modified angular velocity is ω, then

$$F_0 + F = m\omega^2 r$$

Now for this case

$$F = \frac{qvH}{c}$$

and is radially outward if q is reckoned as positive.

$$\therefore F_0 - \frac{Hq\omega r}{c} = m\omega^2 r$$

Hence
$$m\omega_0^2 = m\omega^2 + \frac{Hq\omega}{c}$$

Putting
$$\omega^2 \approx \omega_0^2 + 2\omega_0\, \delta\omega,$$

we have
$$\delta\omega \approx -\frac{qH}{2mc} \quad \text{simply}$$

i.e., this right-handed motion is slowed down if q is positive. Equally it is speeded up if q is negative. For the left-handed rotation we have just an equal and opposite shift. (The above calculation assumes that the orbit radius r remains unchanged when the field H is applied. For an elementary discussion of this point see Appendix II.)

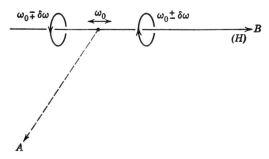

Figure 3.6. The motions that comprise the classical Zeeman effect.

We are now in a position to explain the Zeeman pattern, with the help of Fig. 3.6. (a) Viewing at B along the line of H, we should see two circularly polarized components, shifted equal amounts $\delta\omega$ on either side of the original line. There is no line at the original frequency. (b) Viewing at A perpendicular to H, we see (i) a line of the original frequency, plane-polarized parallel to H, and (ii) the edge view of the two circular components, which thus appear to be plane-polarized perpendicular to H.

We shall see in Chapter 8 that this simple theory has to be modified, and that the Zeeman effect exists in more complicated forms. But two very important pieces of information can be extracted with this simple theory, viz. the value of q/m and the sign of q. We see that the right-handed circular component is slowed down if q is positive, but Zeeman found that in fact it was raised in frequency. Hence the radiating particles have negative charge. Zeeman also found $q'/m = q/mc \approx 10^7$ emu per g.

3.4 DISCOVERY OF THE ELECTRON

The direct observation of this basic charge carrier was finally achieved in experiments with the low-pressure gas-discharge tube. This work had its beginning with the Geissler tube (1854), and many studies were made by Plücker (1858–1862). He found a green fluorescence near the cathode which could be moved about by a magnet (see Fig. 3.7a).

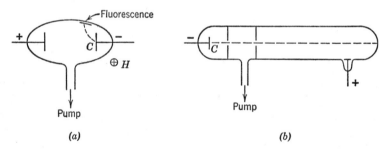

Figure 3.7. (a) Plücker's experiment. (b) A Crookes tube with a collimated beam of cathode rays.

It was found that, if the pressure were made low enough, the fluorescence could be made to appear as a spot at the end of a suitably designed tube. The effect could be ascribed to "cathode rays" which traveled in straight lines through defining apertures (Fig. 3.7b) but could be deflected by a magnet. During the period 1879–1885 Crookes made a series of experiments to show that the cathode rays were a stream of swift particles emitted from the cathode. In 1895 Perrin collected the rays in an insulated cup and showed that their sign was negative. Some people still thought that the cathode rays were something different from these negative particles, and were bothered by the apparent failure of the rays to be deflected by *electric* fields, though they were bent by magnetic fields.

In 1897 J. J. Thomson made the necessary exact experiments, which led to the identification of the rays. Using a pair of long parallel

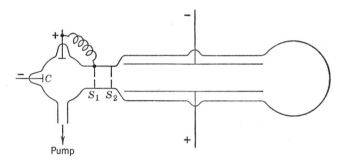

Figure 3.8. A schematic diagram of J. J. Thomson's e/m apparatus.

Figure 3.9. Deflection of an electron beam in crossed electric and magnetic fields.

plates, he demonstrated the deflection of cathode rays by an electric field (Fig. 3.8). He also showed that this deflection could be canceled out by an appropriate magnetic field perpendicular to the electric field. By measuring the deflection in one field alone, and by knowing the values of the electric and magnetic fields, he could obtain quantitative information about charged particles in the rays.

For a potential difference V between plates a distance d apart, we have

$$E = \frac{V}{d}$$

Let the length of path in the crossed electric and magnetic fields be l (Fig. 3.9). Then the time of transit is given by $t = l/v$. The electric force is Ee, giving rise to a transverse acceleration Ee/m.

$$\therefore \quad s = \frac{1}{2} \frac{Ee}{m} t^2 = \frac{1}{2} \frac{Ee}{m} \cdot \frac{l^2}{v^2}$$

For balancing of electric and magnetic deflections,

$$Ee = \frac{Hev}{c}$$

$$\therefore \quad v = \frac{Ec}{H}$$

$$\therefore \quad s = \frac{1}{2}\frac{Ee}{m} \cdot \frac{l^2 H^2}{E^2 c^2}$$

whence $\quad \dfrac{e}{m} = \dfrac{2sc^2 E}{l^2 H^2} = \dfrac{2c^2}{l^2 d} \cdot \dfrac{V}{H^2} s$

V is to be measured in esu (1 esu = 300 volts) and H in gauss.*
Thomson found $e/m \approx 1.7 \cdot 10^7$ emu per g under all conditions—independent of the nature of the gas in the cathode-ray tube and of the nature of the electrodes. This figure may be compared with the specific charge (i.e. ratio of charge to mass) concerned in the transfer of the lightest atoms (i.e. hydrogen) in electrolysis:

$$\frac{q}{M_H} = \frac{F}{M_H} \approx 10^4 \text{ emu per g} \qquad (1 \text{ coulomb} = \tfrac{1}{10} \text{ emu})$$

Thus we must have, in the discharge tube, carriers that are either lighter than hydrogen atoms, or more heavily charged, or perhaps both. These carriers must be constituents of all matter. Lenard (1900) showed that the carriers of photoelectric current had this same e/m value ($\approx 1.2 \cdot 10^7$ emu/g).

3.5 THE MEASUREMENT OF e.

In 1897 Townsend studied the gases and vapors liberated at the electrodes in electrolysis. Not all the charge is given up at the electrodes; some is carried out on small droplets. Townsend assumed that each ion resided on a droplet. He measured (1) the total charge Q in a sample of the vapor; (2) the total mass M of the vapor, (3) the radius a and hence the mass m of a droplet by observing the fall of a cloud of the vapor under gravity, and making use of Stokes's law:

---

* We have preferred, here and elsewhere, to express the magnetic field in gausses, rather than oersteds, since the magnetic induction B is normally implied.

The Atomicity of Electric Charge

$$\tfrac{4}{3}\pi a^3 \rho g = 6\pi a \eta v$$

Then
$$e = \frac{Q}{M/m}$$

and he found a value of about $3 \cdot 10^{-10}$ esu.

In 1898 J. J. Thomson used a method similar to Townsend's. The ions were produced this time by X rays, and an ionization current density I was measured: $I = ne(u + v)E$, where u and v are the mobilities of positive and negative ions (i.e. the velocities in unit electric field) and n is the number of ion pairs per unit volume. Some early experiments by Rutherford had already measured the velocities of ions, and so this gave $ne\ (=Q)$. Thomson obtained various values for e in the range $(5.5\text{ to }8.4) \cdot 10^{-10}$ esu.

In 1903 H. A. Wilson performed the prototype of Millikan's method. Wilson formed a cloud of electrically charged drops by an expansion method. He then measured the rate of fall of the upper edge of the cloud under gravity, and under gravity plus an electric field E.

We have, for a given drop carrying one electronic charge,

$$mg = 6\pi \eta a v_1$$

$$mg + Ee = 6\pi \eta a v_2$$

$$\therefore\ e = \frac{6\pi \eta a (v_2 - v_1)}{E}$$

Also, from the first equation,

$$\tfrac{4}{3}\pi a^3 \rho g = 6\pi \eta a v_1$$

$$\therefore\ a = \left(\frac{9\eta}{2\rho g} \cdot v_1\right)^{1/2}$$

Thus
$$e = \left(\frac{9\eta}{2\rho g}\right)^{1/2} 6\pi \eta \frac{(v_2 - v_1)v_1^{1/2}}{E}$$

By observing the top of the cloud only, the multiply charged drops are not included.

Wilson found $e = 3.1 \cdot 10^{-10}$ esu. In 1910 Millikan, by observation of individual drops, found $e = 4.65 \cdot 10^{-10}$ esu, with good consistency ($\pm 0.1 \times 10^{-10}$ esu). He also observed sudden changes in the speed of drops under an electric field and could verify that the change of charge corresponding to the capture of an individual ion was equal to e or some small multiple of it. Millikan later used oil drops to overcome the difficulty of evaporation and hence change of size of a drop while under observation. (A schematic diagram of the apparatus for this world-famous experiment is shown in Fig. 3.10.)

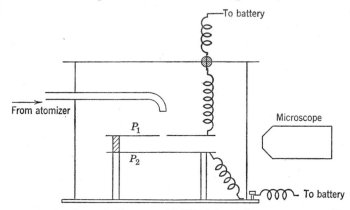

Figure 3.10. A schematic diagram of Millikan's oil-drop apparatus. Motion of the drops is observed between the plates P_1 and P_2 under the combined electric and gravitational fields.

The apparent value of e began to change as very small drops were used. This led to the discovery that η was not strictly a constant. The final result of the experiments was

$$e = (4.807 \pm 0.005) \cdot 10^{-10} \text{ esu}$$

3.6 CONDUCTIVITY OF METALS

The discovery of the electron paved the way for a theory of the behavior of metals as opposed to that of insulators or electrolytes. It was natural to suggest that a metal contained free electrons, by which current could be carried without the transport of mass in the form of atoms or ions. This picture was developed by Drude (1900) in the following way:

Suppose an electron "gas," of n electrons per cm^3, in thermal equilibrium with the metal. Consider first the effect of applying an electric field E which produces a drift velocity small compared with the random agitation. Let the mean free path of the electrons be λ. Then

$$\tfrac{1}{2}mc^2 = \tfrac{1}{2}m\overline{v^2} = \tfrac{3}{2}kT$$

$$\therefore \quad c^2 = \frac{3kT}{m}$$

Force on electron $= Ee$; therefore acceleration $= Ee/m$. The time for which this acceleration continues is $\lambda/\bar{v}$ (where $\bar{v}$ is the mean velocity of agitation), after which a random change of direction occurs as the

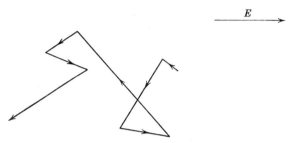

Figure 3.11. Drift motion superimposed on random agitation for an electron in an electric field inside a metal.

result of a collision (Fig. 3.11), and the electron effectively starts again from rest so far as systematic drift is concerned. Then we have

$$\text{Velocity acquired between collisions} = \frac{Ee}{m} \cdot \frac{\lambda}{\bar{v}}$$

$$\therefore \quad \text{Mean drift velocity} = \frac{Ee\lambda}{2m\bar{v}}$$

$$\therefore \quad \text{Mean current density} = ne\frac{Ee\lambda}{2m\bar{v}} = \frac{ne^2\lambda}{2m\bar{v}} E = \sigma E$$

where σ = electrical conductivity.

$$\therefore \quad \sigma = \frac{ne^2\lambda}{2m\bar{v}}$$

Consider now the thermal conductivity of a metal as due to its free electrons. This is a transport phenomenon of the type discussed in Chapter 1, Section 9. The energy transfer by electrons through 1 cm^2 at z (Fig. 3.12) is given by

$$\tfrac{1}{6}n\bar{v} \cdot \tfrac{3}{2}k\left(T + \lambda\frac{dT}{dz}\right) - \tfrac{1}{6}n\bar{v} \cdot \tfrac{3}{2}k\left(T - \lambda\frac{dT}{dz}\right) = \tfrac{1}{2}n\bar{v}k\lambda\frac{dT}{dz} = K\frac{dT}{dz}$$

where K = thermal conductivity.

$$\therefore \quad K = \frac{n\bar{v}k\lambda}{2}$$

Dividing σ by K, we have

$$\frac{\sigma}{K} = \frac{e^2}{m\bar{v}^2 k} = \frac{3\pi}{8} \frac{e^2}{m\overline{v^2} k}$$

$$= \frac{3\pi}{8} \frac{e^2}{mc^2 k} \qquad (c = \text{rms velocity})$$

We have here made use of the relation between $\bar{v}^2$ and $\overline{v^2}$ (see Appendix I). Hence

$$\frac{\sigma}{K} = \frac{\pi}{8} \frac{e^2}{k^2 T} = \frac{\pi}{8}\left(\frac{Ne}{Nk}\right)^2 \frac{1}{T} = \frac{\pi}{8}\left(\frac{F}{R}\right)^2 \frac{1}{T}$$

This accounts for the experimentally observed fact that the ratio σ/K for many metals is nearly the same (Wiedemann–Franz law).

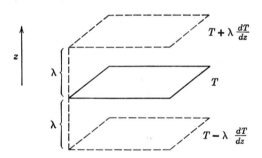

Figure 3.12. Heat conduction by electrons—a transport phenomenon.

It also gives a $1/T$ dependence that seems to be more or less correct (Lorenz, 1872). Furthermore, we see that the absolute magnitude of σ/K is given by Drude's formula. Putting $F = 9650$ emu, $R = 8.31 \cdot 10^7$ erg/°K, $T = 300$°K, we have

$$\frac{\sigma}{K} = \frac{\pi}{8}\left(\frac{9.65 \cdot 10^3}{8.31 \cdot 10^7}\right)^2 \frac{1}{300} = 1.76 \cdot 10^{-11}$$

This is for σ and K measured in absolute units. For practical units σ is in ohm^{-1} cm^{-1} = 10^9 σ_{abs}, K is in cals/cm/°C = $(1/4.2 \cdot 10^7) K_{\text{abs}}$

$$\therefore \left(\frac{\sigma}{K}\right)_{\text{practical}} = 4.2 \cdot 10^{16} (\sigma/K)_{\text{abs}} = 7.4 \cdot 10^5 \text{ at } 300° \text{ K}$$

The Atomicity of Electric Charge

Some examples are given in Table 3.1. We see that the electron

TABLE 3.1

Metal	σ	K	σ/K
Al	$3.54 \cdot 10^5$	0.504	$7.0 \cdot 10^5$
Cu	$5.82 \cdot 10^5$	0.918	$6.3 \cdot 10^5$
Pb	$4.55 \cdot 10^4$	0.083	$5.5 \cdot 10^5$

$$(\sigma/K)_{\text{mean}} = 6.3 \cdot 10^5$$

theory is fairly successful in accounting for the data, although we shall find that this simple kinetic theory of the electrons in a metal is far from adequate in other respects.

3.7 OPTICAL PROPERTIES OF INSULATORS

The electron theory pictures an insulator as a material in which all the electrons are bound to fixed positions, so that the free flow of current and thermal energy is impossible. Lorenz (1880) and Lorentz (1909) developed a definite model of an insulator in which the electrons were regarded as harmonic oscillators that were driven into a forced motion by the passage of a light wave. The object is to find a theoretical expression for the dielectric constant ϵ of the material. We have seen (Chapter 2, Section 6) that the velocity of electromagnetic waves in a medium is given by

$$v = \frac{c}{\sqrt{\epsilon\mu}}$$

Thus the refractive index n is equal to $\sqrt{\epsilon\mu}$. In all insulators $\mu \approx 1$. Hence

$$n^2 = \epsilon$$

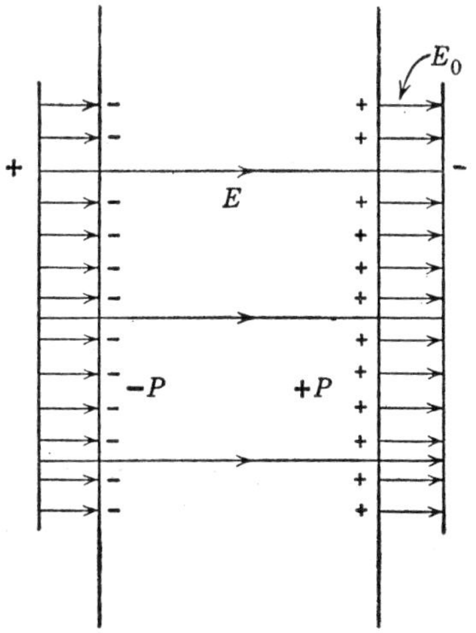

Figure 3.13. The polarization of a dielectric.

Now ϵ represents the ratio by which an electric field E is weakened upon entering a medium. We can describe this effect in terms of a surface charge per square centimeter—the polarization P, which is also equivalent to a certain electric moment per cubic centimeter (see Fig. 3.13).

By Gauss's theorem,
$$E_0 = E + 4\pi P$$
By the definition of ϵ,
$$E_0 = \epsilon E$$
$$\therefore \quad \epsilon = 1 + 4\pi \frac{P}{E}$$

If P/E is small, we then have, approximately,
$$n = 1 + 2\pi \frac{P}{E}$$

We now consider the polarizing field E to be the electric field set up by a plane wave passing through the medium:
$$E = E_0 e^{j\omega t} \qquad (j = \sqrt{-1})$$
The equation of motion of bound electrons, of natural frequency ω_0, is
$$m\ddot{x} + m\omega_0^2 x = Ee$$
$$\therefore \quad \ddot{x} + \omega_0^2 x = \frac{E_0 e}{m} e^{j\omega t}$$
Put
$$x = x_0 e^{j\omega t},$$
giving
$$\ddot{x} = -\omega^2 x_0 e^{j\omega t}$$
Then
$$x_0 = \frac{E_0 e}{m(\omega_0^2 - \omega^2)}$$

The induced electric moment for 1 electron $= ex = \dfrac{E e^2}{m(\omega_0^2 - \omega^2)}$

If there are N_0 electrons per cm^3, we have

Total moment per cm^3 $= P = \dfrac{N_0 e^2}{m(\omega_0^2 - \omega^2)} E$

Hence
$$n(\omega) = 1 + \frac{2\pi N_0 e^2}{m(\omega_0^2 - \omega^2)}$$

When damping is taken into account, we get rid of the catastrophic behavior of n at $\omega = \omega_0$ (see Appendix III). We also find that very strong absorption of the light takes place at this frequency (e.g. the Fraunhofer lines). This theory then gives a very good account of both normal and anomalous dispersion (see Fig. 3.14). A colorless transparent substance (e.g. water or glass) has this quality because it has no electron resonance frequencies in the visible region. The refractive index of water (Fig. 3.15) is observed to fall steadily as

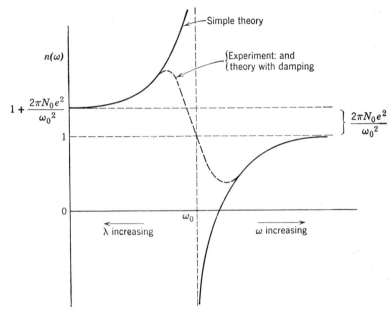

Figure 3.14. Anomalous dispersion.

wavelength increases, and this could be ascribed, on the simplest possible model, to one resonance on the short wavelength side, i.e. in the ultraviolet. We assume

$$n(\omega) = 1 + \frac{B}{\omega_0^2 - \omega^2}$$

where
$$B = \frac{2\pi N_0 e^2}{m}$$

A good fit to the data of Fig. 3.15 is obtained with

$$B = 1.28 \cdot 10^{32}$$
$$\omega_0 = 1.84 \cdot 10^{16} \text{ sec}^{-1}$$

ω_0 corresponds to $\quad \lambda_0 = \dfrac{2\pi c}{\omega_0} = 1.02 \cdot 10^{-5}$ cm

Substituting the value of B, we have

$$N_0 = \frac{mB}{2\pi e^2} = \frac{9.11 \cdot 10^{-28} \times 1.28 \cdot 10^{32}}{2\pi \times 2.30 \cdot 10^{-19}}$$

$$= 8.1 \cdot 10^{22} \text{ oscillator electrons per cm}^3$$

Now: Number of H_2O *molecules* per $cm^3 = \dfrac{6.02 \cdot 10^{23}}{18} = 3.3 \cdot 10^{22}$

Thus, according to our simple description, there would have to be 2 or 3 electrons per molecule of the appropriate natural frequency to

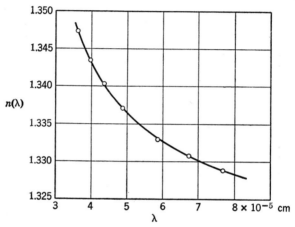

Figure 3.15. Refractive index of water versus wavelength in the visible region, pointing to a resonance in the ultraviolet. (The data are taken from R. A. Houstoun, *A Treatise on Light*, London: Longmans, Green & Co., 1938.)

describe the index of refraction. This is very reasonable, and helps to strengthen our belief that electrons play the major role in optical phenomena.

3.8 POINT CHARGES IN MOTION

We shall now collect together some results of electromagnetic theory that are of importance in describing the behavior of electrons or other charged particles.

(1) Magnetic Field of a Moving Charge

The Biot (Ampère) law (1824) for the magnetic effect of a current tells us that

$$dH = \frac{i\,dl \times n}{r^2} \quad \text{(see Fig. 3.16)}$$

We can apply this to a moving charge in the following simple way:

$$i\,dl = \frac{\text{charge}}{\text{time}} \times \text{distance} \equiv \text{charge} \times \frac{\text{distance}}{\text{time}}$$

$$\equiv \text{charge} \times \text{velocity} = qv$$

Thus
$$H = \frac{qv \sin \theta}{cr^2} \quad \text{(if } q \text{ is in esu)}$$

or
$$H = \frac{q}{cr^2} \mathbf{v} \times \mathbf{n}$$

In 1878 Rowland tested this relation by measuring the magnetic field

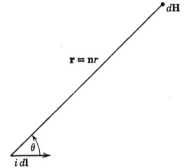
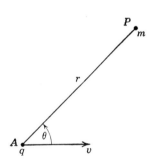

Figure 3.16. Field of a current element.

Figure 3.17. Magnetic interaction between moving charge and magnetic field.

generated by charged brass studs mounted in a rotating disk of insulating material.

(2) Force on Moving Charge in Magnetic Field

A magnetic pole m at P (Fig. 3.17) would experience a force F_m due to the motion of a charge q at A, where

$$F_m = mH = \frac{mqv \sin \theta}{cr^2}$$

Thus the reaction experienced by q is given by

$$F_q = -\frac{mqv \sin \theta}{cr^2}$$

Now m/r^2 represents the magnetic field H_0 at A due to m. We can argue that the force on the charge depends only on the value of the magnetic field it experiences, and not at all on the way that field is set up. Thus we put

$$F_q = -\frac{H_0 q v \sin \theta}{c}$$

or in terms of vectors,

$$\mathbf{F}_q = -\frac{q}{c}\mathbf{H}_0 \times \mathbf{v} = \frac{q}{c}\mathbf{v} \times \mathbf{H}_0$$

(3) Radiation by an Accelerated Charge

Suppose that a charge q is at rest at P up to time $t = 0$, and is then given a uniform acceleration a during a time τ. Let the acceleration be

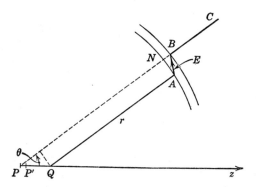

Figure 3.18. Simple picture of a transverse pulse of electric field resulting from acceleration of a point charge.

rapid, so that the charge acquires a certain velocity v in a very short distance PP' (Fig. 3.18). Let us consider the state of the electromagnetic field when q has reached a point Q ($PQ \gg PP'$) at time t. On a naive view we can regard the line $QABC$ as a single line of force:

The portion QA is for $t \geqslant \tau$ ($QA = r \approx ct$).
The portion BC is for $t < 0$.
The portion AB is for the accelerating period τ.

(We tacitly assume that any disturbance caused by the acceleration is propagated with the speed of light c.)

The effects of the acceleration are contained in a shell of thickness

The Atomicity of Electric Charge

$c\tau$. Further, we have

$$v = a\tau$$
$$PQ \approx vt$$

By the geometry of the figure,

$$AN = PQ \sin\theta = vt \sin\theta = a\tau t \sin\theta$$
$$BN = c\tau$$

Thus, if we resolve the electric field E along AB into transverse and radial components E_t, E_r, we have

$$\frac{E_t}{E_r} = \frac{AN}{BN} = \frac{at \sin\theta}{c}$$

But continuity of the component of E measured radially outward from q requires

$$E_r = \frac{q}{r^2}$$

Hence
$$E_t = \frac{q}{r^2} \cdot \frac{at \sin\theta}{c}$$

i.e.,
$$E_t = \frac{qa \sin\theta}{c^2 r}$$

It should be noticed that this transverse field varies as $1/r$, in contrast to the normal radial field which dies away as $1/r^2$. Thus at large distances from the charge only the transverse field will survive.

3.9 ELECTROMAGNETIC MASS

We next wish to mention some purely classical calculations that demonstrate how a moving charge appears to acquire mass as a result of its motion.

(1) Energy in the Field of a Moving Charge

For a charge q of radius a moving with uniform velocity v (Fig. 3.19) the magnetic field at a point P is given by

$$H = \frac{qv \sin\theta}{cr^2}$$

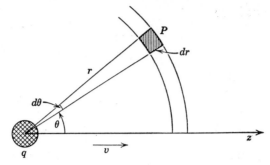

Figure 3.19. Diagram for consideration of magnetic-field energy of a moving charge of finite size.

Therefore energy density of magnetic field at P

$$= \frac{H^2}{8\pi} = \frac{q^2 v^2 \sin^2 \theta}{8\pi c^2 r^4}$$

The volume element between (r, θ) and $(r + dr, \theta + d\theta)$

$$= 2\pi r^2 \sin \theta \, d\theta \, dr$$

The contribution to the total energy of the field is given by

$$dW = \frac{1}{4} \frac{q^2 v^2}{c^2} \cdot \frac{\sin^3 \theta \, d\theta \, dr}{r^2}$$

$\therefore$ Total energy $\quad W = \frac{1}{4} \frac{q^2 v^2}{c^2} \int_{r=a}^{\infty} \frac{dr}{r^2} \int_{\theta=0}^{\pi} \sin^3 \theta \, d\theta$

$$= \frac{q^2 v^2}{3ac^2}$$

Since this energy is proportional to the square of the velocity, we can think of it as the kinetic energy of a certain effective mass m_{eff}, where

$$\tfrac{1}{2} m_{\text{eff}} v^2 = \frac{q^2 v^2}{3ac^2}$$

$$\therefore \quad m_{\text{eff}} = \frac{2}{3} \frac{q^2}{c^2 a}$$

Note that the energy of the *electric* field is

$$\frac{1}{8\pi} \int_{r=a}^{\infty} 4\pi r^2 \, dr \left(\frac{q}{r^2}\right)^2 = \frac{1}{2} \frac{q^2}{a}$$

The Atomicity of Electric Charge 65

(2) Momentum in the Field of a Moving Charge

The starting point of this calculation is the flow of energy in the electromagnetic field as given by the Poynting vector. At any point

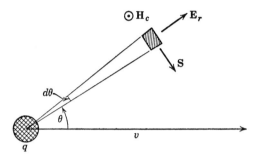

Figure 3.20. Diagram for consideration of momentum of electromagnetic field of a moving charge of finite size.

(Fig. 3.20), we have a radial electric field E_r and a circumferential magnetic field H_c:

$$E_r = \frac{q}{r^2}$$

$$H_c = \frac{qv}{cr^2} \sin \theta$$

Now the Poynting vector **S** is given by

$$\mathbf{S} = \frac{c}{4\pi} \mathbf{E} \times \mathbf{H}$$

Its direction is as shown in the diagram, and we have

$$S = \frac{c}{4\pi} E_r H_c$$

$$= \frac{q^2 v \sin \theta}{4\pi r^4}$$

The momentum density G of the field is given by S/c^2, as we saw earlier (Chapter 2, Section 7). Thus

$$G = \frac{q^2 v \sin \theta}{4\pi c^2 r^4}$$

Evidently, from the symmetry of the problem, the components of **G** perpendicular to the direction of v will cancel out, and we have

$$\text{Total momentum of field} = \int G \sin\theta \cdot 2\pi r^2 \sin\theta \, d\theta \, dr$$

$$\text{Momentum} = \frac{q^2 v}{2c^2} \int_{r=a}^{\infty} \frac{dr}{r^2} \int_{\theta=0}^{\pi} \sin^3\theta \, d\theta$$

$$= \frac{2}{3} \frac{q^2 v}{c^2 a}$$

We see that this can be written as $m_{\text{eff}} \times v$, where m_{eff} is the same effective mass as we obtained from the kinetic energy calculation, and so we have a self-consistent description of the mechanical properties of an electromagnetic field.

3.10 THE ELECTRON RADIUS

The very small value of the electron mass led physicists to suggest that the whole of this mass was electromagnetic in origin. This led to a picture of the electron having a certain radius a_1, such that $m = \frac{2}{3}(e^2/c^2 a_1)$, or $a_1 = \frac{2}{3}(e^2/mc^2)$. Did this picture have any meaning? It certainly appeared so, when the scattering of light by an electron was investigated with the help of classical electromagnetic theory. The problem was treated as follows (J. J. Thomson):

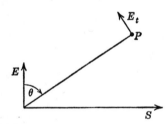

Figure 3.21. Production of a scattered electromagnetic wave.

Consider a plane wave of plane-polarized light falling on an electron, as shown in Fig. 3.21. We have

$$E = E_0 e^{j\omega t}$$

and an energy flux

$$S = \frac{cE^2}{4\pi} \text{ per cm}^2/\text{sec}$$

The electron acceleration is Ee/m. Hence at point P the electric field is

$$E_t(\theta) = \frac{e(Ee/m)\sin\theta}{c^2 r} = \frac{Ee^2 \sin\theta}{mc^2 r}$$

The Atomicity of Electric Charge

This is an oscillating field, propagating with speed c (in vacuo), and so represents an outward flow of energy $S'(\theta)$:

$$S'(\theta) = \frac{c}{4\pi} E_t^2(\theta) \text{ per cm}^2/\text{sec}$$

$\therefore$ Total radiated energy $= \displaystyle\int_{\theta=0}^{\pi} S'(\theta) \cdot 2\pi r^2 \sin\theta \, d\theta$ per sec

$$= \int_{\theta=0}^{\pi} \frac{c}{4\pi} \cdot \frac{E^2 e^4 \sin^2\theta}{m^2 c^4 r^2} 2\pi r^2 \sin\theta \, d\theta$$

$$= \frac{cE^2}{4\pi} 2\pi \left(\frac{e^2}{mc^2}\right)^2 \int_{\theta=\pi}^{0} (1 - \cos^2\theta) \, d(\cos\theta)$$

$$= \frac{cE^2}{4\pi} \cdot \frac{8\pi}{3} \left(\frac{e^2}{mc^2}\right)^2 \text{ per sec}$$

Now if we remember that the incident energy flow is $cE^2/4\pi$ per cm^2 per sec, we see that the electron behaves as though it presents an area A given by

$$A = \frac{8\pi}{3} \left(\frac{e^2}{mc^2}\right)^2$$

and scatters all the radiant energy falling on this area. But, if the electron is regarded as being a sphere of radius a_2, we shall have

$$A = \pi a_2^2$$

Hence $$a_2 = \sqrt{\frac{8}{3}} \cdot \frac{e^2}{mc^2}$$

This is nearly but not quite the same as a_1 deduced from the electron mass, and, without some idea concerning the precise distribution of charge within the electron volume, we cannot be sure about numerical factors of order unity. It was therefore decided to define a so-called "classical electron radius" r_0 through the equation

$$r_0 = \frac{e^2}{mc^2} = 2.82 \cdot 10^{-13} \text{ cm}$$

It may be seen that r_0 lies between a_1 and a_2, and is a convenient unit of length for describing processes in which electrons are concerned.

Problems

3.1. What would be seen, according to classical physics, if a source of light emitting a pure spectral line were placed in a uniform magnetic field and viewed in a direction at 45° to the field? Include a consideration of the state of polarization of the spectral components.

3.2. A spectrometer has a resolving power $(\lambda/\delta\lambda)$ of 10^4. Design an experiment to measure the Zeeman effect with this spectrometer for light of wavelength 10,000 A.

3.3. In one of J. J. Thomson's early experiments to determine e/m for electrons, the electron beam passed through crossed electric and magnetic fields for a distance of 5 cm. The beam was undeflected when the electric field was 100 volts per cm, and the magnetic field was 3.6 gausses. The magnetic field alone produced a deflection of 3°40'. Find e/m and the velocity of the electrons from these data. 2.78×10^8 cm/sec

3.4. The following data were recorded during an oil drop experiment: 3000 volts applied between horizontal plates 1 cm apart. Drop observed over a vertical distance of 0.5 cm. Density of oil = 0.9 g/cm³; Viscosity of air = $1.8 \cdot 10^{-4}$ cgs. Average time of fall of a given drop = 12.5 sec. Successive times of *rise* = 25 sec, 40 sec, 18 sec, 10 sec, 14 sec. Present these results *graphically* to identify the various states of charge and to find a best value of e.

3.5. (a) The density of copper is 8.9 g per cm³, and the atomic weight is 63.5. What electron drift velocity is needed to carry a current density of 1000 amperes/cm², assuming that there is one free conduction electron per atom?

(b) What would be the root-mean-square agitation velocity of these conduction electrons, assuming them to be in thermal equilibrium with the mass of the copper at 27° C?

(c) Taking the result of (b) above, and with the help of Table 3.1, calculate the mean free path of the conduction electrons from the values of σ and K. (Remember that, with σ, one must be consistent in converting both σ and e into esu or emu.)

(d) Compare this mean free path with the approximate spacing between copper ions in the lattice, using the data in (a) above. (The free path so obtained is much shorter than the true value; the discrepancy represents a failure of this classical approach.)

3.6. An electron is caused to move in a circular path by application of a magnetic field of 10^4 gausses. If the radius of the path is 0.1 mm, what is the total rate of radiation of energy? What would be the initial rate of collapse of the orbit radius if no energy were supplied to maintain the motion?

4 | Thermal Radiation and the Quantum Theory

4.1 THERMAL RADIATION

The study of radiation is usually introduced as part of a course in heat, and this is justified to the extent that thermodynamics can tell us a good deal about its general features. But the attempt to describe radiation phenomena in detail took physicists into electromagnetic theory, and finally uncovered a fundamental breakdown of classical physics, from which the quantum theory was born.

By radiation we understand the transfer of energy from one body to another in the absence of anything that we can recognize as a material medium connecting them. The essential features of radiation are precisely as we have listed them for the special case of visible light, for, as we have seen, light is merely a very tiny part of an immense range of radiations of the kind we call electromagnetic. A given radiation is characterized by its wavelength λ, and the approximate extremes known to us at present are

(a) Very "hard" gamma rays: $\lambda \approx 10^{-12}$ cm
(b) Long radio waves: $\lambda \approx 10^{+6}$ cm

Visible light, with $\lambda \approx 5 \cdot 10^{-5}$ cm, thus lies somewhere near the middle of the range on a logarithmic scale. The whole range is covered, without any breaks, by radiations known to us.

When a given body is examined, it is found in general to be emitting radiations of various wavelengths. If it emits a quantity dE_λ per second of radiation energy with wavelengths between λ and $\lambda + d\lambda$, then a plot of $dE_\lambda/d\lambda$ against λ is said to represent the *spectrum* of the radiation. It is a matter of common knowledge that the spectrum for a given body depends on its temperature. As it is heated, it emits first heat, and then heat and red light, and then it becomes brighter and brighter, passing through white toward blue. These are the qualitative features; we must now discuss the problem in more detail.

4.2 PRÉVOST'S THEORY OF EXCHANGES

At one time people used to speak of hot and cold radiations. A block of ice, for example, was thought to radiate cold to any body placed beside it, and this idea is certainly in keeping with one's subjective impressions. But, if we examine this idea, we see that it is logically unsound. What we in fact observe is a tendency for two bodies at different temperatures to come to the same temperature, even though there is a vacuum between them. A block of ice appears to radiate cold to a human being, but it would appear to radiate heat to a block of Dry Ice (solid CO_2) at a temperature of $-78°$ C. All that we can truly observe is an exchange of energy by radiation until two bodies reach the same temperature, after which nothing seems to happen. We could assume that the transfer of radiant energy ceases altogether when the bodies have reached the same temperature. We do not like to accept this idea, however, for it implies that the bodies, although entirely separated from each other physically, know when they ought to stop radiating, through an awareness of each other's temperatures.

To make the situation logically satisfactory, Prévost in 1792 put forward the theory of exchanges. This asserts that a body emits radiant energy at all temperatures, the amount increasing with the temperature. What is observed is simply the difference between what it emits and what it receives. When a body has settled down to a constant temperature, it has not stopped radiating, but receives from its surroundings just as much energy as it emits in a given time. In other words the equilibrium is dynamic, not static.

4.3 EMISSIVE AND ABSORPTIVE POWERS

The well-known experiment of Leslie's cube shows that the rate of radiation of energy from a body at a given temperature depends on the nature of its surface. A very simple and important experiment by

Thermal Radiation and the Quantum Theory

Ritchie (1833), also well known, showed that there is an essential connection between the abilities of a given kind of surface to radiate and to absorb. The experiment (see Fig. 4.1) consisted in showing that no temperature difference developed between A' and B' when exposed to the radiations from B and A, respectively.

Let us define an *emissive power* e for a surface as the ergs per square centimeter per second emitted by it as radiant energy. Let us also define the *absorption coefficient* a of the surface as the fraction of incident energy falling on it that is absorbed. If A,A' are described by α, and B,B' by β, then Ritchie's experiment shows

$$e_\alpha a_\beta = e_\beta a_\alpha$$

$$\therefore \quad \frac{e_\alpha}{a_\alpha} = \frac{e_\beta}{a_\beta}$$

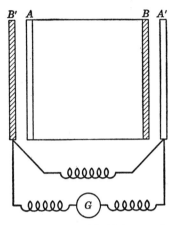

Figure 4.1. Ritchie's experiment to show that good emitters are good absorbers. A,A' have one type of surface (e.g. polished), and B,B' a different type (e.g. blackened).

We shall see that this result comes from a thermodynamic argument and can be extended to deal with radiations at particular wavelengths.

We can imagine a surface for which $a = 1$. A body with a fully absorbing surface of this kind is called a *black body*. All other surfaces have $a < 1$. Let us denote the emissive power of a black body by e_B. Then we have

$$\frac{e_B}{1} = \frac{e_\alpha}{a_\alpha}$$

or $e_B > e_\alpha$; i.e., the surface of a black body emits more radiation per square centimeter per second than any other surface at the same temperature.

4.4 UNIFORM-TEMPERATURE ENCLOSURES

As a first step toward considering the thermodynamics of radiation, we introduce the idea of a constant-temperature enclosure. This is defined as an enclosure whose walls are impervious to radiation and are maintained at a uniform temperature. Then we can show that both the quantity (energy density) and quality (spectrum) of the radiation within the enclosure depend only on its temperature, and are entirely independent of its size and shape, of the nature of the walls, and of the nature and shape of any bodies placed inside it.

Suppose two enclosures, A and B (Fig. 4.2), which are at the same temperature but may have arbitrary sizes, shapes and surfaces. We assume, however, that the enclosures are connected by a tube in which is a shutter C impervious to radiation. If C is replaced by a window C' transparent to all radiation, then there will be a net transfer of radiation from A to B if A initially has a greater radiation energy density than B. After a time replace C' by C; then the enclosures settle down to new equilibrium conditions. A is at a higher temperature than that corresponding to the radiation density in it; so it will cool down, giving radiant energy to the cavity, until equilibrium is reached. Correspondingly, B will rise in temperature. But this means that we have effected a violation of the second law of thermodynamics, for the temperature difference so produced could be used to drive a heat engine, thereby carrying out a "transformation whose only final result is to transform into work heat extracted from a source which is (initially) at the same temperature throughout." (Kelvin's statement of the second law.)

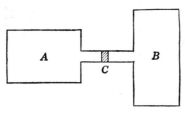

Figure 4.2. Equilibrium of two uniform temperature enclosures.

If we go through the same argument, but postulating C' to be a filter that passes only a narrow band of wavelengths, we can prove that the energy density per unit wavelength interval at any wavelength in a uniform-temperature enclosure is a function only of temperature.

The same type of consideration holds for a body placed inside a uniform-temperature enclosure. The thermodynamic argument shows that any body placed in such an enclosure must take up the temperature of the enclosure and remain in equilibrium. The theory of exchanges tells us that the body then emits just as much energy as it absorbs, thus leaving the radiation density in the enclosure unaffected.

4.5 KIRCHHOFF'S LAW. CAVITY RADIATION

Let us now suppose that we place a body in an enclosure of temperature T. Then the amount of radiation falling on the body per square centimeter per second depends only on T. Let the amount of energy per square centimeter per second between λ and $\lambda + d\lambda$ be dQ_λ. Then the amount *absorbed* per second $= a_\lambda \, dQ_\lambda$, where a_λ is a new absorption coefficient defined at a particular wavelength. If now we introduce a new emissive power e_λ, such that the emitted energy per square centimeter per second between λ and $\lambda + d\lambda$ is $e_\lambda \, d\lambda$, we have

$$dQ_\lambda = (1 - a_\lambda) \, dQ_\lambda + e_\lambda \, d\lambda$$

$$\therefore \quad \frac{e_\lambda}{a_\lambda} = \frac{dQ_\lambda}{d\lambda} = E_\lambda = f(\lambda, T)$$

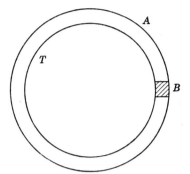

Figure 4.3. To illustrate the principle for constructing a source of black-body radiation. If plug B is removed, the emergent radiation will approximate to the cavity radiation at temperature T.

where E_λ is the emissive power at temperature T and wavelength λ for a black body, and depends on λ and T only. The above result is Kirchhoff's law, and it explains such phenomena as the Fraunhofer lines, since it implies that high emission and high absorption go hand in hand.

Let us suppose that we make a constant-temperature enclosure (Fig. 4.3), whose walls A are of quite arbitrary material except for a small perfectly black section B. Then B absorbs all the radiation falling on it. Now let us remove B altogether. The hole left in its place continues to act like a black body so far as absorption is concerned. The only change is the loss to the enclosure of the radiation from B. We can reduce this loss as much as we please by making the hole small enough. Thus the radiation intensity passing out of the hole can be made equal to $dQ_\lambda/d\lambda$, which as we have seen is the emissive power of a perfectly black body. We thus have a practical means of constructing a black body from ordinary materials that are not black for all wavelengths.

The perfect blackness of a small hole in a large enclosure can also be seen by considering the reverse process of radiation entering the hole from outside. Its chance of escaping again is very small if it is forced to undergo multiple reflections and scatterings before arriving once again at the aperture. Because of the way in which it can be produced, black-body radiation is also given the name *cavity radiation*.

4.6 TOTAL RADIATION AND TEMPERATURE

Measurements of the rate of cooling of a body as a function of temperature were made by Dulong and Petit (1817) and by Tyndall (1865).

It was found that the results could be expressed in the form

$$-\frac{d\theta}{dt} = f(\theta) - f(\theta_0)$$

where θ is the temperature of the body and θ_0 the temperature of the surroundings. This supports Prévost's theory of exchanges (Section 2 of this chapter) by its analytical form. In 1879 Stefan showed that $f(\theta)$ was proportional to the fourth power of the absolute temperature, and his findings, later justified theoretically by Boltzmann (1884)

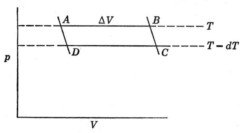

Figure 4.4. A Carnot cycle with cavity radiation as the working substance.

were expressed in the Stefan–Boltzmann law: If a black body at absolute temperature T is surrounded by another black body at temperature T_0, the amount of energy E lost per second per square centimeter of the former is given by

$$E = \sigma(T^4 - T_0^4)$$

where σ is Stefan's constant.

Stefan's law can be established by thermodynamical arguments. We have seen that the energy density of radiation in a uniform-temperature enclosure depends only on the temperature. Let us denote it by $u(T)$. We have also seen that a unidirectional plane wave, whose Poynting vector is **S**, exerts a pressure given by

$$p = \frac{S}{c} = \frac{(E^2)_{\text{avg}}}{4\pi} = W$$

where W is the energy per cubic centimeter in the electromagnetic field. For an enclosure containing radiation moving in all directions, we can say that, on the average, only one third of the radiation is moving normal to a given element of the containing vessel. Hence in this case we have

$$p = \tfrac{1}{3}u$$

We now imagine a cylinder with a movable piston used as a heat engine, but having as its working substance the thermal radiation, not a gas. In the cycle $ABCD$ (Fig. 4.4), we have

Net work done *by* working substance

$$= \Delta V \cdot p(T) - \Delta V \cdot p(T - dT)$$
$$= \tfrac{1}{3} du \cdot \Delta V$$

Net heat put *into* working substance along AB

$$= \Delta(uV) + \Delta V \cdot p(T)$$
$$= u \cdot \Delta V + \Delta V \cdot \tfrac{1}{3}u$$
$$= \tfrac{4}{3} u \cdot \Delta V$$

Hence, regarding this as a reversible heat engine, we have its efficiency given as

$$\frac{dT}{T} = \frac{\tfrac{1}{3} du \cdot \Delta V}{\tfrac{4}{3} u \cdot \Delta V} = \frac{1}{4} \frac{du}{u}$$

$$\therefore \quad u(T) = aT^4$$

where a is a constant. If we consider the efflux of energy from a hole in an enclosure containing such radiation, we have (by analogy with kinetic theory*) that the energy emitted per square centimeter per second is given by

$$E(T) = \frac{c}{4} u(T) = \frac{ac}{4} T^4$$

This is the Stefan–Boltzmann law if we identify $ac/4$ with σ.

Various direct experimental determinations have led to the result

$$\sigma = 5.670 \cdot 10^{-5} \text{ ergs/sec/cm}^2/°K^4$$

Thus, for example, 1 cm^2 of a perfectly black body at a temperature of 2000° K would emit radiant energy at a rate given by

$$P \approx 6 \cdot 10^{-5} \times 16 \cdot 10^{12} \text{ ergs/cm}^2/\text{sec} \approx 100 \text{ watts}$$

4.7 THE BLACK-BODY SPECTRUM

If the radiation from a black body is focused on a spectrometer slit, analyzed by a prism of fluorite or quartz, and allowed to fall on a sensitive detector which can be moved across the spectrum, a distribu-

---

* See Problem 6 of Chapter 1.

tion curve with a well-defined shape is obtained. It has a single maximum, occurring at some wavelength λ_m. If measurements of the spectrum are made for a series of different temperatures, and if the detector response (which will be proportional to e_λ) is plotted against λ for each temperature, then it is observed (see Fig. 4.5) (a) that the

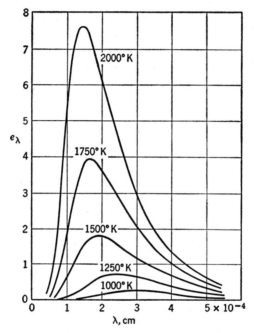

Figure 4.5. The black-body spectrum at several temperatures.

curves get rapidly higher as T increases, and in accordance, in fact, with the Stefan–Boltzmann law, i.e.,

$$\int_0^\infty e_\lambda \, d\lambda = \sigma T^4$$

(b) that the position of the maximum of the curve shifts toward shorter wavelengths as T increases. This is already obvious to us if we remember how a hot wire, for example, passes from red toward blue heat as its temperature is raised. The exact way in which the maximum shifts is very simple; viz.,

$$\lambda_m T = \text{const}$$

The experimental results by which this formula was confirmed were obtained by Lummer and Pringsheim (1899), but the formula itself

Thermal Radiation and the Quantum Theory

had been discovered by Wien (1893) as one result of a theoretical analysis.

Wien considered a reversible adiabatic expansion of a cavity containing radiation. Such an expansion reduces the total energy of the cavity (because external work has been done) and also shifts its radiation spectrum toward longer wavelengths. If we imagine the radiation (in equilibrium with its enclosure) as constituted of standing waves, it is plausible that the increase of linear dimension l of the cavity should be accompanied by a "stretching" of the radiation wavelengths λ in such a way that the general pattern of standing waves in the cavity remains unaffected. This will be true if we put $\lambda_1/\lambda_2 = l_1/l_2$. Corresponding wavelength *intervals* are similarly defined by $d\lambda_1/d\lambda_2 = l_1/l_2$. That this relation between corresponding wavelengths does in fact hold good was shown by Wien to follow from the Doppler shift for reflection of the radiation at the walls of the expanding enclosure. The amount of the shift is shown to remain the same, even when the expansion is made to take place infinitely slowly. It thus becomes possible to discuss the expansion as a reversible thermodynamic process, in which the temperature of the radiation falls from T_1 to T_2.

This time we restrict attention to the radiation within a small wavelength range. We introduce the energy density *per unit wavelength* $u(\lambda)$. Then we are interested in an amount of energy $u(\lambda_1)\,d\lambda_1$ in the initial state and the corresponding (but not equal) amount $u(\lambda_2)\,d\lambda_2$ in the final state. We first apply to our small band of wavelengths the argument that led to Stefan's law for the total radiation:

$$T_2{}^4 u(\lambda_1)\,d\lambda_1 = T_1{}^4 u(\lambda_2)\,d\lambda_2$$

We next write down the equation for the adiabatic process:

$$d[V\,u(\lambda)\,d\lambda] + p(\lambda)\,dV = 0$$

Here $p(\lambda)$ represents that part of the pressure due to the wavelength we are considering; i.e.,

$$p(\lambda) = \tfrac{1}{3} u(\lambda)\,d\lambda$$

Thus
$$d[3V\,p(\lambda)] + p(\lambda)\,dV = 0$$

$$\therefore\ 3V\,dp(\lambda) + 4p(\lambda)\,dV = 0$$

and so
$$p(\lambda_1) V_1{}^{4/3} = p(\lambda_2) V_2{}^{4/3}$$

describes this adiabatic expansion. Rewriting this in terms of energy

densities and linear dimensions, we have
$$u(\lambda_1)\, d\lambda_1 \cdot l_1^4 = u(\lambda_2)\, d\lambda_2 \cdot l_2^4$$
i.e.,
$$u(\lambda_1) \cdot \lambda_1^5 = u(\lambda_2) \cdot \lambda_2^5 \tag{4.1}$$

Now Stefan's law can be rewritten as
$$p(\lambda_1) T_2^4 = p(\lambda_2) T_1^4$$

Eliminating p between this and the equation to the adiabatic, we have
$$T_1^4 V_1^{4/3} = T_2^4 V_2^{4/3}$$
i.e.,
$$T_1 l_1 = T_2 l_2$$
or
$$T_1 \lambda_1 = T_2 \lambda_2 \tag{4.2}$$

We require equations (4.1) and (4.2) to be satisfied simultaneously, and, if this is so, we may put
$$\frac{u(\lambda_1) \cdot \lambda_1^5}{f(\lambda_1 T_1)} = \frac{u(\lambda_2) \cdot \lambda_2^5}{f(\lambda_2 T_2)} = \text{const}$$

where $f(\lambda T)$ is an unspecified function of λT. It follows, then, from thermodynamic reasoning alone, that we shall have
$$u(\lambda) = \frac{A}{\lambda^5} f(\lambda T)$$

where A is a constant. It will appear later that it is often more convenient to discuss the spectrum in terms of frequency ν than in terms of wavelength λ. We therefore introduce the energy per unit frequency interval, $u(\nu)$. By definition,
$$u(\nu)\, d\nu \equiv u(\lambda)\, d\lambda$$
But
$$\lambda = \frac{c}{\nu}$$
$$\therefore\ |d\lambda| = c\, \frac{d\nu}{\nu^2}$$
$$\therefore\ u(\nu) = B\nu^3 g\!\left(\frac{\nu}{T}\right)$$

where B is another constant and $g(\nu/T) \equiv f(\lambda T)$.

Wien's displacement law follows as a particular consequence of our expression for $u(\lambda)$. For suppose that we measure the black-body

Thermal Radiation and the Quantum Theory

spectrum at two different temperatures, T_1 and T_2. Then we have

$$u_1(\lambda) = \frac{A}{\lambda^5} f(\lambda T_1)$$

$$u_2(\lambda) = \frac{A}{\lambda^5} f(\lambda T_2)$$

Let us choose not λ but $x = \lambda T$ as the independent variable. Then

$$u_1(x_1) = \frac{A T_1^5}{x_1^5} f(x_1)$$

$$u_2(x_2) = \frac{A T_2^5}{x_2^5} f(x_2)$$

Let us further choose not u but $y = u/T^5$ as the dependent variable. Then

$$y_1(x_1) = \frac{A}{x_1^5} f(x_1)$$

$$y_2(x_2) = \frac{A}{x_2^5} f(x_2)$$

We have only to choose $x_2 = x_1$ to make $y_2 = y_1$, whatever the function f may be. This means that a plot of $u(\lambda)/T^5$ against λT will be a single curve (Fig. 4.6) and will represent the black-body spectrum at all temperatures. If the curve has a maximum, this will occur at some value x_m which is the same for all T. Hence we have $\lambda_m T = $ constant,

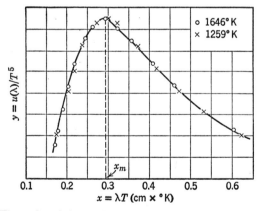

Figure 4.6. The reduced form of the cavity radiation spectrum, representing energy density versus wavelength for all temperatures. The experimental points are from Lummer and Pringsheim (1899). The peak occurs for $\lambda T \approx 0.29$ cm $\times$ °K.

which is the displacement law. It may be seen from the figure that the agreement between theory and experiment is excellent.

We have now gone as far as it is possible to go in describing the spectrum of black-body radiation without recourse to some definite picture, or model, of the radiation process. Our next step is to consider the first significant attempt that was made to explain the spectrum in detail.

4.8 OSCILLATORS AND RADIATION FIELDS

With the development of Maxwell's electromagnetic theory, and the discovery of the electrical structure of neutral matter, it was natural to describe thermal radiation in terms of the generation of electromagnetic waves by atomic oscillators. For the problem of cavity radiation it was necessary to consider the equilibrium between these oscillators and the radiation produced by them at some specified temperature T. Let us therefore consider the emission and absorption of radiation by an electron elastically bound to a center of force. This resembles the problem of optical dispersion as we have already treated it (Chapter 3, Section 7), but now we shall also consider a small viscous damping term in the equation of motion.

(1) Emission

For free oscillations, leading to spontaneous emission of radiation, we have
$$m\ddot{x} + m \cdot 2k\dot{x} + m\omega_0^2 x = 0$$

Let us put $\quad x = Re(z) \quad$ where $\quad z = x_0 e^{j\omega t}$

Then $\quad (\omega_0^2 - \omega^2)z + j \cdot 2kz\omega = 0$

Putting $\quad \omega = n + js$

we find $\quad s = k$

$$n = \sqrt{\omega_0^2 - k^2} \approx \omega_0 \quad \text{if } k \text{ is small}$$

Hence $\quad x \approx x_0 e^{-kt} \cos \omega_0 t$

or, to take a further approximation for $k \ll \omega_0$,
$$x \approx x_0 \cos \omega_0 t$$
$$\therefore \ddot{x} = -\omega_0^2 x_0 \cos \omega_0 t$$

From the theory of radiation fields due to accelerated charges [Chapter

Thermal Radiation and the Quantum Theory

3, Section 8(3)], we then find

$$E_t(r, \theta) = \frac{e\omega_0^2 x_0 \cos \omega_0 t}{c^2 r} \sin \theta$$

Radiated energy per sec $= \dfrac{c}{4\pi} \displaystyle\int E_t^2 \cdot 2\pi r^2 \sin \theta \, d\theta$

$$= \frac{c}{4\pi} \cdot \frac{2\pi e^2 \omega_0^4 x_0^2 \cos^2 \omega_0 t}{c^4} \int_{-1}^{+1} (1 - \cos^2 \theta) \, d(\cos \theta)$$

$$= \frac{2}{3} \frac{e^2 \omega_0^4 x_0^2 \cos^2 \omega_0 t}{c^3}$$

∴ Mean rate of radiating energy $= \dfrac{e^2 \cdot \omega_0^4 x_0^2}{3c^3}$ since $\overline{\cos^2 \omega_0 t} = \tfrac{1}{2}$

The total energy of the oscillating system is given by

$$\varepsilon = \tfrac{1}{2} m \omega_0^2 x_0^2 \quad \text{at all times}$$

∴ Rate of radiation $= \dfrac{2}{3} \cdot \dfrac{e^2 \omega_0^2}{mc^3} \varepsilon$ \hfill (4.3)

(2) Absorption

We now consider the same oscillator exposed to a plane wave of radiation, with an electric field given by

$$E = E_0 \cos \omega t$$

This time we put $x = \mathrm{Re}(z)$

where

$$z = x_0 e^{j(\omega t - \delta)}$$

and

$$m\ddot{z} + m \cdot 2k\dot{z} + m\omega_0^2 z = eE_0 e^{j\omega t}$$

i.e.,

$$(\omega_0^2 - \omega^2)x_0 + j \cdot 2k\omega x_0 = \frac{eE_0}{m} e^{j\delta}$$

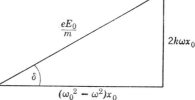

Figure 4.7. The vector diagram for forced vibrations of an elastically bound charge with damping.

This can be exhibited as a vector sum (see Fig. 4.7).

We thus see that the displacement maintains a certain definite phase relationship with the driving field. Completing the solution of the equation, we have

$$x = \frac{eE_0/m}{[(\omega_0^2 - \omega^2)^2 + (2k\omega)^2]^{1/2}} \cos(\omega t - \delta)$$

where $\quad \tan \delta = \dfrac{2k\omega}{\omega_0^2 - \omega^2}$

This can be rewritten in the form

$$x = A_1 \cos \omega t + A_2 \sin \omega t$$

Now the power absorbed by the oscillator is the mean rate at which work is done on it, and for the work in time dt we have

$$dW = F\,dx = eE_0 \cos \omega t (-A_1 \omega \sin \omega t + A_2 \omega \cos \omega t)\,dt$$

$$\therefore \frac{dW}{dt} = -eE_0 A_1 \omega \sin \omega t \cos \omega t + eE_0 A_2 \omega \cos^2 \omega t$$

When we come to evaluate the average of dW/dt over a cycle, the term containing A_1 contributes nothing. (Note however that A_1 is responsible for the anomalous dispersion curve, Chapter 3, Section 7.) We thus have

$$\text{Mean power absorbed} = eE_0 A_2 \omega \,\overline{\cos^2 \omega t} = \tfrac{1}{2} eE_0 A_2 \omega$$

But $\quad A_2 = \dfrac{eE_0/m}{[(\omega_0^2 - \omega^2)^2 + (2k\omega)^2]^{1/2}} \sin \delta$

$$= \frac{eE_0}{m} \cdot \frac{2k\omega}{(\omega_0^2 - \omega^2)^2 + (2k\omega)^2}$$

$$\therefore \text{Mean power} = \frac{ke^2 E_0^2}{m} \cdot \frac{\omega^2}{(\omega_0^2 - \omega^2)^2 + (2k\omega)^2}$$

This would represent the absorption of energy by the oscillator if the incident radiation were of a pure frequency ω. We are, however, concerned with a continuous spectrum of radiation. To take account of this we note that $E_0^2/8\pi$ ($= \overline{E^2}/4\pi$) would represent the energy per cubic centimeter of the field of the plane wave if it were monochromatic, and would be equivalent to one third of the total energy of the field, allowing all possible directions of propagation. But, if we consider an energy spectrum $u(\nu)$ ($\nu = \omega/2\pi$) we say that the energy in frequency range $d\nu$ is $u(\nu)\,d\nu$. In our formula for the power absorbed we therefore make the substitution

$$\frac{E_0^2}{8\pi} \to \tfrac{1}{3} u(\nu)\,d\nu$$

Thermal Radiation and the Quantum Theory

Thus, for the energy absorbed per second in frequency range $d\nu$ we have

$$dP(\nu) = \frac{ke^2}{m} \cdot \frac{8\pi}{3} u(\nu) \, d\nu \, \frac{\omega^2}{(\omega_0^2 - \omega^2)^2 + (2k\omega)^2}$$

or

$$dP(\omega) = \frac{4ke^2}{3m} \cdot \frac{u(\nu)\omega^2 \, d\omega}{(\omega_0^2 - \omega^2)^2 + (2k\omega)^2}$$

This represents the characteristic resonance absorption by an oscillator (see Fig. 4.8 and Appendix III). If we suppose $k \ll \omega_0$, as is typical for the very sharp absorption lines found in spectroscopy, we

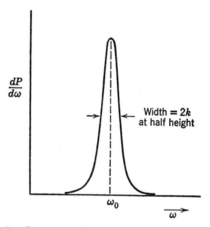

Figure 4.8. Resonance absorption by a damped oscillator.

see that nearly all the absorption takes place in the immediate neighborhood of ω_0. This enables us to simplify the evaluation of the total power absorbed by putting

$$\int_0^\infty dP(\omega) \approx \frac{2ke^2}{3m} u(\nu_0)\omega_0 \int_0^\infty \frac{d(\omega^2)}{(\omega_0^2 - \omega^2)^2 + (2k\omega_0)^2}$$

$$\approx \frac{2ke^2}{3m} u(\nu_0)\omega_0 \int_{-\infty}^\infty \frac{d(\omega^2 - \omega_0^2)}{(\omega^2 - \omega_0^2)^2 + (2k\omega_0)^2}$$

$$= \frac{2ke^2}{3m} u(\nu_0)\omega_0 \frac{\pi}{2k\omega_0}$$

$$\therefore \text{Total power absorbed} = \frac{\pi e^2}{3m} u(\nu_0) \qquad (4.4)$$

(3) Equilibrium

It remains only to equate the emitted and absorbed powers (equations 4.3 and 4.4). This gives the identity

$$\frac{2}{3}\frac{e^2\omega_0^2}{mc^3}\varepsilon = \frac{\pi e^2}{3m}u(\nu_0)$$

and so
$$u(\nu_0) = \frac{2\omega_0^2}{\pi c^3}\varepsilon = \frac{8\pi\nu_0^2}{c^3}\varepsilon \qquad (4.5)$$

We proceed to use this result to deduce the theoretical black-body spectrum.

4.9 THE RAYLEIGH—JEANS LAW

We now have an expression for the energy density of radiation in dynamic equilibrium with a typical oscillator. The walls of our uniform-temperature enclosure are, however, assumed to contain large numbers of oscillators, the connection between them being the condition that the temperature of the enclosure has a certain value T. It is not required that the oscillators should all possess the same energy, but simply that they should be in thermal equilibrium with each other; this is essentially the same problem as in the distribution of energy among the molecules of a gas at a given temperature. We must, in fact, apply the principles of *statistical mechanics*, which governs the behavior of systems of many particles, and of which the kinetic theory of gases is a good example.

Now we have seen that the kinetic theory of gases leads us to the result

$$p = \tfrac{1}{3}nmc^2 = \rho\frac{RT}{M}$$

whence
$$\tfrac{1}{2}mc^2 = \frac{3}{2}\frac{RT}{N} = \tfrac{3}{2}kT$$

Here $\tfrac{1}{2}mc^2$ is the mean value of the total kinetic energy of a molecule. We can, however, analyze this into three equal parts, corresponding to the mutually independent components of the motion along three axes at right angles. Thus

$$(\tfrac{1}{2}mv_x^2)_{\text{avg}} = (\tfrac{1}{2}mv_y^2)_{\text{avg}} = (\tfrac{1}{2}mv_z^2)_{\text{avg}} = \tfrac{1}{2}kT$$

Thermal Radiation and the Quantum Theory 85

We speak of these independent modes of motion as *degrees of freedom*, and the kinetic theory then tells us that the mean energy per degree of freedom in this translational motion is $\frac{1}{2}kT$. Now, unless molecules are geometrical points, we can expect that they will be able to take up energy in internal motions—rotation and vibration—as well as in translation, and it appeared in the development of statistical mechanics by Boltzmann and Maxwell that there was a far-reaching principle governing the distribution of energy among the various modes of motion. This was the principle of equipartition of energy and stated

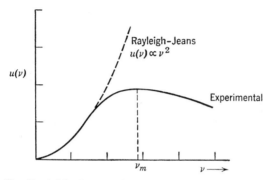

Figure 4.9. The Rayleigh–Jeans radiation spectrum (showing the "ultraviolet catastrophe") with an experimental spectrum for comparison.

that in thermal equilibrium a mean energy $\frac{1}{2}kT$ was associated with each distinct degree of freedom.

We can try to apply this principle to our assembly of oscillators. They are assumed to be fixed in position, and so to have no translational degrees of freedom. As linear oscillators, however, they have two degrees of vibrational freedom—one for the kinetic energy, and the other for the potential energy. The occurrence of two degrees of freedom, rather than one, may be seen from the fact that the velocity $\dot{x}$ of the oscillating particle may be given quite independently of the displacement x at the same instant. The two quantities together determine the amplitude and the total energy of the motion. Thus the average total energy $\bar{\varepsilon}$ of an oscillator at temperature T is kT. We therefore conclude from equation (4.5), Section 8 of this chapter, that the energy density of black-body radiation should be given by the formula

$$u(\nu) = \frac{8\pi\nu^2}{c^3} kT$$

This result was obtained by Rayleigh (1900) and Jeans (1905), and was found to provide a good fit to the experimental results in the region of long wavelengths (see Fig. 4.9). It is, however, open to a fatal objection, an objection that has become famous as the "ultra-violet catastrophe." As the wavelength λ tends to zero, and so ν tends to infinity, the associated energy density tends to infinity; moreover, the energy density integrated over all frequencies becomes infinite at any temperature. This is in total disagreement with experiment which plainly shows that $u(\nu)$ passes through a maximum at some frequency ν_m and falls steadily for all further increase of ν. To show that this is a fundamental breakdown of classical physics, and not merely a consequence of our particular model, we shall obtain this

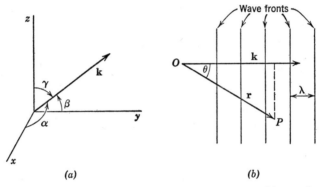

Figure 4.10. (a) The wave vector **k** of a standing wave, making angles (α, β, γ) with the co-ordinate axes. (b) A view parallel to the wave fronts. The phase of the wave at P relative to O is $2\pi(r \cos \theta)/\lambda = 2\pi(\mathbf{k}\cdot\mathbf{r})$.

same result by another method—in fact by the method that was originally used by Rayleigh and Jeans. This consists in studying the pattern of standing waves set up in an enclosure.

For simplicity we consider a cube, although this is not necessary to the result. For a vibration of any given wavelength to fit into the cube as a stationary wave it must have nodes at the walls.

If we were dealing with a one-dimensional problem, the equation to a plane wave could be written

$$\psi = A \sin(2\pi k x)$$

where k ($=1/\lambda$) is the wave number. In the three-dimensional problem, k is replaced by a corresponding vector **k** whose direction (normal to the wave front) can be characterized by the angles α, β, γ made between **k** and the $x, y,$ and z axes, respectively (see Fig. 4.10a).

Thermal Radiation and the Quantum Theory

The equation to the wave thus becomes

$$\psi = A \sin 2\pi (\mathbf{k} \cdot \mathbf{r})$$
$$= A \sin 2\pi (k_x x + k_y y + k_z z)$$
$$= A \sin 2\pi (kx \cos \alpha + ky \cos \beta + kz \cos \gamma)$$

To have the desired nodes we then require that ψ should be zero whenever x, y, or z is equal to 0 or a, where a is the length of a cube edge. These conditions are all satisfied if we put

$$2\pi ka \cos \alpha = n_1 \pi$$
$$2\pi ka \cos \beta = n_2 \pi$$
$$2\pi ka \cos \gamma = n_3 \pi$$

where (n_1, n_2, n_3) is a set of positive integers. Squaring and adding these equations, we have (since $\cos^2 \alpha + \cos^2 \beta + \cos^2 \gamma = 1$)

$$n_1^2 + n_2^2 + n_3^2 = 4k^2 a^2 = \frac{4a^2}{\lambda^2} = \frac{4a^2 \nu^2}{c^2}$$

Each possible set (n_1, n_2, n_3) defines one possible stationary vibration inside the enclosure, and we can depict this by constructing a three-dimensional plot in which the integers are used as the three co-ordinates, fixing the position of a point by which the particular vibration may be identified. All admissible vibrations will then be represented by separate points on a cubic lattice, and the total number of such vibrations in the range of frequencies between zero and ν may be found by counting the lattice points contained in the positive octant of a sphere of radius $2a\nu/c$. Since there is one such lattice point per unit volume, this procedure is almost equivalent to finding the volume of the octant—not exactly so, in general, because the number of vibrations must, by definition, be an integer, whereas the volume of the octant is a continuous function of ν. It is, however, plain that, as ν becomes large, the number of different stationary vibrations $N(\nu)$ between 0 and ν approximates more and more closely to the formula

$$N(\nu) = \frac{1}{8} \cdot \frac{4\pi}{3} \left(\frac{2a\nu}{c} \right)^3 = \frac{4\pi a^3}{3c^3} \nu^3$$

The number of vibrations in a range $d\nu$ at ν is thus given by

$$dN(\nu) = \frac{4\pi a^3}{c^3} \nu^2 \, d\nu$$

Now a^3 is the volume V of the cube. Thus the number of vibrations per cubic centimeter is simply

$$\frac{1}{V} dN(\nu) = \frac{4\pi}{c^3} \nu^2 \, d\nu$$

This result would be the final one if our vibrations were longitudinal (i.e. sound waves). But we are concerned with transverse electromagnetic vibrations, and in this case we need to specify not only the wave propagation vector **k** but also the state of polarization. To describe the polarization we need to combine two vibrations at right angles, of identical frequency, but with no necessary relationship in amplitude or phase. Thus to each value of **k** there are in fact *two* independent transverse vibrations. If to each such vibration we ascribe a mean energy kT, according to the principle of equipartition, we have the energy density within the interval $d\nu$ given by

$$u(\nu) \, d\nu = 2 \cdot \frac{4\pi}{c^3} \nu^2 \, d\nu \cdot kT$$

i.e., $$u(\nu) = \frac{8\pi \nu^2}{c^3} kT$$

exactly as in the other treatment. We see, therefore, that it does not matter whether the classical equipartition of energy is applied to the oscillators in the walls of the cavity, or to the vibrations of the radiation itself; we arrive at the same impossible result. Our next step is to see how the difficulty was resolved, and for this purpose we must examine more closely the principles of statistical mechanics.

4.10 BOLTZMANN'S DISTRIBUTION

The basic problem of the kinetic theory of gases is to discover how a certain fixed amount of energy is to be shared among a certain fixed number of molecules, so as to fulfill the condition that the mean energy per molecule has a well-defined value. The same type of problem appears to be represented by the oscillators generating cavity radiation or by the characteristic vibrations of the cavity itself. To see how such problems are approached in statistical mechanics, consider a very simple situation. Suppose we have four molecules of a total energy $4E$, and suppose that the energy of each one is restricted to the possible values $0, E, 2E, 3E, 4E$, but that the distribution of energy amongst the molecules is otherwise a matter of pure chance. Then we can work out the possible groupings, and we find that we are limited to those shown in Fig. 4.11. In any one of these groupings we shall have a new and distinct arrangement of the molecules if molecules at different levels are interchanged, but no recognizably different system if molecules at the same level are permuted in any way. We can thus set up Table 4.1 showing the number of different arrangements belonging to

TABLE 4.1

Type of grouping	(a)	(b)	(c)	(d)	(e)	
Number of arrangements	1	12	12	6	4	Total = 35

each type of grouping. Now in a state of pure randomness, we can assume that any one of these arrangements is equally likely. Thus,

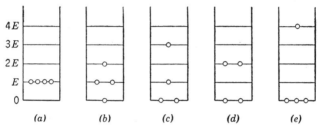

Figure 4.11. The possible groupings of four molecules with a total energy $4E$ among levels spaced E apart.

for example, in 35 such samples of four molecules we should expect to find 12 cases of grouping b, having one molecule of zero energy, two of E, and one of $2E$. We can say that the *probability* p of grouping b is $12/35$. It is interesting to note that the grouping in which the molecules all have equal energies is by far the least likely. Now if a grouping of probability p has n molecules at a certain level, we can say that it contributes the amount pn to the average number of molecules at that level. Products such as pn must then be summed over all the possible groupings. Or, to put it another way, the average number of molecules at a given level may be found by adding up the number of molecules found at that level in all the possible arrangements and then dividing by the total number of arrangements. Applying this to the above problem, we find the values shown in Table 4.2. We see that the lowest level

TABLE 4.2

Energy of level, ε	0	E	$2E$	$3E$	$4E$	
Mean number of molecules	1.72	1.14	0.69	0.34	0.11	Total = 4.00

is the most highly populated, on the average.

To apply this analysis to the kinetic theory of gases we must allow the number of molecules in our sample to become enormously great. We suppose that a typical energy level ε_i contains n_i molecules. Let the total energy be E, and the total number of molecules be N.

Then
$$\sum_i n_i = N \qquad (4.6)$$

$$\sum_i n_i \varepsilon_i = E \qquad (4.7)$$

A typical grouping is characterized by a certain set of numbers $(n_1, n_2, \cdots, n_i \cdots)$, and the number of distinct arrangements belonging to this grouping is given by

$$P = \frac{N!}{n_1! n_2! \cdots n_i! \cdots} = \frac{N!}{\prod_i n_i!} \qquad (4.8)$$

The grouping for which P is largest, subject to conditions (4.6) and (4.7), will be the most probable distribution. It is convenient to consider $\log P$ rather than P itself. A maximum, and therefore stationary, value of P is then defined by the simultaneous conditions

$$\sum_i \delta n_i = 0, \quad \sum_i \varepsilon_i \delta n_i = 0, \quad \sum_i \delta(\log n_i!) = 0 \qquad (4.9)$$

The third condition we express in another way by using Stirling's formula for the factorial of a large number:

$$n! \approx \sqrt{2\pi}\, e^{-n} n^{n+\frac{1}{2}}$$

i.e.,
$$\log n! \approx n \log n - n$$

$\therefore \delta(\log n!) = \delta n \cdot \log n$ and therefore $\displaystyle\sum_i \delta(\log n_i!) = \sum_i \delta n_i \cdot \log n_i$

(It should be understood that logarithms to base e are used throughout.) According to Lagrange's theory of undetermined multipliers, the three conditions (4.9) are simultaneously satisfied if, for all i,

$\alpha\, \delta n_i + \beta \varepsilon_i\, \delta n_i + \delta n_i \cdot \log n_i = 0$ where α and β are constants

$$\therefore n_i = A e^{-\beta \varepsilon_i}$$

or, in general $n(\varepsilon) = A e^{-\beta \varepsilon}$, where the constants A and β are to be determined.

Thermal Radiation and the Quantum Theory

Suppose that we apply this result to a collection of harmonic oscillators. There is no restriction, according to classical ideas, on the amount of energy that an oscillator may have. Thus, in place of a series of distinct values of $n(\varepsilon)$ at discrete values of ε we have a continuous curve (Fig. 4.12). The number of oscillators in a small energy range $d\varepsilon$ at ε is proportional to $e^{-\beta\varepsilon}\,d\varepsilon$. The mean energy $\bar{\varepsilon}$ per oscillator is then given by

$$\bar{\varepsilon} = \frac{\int_0^\infty \varepsilon e^{-\beta\varepsilon}\,d\varepsilon}{\int_0^\infty e^{-\beta\varepsilon}\,d\varepsilon} = \frac{1}{\beta}$$

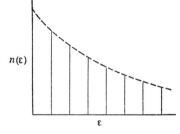

Figure 4.12. Boltzmann's distribution for discrete or continuous energy values.

(Cf. the mean free path problem, Chapter 1, Section 7.)

By classical equipartition theory, however, we have $\bar{\varepsilon} = kT$ for this case. Hence $\beta = 1/kT$, and so we arrive at the *Boltzmann distribution function*:

$$dn(\varepsilon) = A \exp(-\varepsilon/kT)\,d\varepsilon \qquad (4.10)$$

where $dn(\varepsilon)$ may be taken to mean the number of entities (oscillators, molecules, etc.) per energy interval $d\varepsilon$ having energy ε. The constant A is readily obtained from the condition

$$N = \int_0^\infty dn(\varepsilon)$$

and we find
$$A = N\beta = \frac{N}{kT}$$

It is to be noted that in our calculations we have assumed that an individual molecule has the same chance of finding itself at one energy level as at any other (the Boltzmann distribution arises from the limitation of the total energy available). But we might, for example, in our problem of four molecules provide *two* available levels at some particular energy, in which case the chance that a molecule will fall into that level is at once doubled. In general it is necessary to take account of this kind of situation by associating a certain *statistical weight* $g(\varepsilon)$ with the level ε. We then put

$$n(\varepsilon) = A\,g(\varepsilon)\exp(-\varepsilon/kT)$$

In the Maxwell velocity distribution, for example, we have
$$dn(\varepsilon) = A\varepsilon^{1/2} \exp(-\varepsilon/kT)\, d\varepsilon$$
i.e.,
$$g(\varepsilon) \propto \varepsilon^{1/2}\, d\varepsilon$$

(This may be seen by equating $f(v)\, dv$ (Chapter 1, Section 5) to $dn(\varepsilon)$ as given above.)

4.11 PLANCK'S QUANTUM HYPOTHESIS

The failure of classical physics where black-body radiation was concerned led Max Planck to re-examine the basis of this subject, and in 1900 he discovered a theoretical means of describing the experimental

TABLE 4.3

ε	$n(\varepsilon)$
0	A
ε_0	$Ae^{-\beta\varepsilon_0}$
$2\varepsilon_0$	$Ae^{-2\beta\varepsilon_0}$
etc, to infinity	

results with very great accuracy. Instead of allowing the atomic oscillators producing the radiation to have any arbitrary energy, he supposed that they were restricted to integral multiples of some unit ε_0, just as in our four-molecule problem. Thus the numbers of oscillators in successive levels are as shown in Table 4.3. We then evaluate the total number of oscillators and their total energy as follows:

$$N = A + Ae^{-\beta\varepsilon_0} + Ae^{-2\beta\varepsilon_0} + \cdots$$
$$= \frac{A}{1 - e^{-\beta\varepsilon_0}}$$
$$E = 0 \cdot A + \varepsilon_0 \cdot Ae^{-\beta\varepsilon_0} + 2\varepsilon_0 \cdot Ae^{-2\beta\varepsilon_0} + \cdots$$
$$= \varepsilon_0 A e^{-\beta\varepsilon_0}(1 + 2e^{-\beta\varepsilon_0} + 3e^{-2\beta\varepsilon_0} + \cdots)$$
$$= \frac{\varepsilon_0 A e^{-\beta\varepsilon_0}}{(1 - e^{-\beta\varepsilon_0})^2}$$

$$\left[\frac{1}{(1-x)^2} = 1 + 2x + 3x^2 + \cdots ; \text{ or notice that } E = -\frac{\partial N}{\partial \beta}\right]$$

Hence
$$\bar{\varepsilon} = \frac{\varepsilon_0 e^{-\beta\varepsilon_0}}{1 - e^{-\beta\varepsilon_0}} = \frac{\varepsilon_0}{e^{\beta\varepsilon_0} - 1} \qquad (4.11)$$

Thermal Radiation and the Quantum Theory

We take this new value of $\bar{\varepsilon}$ and insert it in the expression for the energy density of radiation (equation 4.5, Section 8 of this chapter). We thus have

$$u(\nu)\, d\nu = \frac{8\pi\nu^2\, d\nu}{c^3} \cdot \frac{\varepsilon_0}{e^{\beta\varepsilon_0} - 1}$$

Now, for $\beta\varepsilon_0 \ll 1$, this simplifies to the form

$$u(\nu)\, d\nu = \frac{8\pi\nu^2\, d\nu}{c^3} \cdot \frac{1}{\beta}$$

and we notice that, if we put $1/\beta = kT$, as we did before, we have reproduced the Rayleigh–Jeans formula. Furthermore, it is necessary that β should have this form in order that we should satisfy the form of $u(\nu)$ demanded by thermodynamics:

$$u(\nu) = B\nu^3\, g\left(\frac{\nu}{T}\right)$$

Thus we arrive at the expression

$$u(\nu) = \frac{8\pi\nu^2}{c^3} \cdot \frac{\varepsilon_0}{\exp(\varepsilon_0/kT) - 1}$$

But this, too, we must be able to cast into the form required by the general thermodynamic arguments. This is possible only if ε_0 is proportional to ν, so as to give the ratio ν/T in the exponential term. We therefore put $\varepsilon_0 = h\nu$ where h is some constant. By doing this we arrive at the famous formula discovered by Planck:

$$u(\nu) = \frac{8\pi h\nu^3}{c^3} \cdot \frac{1}{\exp(h\nu/kT) - 1} \qquad (4.12)$$

It was at once apparent that Planck's expression gave a very close fit to the experimental curves then available, provided a suitable value of the constant h was chosen. A number of subsequent studies of the black-body spectrum confirmed this agreement, and such work culminated in the work of Rubens and Michel, who in 1919 showed that Planck's formula accurately represents the spectrum from liquid air temperature ($-160°$ C) up to about $1800°$ C. Over this range the total radiation changes, according to the Stefan–Boltzmann law, by a factor of about $2.5 \cdot 10^5$.

We have seen in Section 6 of this chapter that Stefan's constant

σ is given by the equation

$$\sigma T^4 = \frac{c}{4}\int_0^\infty u(\nu)\,d\nu$$

Thus, from equation (4.12) we have

$$\sigma = \frac{1}{T^4}\cdot\frac{c}{4}\cdot\frac{8\pi h}{c^3}\int_0^\infty \frac{\nu^3\,d\nu}{\exp(h\nu/kT)-1}$$

We make the substitution $x = h\nu/kT$, and so obtain

$$\sigma = \frac{2\pi k^4}{h^3 c^2}\int_0^\infty \frac{x^3\,dx}{e^x - 1}$$

The value of this integral is $\pi^4/15 = 6.49$, as we find after rather lengthy analysis (Appendix IV).

Thus $$\sigma = \frac{2\pi k^4}{h^3 c^2}\times 6.49$$

It is interesting to compare this result with that obtained by an approximation. We note that the integrand tends rapidly to zero as $x \to 0$, and so over most of the integration we may neglect unity in comparison with e^x. (This simplified formula corresponds to an expression put forward by Wien in 1896 in an attempt to explain the radiation spectrum.) We then have

$$\sigma \approx \frac{2\pi k^4}{h^3 c^2}\int_0^\infty x^3 e^{-x}\,dx = \frac{2\pi k^4}{h^3 c^2}\times 6$$

and we see that our approximation leads to a value of σ that is in error by about 8%.

If we express Planck's law in terms of wavelengths, we have

$$u(\lambda) = \frac{8\pi hc}{\lambda^5}\cdot\frac{1}{\exp(hc/\lambda kT)-1} = \frac{8\pi k^5 T^5}{h^4 c^4}\cdot\frac{x^5}{e^x - 1}$$

The wavelength at which this has a maximum is then defined by

$$\frac{du(\lambda)}{d\lambda} = 0$$

When this equation is solved numerically, we find $x_m = 4.965$. Again it is interesting to notice that Wien's formula gives a quite good answer. If we put $u(\lambda) \sim x^5 e^{-x}$, we see that the maximum of the spectrum is given approximately by the condition

$$0 = 5x^4 e^{-x} - x^5 e^{-x}$$

i.e., $\qquad x_m = 5.000$

The relationship of the Planck and Wien treatments is illustrated in

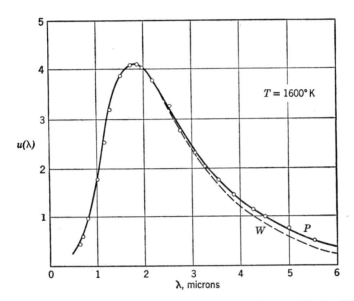

Figure 4.13. The thermal radiation spectrum according to Planck (P) and Wien (W).

Fig. 4.13. In the exact form we have Planck's statement of Wien's displacement law, viz:

$$\lambda_m T = \frac{hc}{4.965k}$$

Since the measurements on black-body radiation provide the values both of Stefan's constant and of $\lambda_m T$,[†] we may treat h and k as unknowns and solve for both. When this is done, it is confirmed that k is identical with Boltzmann's constant as deduced from kinetic theory, and the value of h, according to the latest data, is found to be

$$h = 6.625 \cdot 10^{-27} \text{ erg sec}$$

[†] See Fig. 4.6, p. 79. But note that the abscissa $x(=\lambda T)$ of that figure is not the same as the $x(= hc/\lambda kT)$ of the present section.

4.12 DISCUSSION OF PLANCK'S HYPOTHESIS

The success of Planck's formula left no doubt that in some way the process of emission of radiant energy is discontinuous. According to Planck's ideas, the energy of an oscillator of frequency ν was limited to multiples of $h\nu$. Thus $h\nu$ represented a basic unit of energy, which Planck named the "quantum." The new constant h has the physical dimensions of energy × time, or, what is exactly equivalent to this, of momentum × distance. A quantity of these physical dimensions was already familiar in classical mechanics under the name of "action," and was embodied in Maupertuis's "Principle of Least Action" (1740), according to which the path taken by a particle between two fixed points is that for which the integral of the action along the path is a minimum (or at any rate has a stationary value);

i.e., $$\delta \int mv \, ds = 0$$

We shall see later how this aspect of h was very relevant to the further development of atomic mechanics.

Although the original formulation of Planck's hypothesis was a startling advance in physics, there was nothing to sustain it outside the field of black-body radiation, and before long it was modified and then assimilated in a still more far-reaching approach to phenomena at the atomic level. Nevertheless, the basic idea of the quantization of energy has stood firm, and we must next examine the discoveries by which the breakaway from classical physics was completed.

Problems

4.1. The equation for free oscillation of a damped harmonic oscillator (Section 4.8) is

$$x = x_0 e^{-kt} \cos \omega_0 t$$

Assuming that $k \ll \omega_0$ and that the oscillator is a particle of mass m and charge e, the work of Section 8 of this chapter shows that the equation for the rate of loss of energy is

$$\frac{d\varepsilon}{dt} = -\frac{2}{3} \frac{e^2 \omega_0^2}{mc^3} \varepsilon$$

where $\varepsilon = \frac{1}{2}m\omega_0^2 x_0^2$. From the above equations deduce that

$$k = \frac{1}{3}\frac{e^2\omega_0^2}{mc^3}$$

4.2. According to Fourier analysis, an arbitrary disturbance $x = f(t)$ as a function of time can be resolved into a continuous spectrum of frequencies ω as follows:

$$x = f(t) = \frac{1}{\sqrt{2\pi}}\int_{-\infty}^{\infty} g(\omega)e^{+i\omega t}\,d\omega$$

where
$$g(\omega) = \frac{1}{\sqrt{2\pi}}\int_{-\infty}^{\infty} f(t)e^{-i\omega t}\,dt$$

$g(\omega)$ is thus a characteristic amplitude associated with frequency ω. By substituting $f(t) = x_0 e^{-kt}\cos\omega_0 t$ for $t \geq 0$ [and $f(t) = 0$ for $t < 0$], show that, for frequencies ω near to ω_0, the value of $g(\omega)$ is given approximately by

$$g(\omega) = \frac{ix_0}{2\sqrt{2\pi}} \cdot \frac{1}{\omega_0 - \omega + ik}$$

The square of the modulus of $g(\omega)$ gives the intensity associated with the component frequency ω in the radiation from the oscillator.

4.3. From Problem 2 the intensity distribution for the radiation from an exponentially damped oscillator can be written as

$$I(\omega) \propto \frac{1}{(\omega - \omega_0)^2 + k^2}$$

(a) Show that the intensity falls to half maximum at $\omega = \omega_0 \pm k$. This defines the "natural width" of a damped oscillator.

(b) Show that the natural width expressed on a scale of wavelength instead of frequency is independent of the natural frequency ω_0 and is given by

$$\delta\lambda = \frac{4\pi ck}{\omega_0^2} = \frac{4\pi}{3}\cdot\frac{e^2}{mc^2}$$

(c) Evaluate $\delta\lambda$ (which is a multiple of the classical electron radius, Chapter 3, Section 10), and express it as a fraction of λ_0 for $\lambda_0 = 6000$ A. One sees that the natural width represents an exceedingly small fraction of λ_0 for such wavelengths.

4.4. A group of seven particles has a total energy of $5E$; any individual particle is allowed to have energy $0, E, \cdots, 5E$. Draw the possible groupings of the particles amongst the energy levels, calculate the number of possible arrangements for each grouping, and hence find the average number of particles at each level. What would happen to the distribution if the number of particles were greatly increased, keeping the total energy fixed at $5E$?

4.5. According to Planck's quantum theory, the value of Stefan's constant is given by

$$\sigma = 6.49 \times \frac{2\pi k^4}{h^3 c^2}$$

Using the known values of k (Boltzmann's constant), h (Planck's constant), and c (velocity of light), evaluate σ in ergs, joules, and calories per square centimeter per °K^4 per second.

4.6. The earth receives energy from the sun at the rate of 1.94 cal/cm²/min (normal to the rays). (*a*) If the sun were effectively a black body, what would its surface temperature be? (Angular diameter of the sun as seen from the earth = 0.53°.) (*b*) What would be the temperature of a black body that radiates energy into space at the same rate at which the earth receives energy from the sun?

5 | Quanta and Atoms

5.1 THE PHOTOELECTRIC EFFECT

The central feature of Planck's hypothesis is that any change in the energy of a radiation field must take place in discrete quanta; i.e. at the instant of emission or absorption of radiation of frequency ν, the energy of some atomic oscillator must change by an integral multiple of $h\nu$. This does not automatically imply that the radiation inside a cavity is composed of discrete packets, and Planck himself did not make this inference. On the other hand, when we consider the discussion of cavity radiation according to Rayleigh and Jeans, it is tempting to suppose that the stationary waves themselves are in fact quantized. For, as we have seen, the classical equipartition of energy among these vibrations leads to a wrong result—the ultraviolet catastrophe—but if we impose on the vibrations the same restricted energies as for the atomic oscillators, then we shall arrive once more at Planck's formula. From this point of view the very condition of equilibrium between oscillators and radiation seems to demand that the radiant energy may exist only in amounts $nh\nu$, and it was Einstein who (in 1905) postulated that radiation was always carried in the form of light quanta, or *photons*. He suggested as an immediate consequence that, if photons of frequency ν fall on a solid material, the photoelectrons are emitted with a speed v such that $h\nu = \frac{1}{2}mv^2 + e\phi$, where ϕ represents an electric potential (the work function) that must be overcome in extracting an electron from the material.

We have previously mentioned (Chapter 3, Section 2) the work of Lenard and others on the photoelectric effect, which identified the carriers of the photoelectric current as electrons, but did little more. The stimulus of Einstein's suggestion led to many detailed studies of the process, largely concerned with the variation of electron energy with the wavelength of the light. The first reliable experiment of this kind was made by Hughes in 1912 (see Fig. 5.1). The technique was to irradiate a freshly cleaned metal surface M with monochromatic

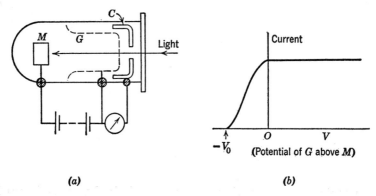

Figure 5.1. (a) Schematic diagram of apparatus used by Hughes to study the photoelectric effect. (b) Typical curve of collector current against collecting volts, with cut-off at $V = -V_0$.

light and measure the current reaching a collector C as a function of the voltage applied between M and a grid G. For positive voltage on G (with respect to M) the current is nearly constant, but for negative voltage the current falls sharply and cuts off altogether at some value $-V_0$. Hughes demonstrated that V_0 was a linear function of ν, and Millikan in 1916 showed that the slope of the line of V_0 against ν was accurately equal to h/e (Fig. 5.2). This is what is required by Einstein's equation, since we measure the kinetic energy $\tfrac{1}{2}mv^2$ in terms of the potential energy eV_0, so that, when the electron is just stopped, we have

$$eV_0 = \tfrac{1}{2}mv^2$$

and therefore

$$h\nu = eV_0 + e\phi$$

It was already known as the result of work by Lenard that the maximum velocity of photoelectrons did not depend on the intensity of the light, although the magnitude of the photoelectric current was proportional to the intensity of the light. The latter result was shown by Elster and Geitel (1913–1914) to hold good for variation of the light intensity by a factor of about $5 \cdot 10^7$. These results are strictly in accord with Einstein's relation, but the constancy of v at all intensities for a given wavelength of light could not be reconciled with classical ideas. So long as light was to be regarded as a wave motion, spreading continually from its source, it would appear that the chance of concentrating a given amount of energy on the target area presented by one electron should be proportional to the luminous intensity. A still more serious difficulty was the question of how the requisite

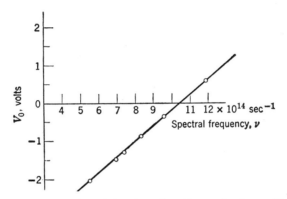

Figure 5.2. Verification of the photoelectric equation $h\nu = eV_0 + e\phi$.

amount of energy could be concentrated at all, between the instant at which a surface is first exposed to light and the instant at which the photoelectron emission begins. Indeed, this question is a crucial one, and we shall consider it carefully.

Suppose that we have a surface irradiated by a luminous intensity of 10^{-7} erg per cm² per sec ($= 10^{-14}$ watt/cm²). If this is taken to be visible light, with a wavelength of about $5 \cdot 10^{-5}$ cm, the number of quanta falling on 1 cm² per second is of the order of 10^4, and the energy per quantum (and so the energy given to the photoelectron) is about 10^{-11} erg. Now we might start by assuming that an electron, initially bound to an atom, has access only to the energy falling upon that atom. Taking the radius of an atom to be about 10^{-8} cm, and therefore the target area presented by the atom to be about 10^{-16} cm², the energy received per second by the atom would be about 10^{-23} erg per sec. It would thus take about 10^{12} sec, or something like 10,000 years, to accumulate the amount of energy needed. It turns out that this argument is fallacious because, as Rayleigh showed (1916), an atomic oscillator presents an effective area of about λ^2 to light of wavelength λ corresponding to its resonant frequency. This would mean an area of about 10^{-9} cm^2, and hence an accumulation time of about 10^5 sec: i.e. in the neighborhood of 10 hours. This is still far too large, however, to correspond with experience. Using light of a smaller intensity than we have assumed in this calculation, Elster and Geitel detected no appreciable time lag in the emission process, and an experiment performed much later (in 1928) by Lawrence and Beams established that any such lag was less than $3 \cdot 10^{-9}$ sec. In this beautiful experiment, pulses of light were generated with the aid of a Kerr cell. (This is a device in which a liquid that becomes doubly refracting under electric stress is placed between crossed Nicol prisms. When the liquid is subjected to an alternating electric field of very high frequency, the system transmits short bursts of light.) By illuminating the cathode of a photoelectric cell with this pulsed beam, and by pulsing the collecting voltage of this cell in a controlled way, it was possible to define intervals of about 10^{-8} sec, and to conclude that the photoelectron emission started and stopped within a time short compared to this.

The photoelectric effect left no room for doubt that the quantum of radiant energy was a real thing. Despite the triumphs of electromagnetic wave theory, and the abundant evidence of interference and diffraction with these waves, the fact remained that a quantum of this electromagnetic energy could somehow collect itself in such a way as to enter an individual atom and become entirely transferred to a single electron. We cannot overstate the truly revolutionary nature of this discovery. Instead of considering the relative merits of the corpuscular and wave theories of light, and making a choice between them, the physicist was suddenly required to accept both at once. The paradox has been dramatically stated by Sir Arthur Eddington in the following words:*

Consider the light waves which are the result of a single emission by a single atom on the star Sirius. These bear away a certain amount of energy endowed with a certain period, and the product of the two is h. The period is carried by the waves without change, but the energy spreads out in an ever widening circle. Eight years and nine months after the emission the wave front is due to reach the earth. A few minutes before the arrival some person takes it into his head to go out and admire the glories of the heavens and—in short—to stick his eye in the way. The light waves when they started could have had no notion what they were going to hit $\cdots$. Their energy would seem to be dissipated beyond recovery over a sphere of 50 billion ($5 \cdot 10^{13}$) miles' radius. And yet if that energy is ever to enter matter again, if it is to work those changes in the retina which give rise to the sensation of light, it must enter as a single quantum $\cdots$.

* Quoted, by permission, from A. S. Eddington, *The Nature of the Physical World*, Cambridge University Press, 1928.

Quanta and Atoms

The physicist of today has simply learned to accept this duality of corpuscular and wave aspects in radiant energy, and, as we shall see, it is very much in line with the enlargement of our concept of the nature of matter itself.

A more prosaic, but important, consequence of the photoelectric equation must be mentioned here. This is that a quantum of frequency ν always has the energy $h\nu$, never a multiple of this. There is therefore something wrong with our derivation of Planck's law from a consideration of the standing waves in a cavity, since to arrive at the correct result we had to allow the vibration of frequency ν to possess energy in any integral multiple of $h\nu$. We shall consider this difficulty more fully later (Chapter 8, Section 13) when we come to examine our assumptions about probabilities based on classical statistics, and consider the modifications that emerge from the quantum theory.

5.2 LINE SPECTRA

The subject of spectroscopy is quite an old one, and the knowledge that an incandescent vapor or flame emits a spectrum of distinct lines rather than a continuous range of colors is of long standing. It was recognized that the set of lines in the spectrum of a pure element was characteristic of that element, and that one could make use of this fact in chemical analysis, or even in the discovery of new elements. The first successful quantitative result of studying spectra, in the sense of finding significant relations between the various lines due to one element, was achieved by Hartley (1883). He noticed that, if he used frequencies instead of wavelengths, the frequency differences between members of certain characteristic groupings of spectral lines were the same for all such groupings at different regions of the spectrum of a given element.

The search for a more far-reaching relationship among the various lines of a spectrum was impeded by the idea, which proved to be quite wrong, that a vibrating optical system would exhibit a set of harmonics, like an acoustic resonator or a vibrating string. But in 1885 Balmer showed that a set of lines in the spectrum of atomic hydrogen (see Fig. 5.3) could be represented very accurately by the equation

$$\lambda_n = 3645.6 \frac{n^2}{n^2 - 4} 10^{-8} \text{ cm} \qquad (n = 3, 4, \cdots, 11)$$

Rydberg (1890) suggested that this expression should be rewritten in terms of frequencies, or rather of wave numbers k ($=1/\lambda$), giving

$$k_n = 27{,}430\left(1 - \frac{4}{n^2}\right) \text{ cm}^{-1}$$

or $\quad k_n = R\left(\dfrac{1}{2^2} - \dfrac{1}{n^2}\right) \quad (n = 3, 4, \cdots, 11)$

where $R = 109{,}720$ cm^{-1} and is called Rydberg's constant. In this last form the description of the Balmer series is very suggestive,

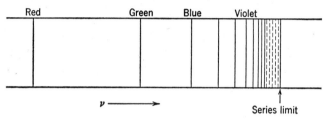

Figure 5.3. The sequence of lines composing the Balmer series, as they appear on a frequency or wave-number scale.

because we at once have the possibility that other spectral sequences exist, defined in general by

$$k(n_0, n) = R\left(\frac{1}{n_0^2} - \frac{1}{n^2}\right),$$

where n_0 and n are both integers. This was not, however, realized at the time. The difficulty was not merely the complexity of spectra, out of which one must try to select the lines belonging to a single sequence, but also the fact that only a few series can be represented in this simple way. When the attempt was made to analyze the spectra of the alkali metals, it was found by Rydberg that the wave numbers were given by expressions of the form

$$k_n = k_\infty - \frac{R}{(n + \mu)^2}$$

where k_∞ represents the series limit for $n = \infty$ (it is evident that the lines crowd more and more closely together as n increases), R is the same constant—the Rydberg constant—that appears in the description of the Balmer series, n is an integer and μ is an appropriately chosen fraction. Certain connections were also found between different spectral series for the same element, but it is more profitable to defer consideration of these until after we have developed the theory of the atom according to wave mechanics (Chapter 8). It should, however, be mentioned that in 1908 Lyman discovered a hydrogen series (in the ultraviolet) described by our formula

$$k(n_0, n) = R\left(\frac{1}{n_0^2} - \frac{1}{n^2}\right)$$

with $n_0 = 1$, and in the same year Paschen found a similar series in the infrared with $n_0 = 3$. These, together with the Balmer series ($n_0 = 2$) were powerful evidence of a great simplicity in the problem, if only the solution could be found.

5.3 THE NUCLEAR ATOM

The discovery of the electron, together with the knowledge that ordinary matter is electrically neutral, implied that matter was built up of positively charged material (carrying most of the mass) together with a requisite number of electrons. The radius of an electron could be estimated, as we have seen, to be a few times 10^{-13} cm, whereas the radius of a complete atom was of the order of 10^{-8} cm, as inferred from the density of matter in liquid or solid form. Thus the electrons could be regarded as effectively point charges, somehow embedded or bound in the atom. Lenard had suggested in 1903 that most of the atom was empty space, because cathode rays, i.e. fast electrons, were observed to penetrate appreciable thicknesses of solid matter, and he proposed a model in which positive charges and electrons paired up in the form of neutral doublets. This description did not provide any clear idea of what characterized the atom of a particular element—and chemistry showed that this was an exceedingly important unit. In the following year J. J. Thomson developed the picture of an atom as a sphere of positive charge, occupying the whole volume of the atom, with electrons embedded in it. Thomson showed that a ring of rotating electrons is stable until the number of electrons in the ring exceeds a certain limit; beyond this point a second ring begins to form, so that one can hope to account for the periodicity of chemical properties as described by Mendeléeff's periodic table.

Atomic models of this sort had to be abandoned, however, when Geiger and Marsden, in 1909, made a series of measurements on the scattering of alpha particles by thin metal foils. It had been discovered earlier that the alpha particles were swiftly moving positive particles which when neutralized by capture of electrons became chemically identical with helium. These alpha particles, ejected in large numbers by certain radioactive substances, provided an ideal means of bombarding other atoms. The most important result found by Geiger and Marsden was that, although most of the alpha particles readily penetrated a thin foil of matter (say 10^{-4} cm thick) about one in every ten thousand was apparently reflected back. Now on Thomson's model an alpha particle would be deflected only gently in passing through the distributed cloud of positive charge in an atom, and it would be likely to have this deflection partly counteracted in passing through successive atoms. (The effect of the electrons could be ignored, because with their very small mass they could not deviate the alpha particle appreciably under any conditions.) The over-all result was what is called a multiple scattering process, and, because of the tendency to accidental cancelation of successive deflections, it is very improbable indeed (i.e. much less than 1 chance in 10,000) that the combined effect should be a large angle of scattering. Furthermore, in a combination of random events such as this, the probability of deflection through a given amount should vary as the square root of the number of encounters [it is equivalent to the random walk problem in Brownian motion, Chapter 1, Section 10(2)]. But the frequency of the large angle deflections caused by a foil of thickness t was found to be proportional to t, not to $\sqrt{t}$. It became clear that these large deflections were the result of one or two violent encounters, rather than of a multitude of small deflections.

In 1911 Rutherford (*Phil. Mag.* **21,** 669) put forward the hypothesis that the positive charge of the atom was concentrated in a very small volume, so that the scattering of alpha particles by an atom could be treated as the relative motion of two point charges under their mutual repulsion through the Coulomb law of force. To account for the observed scattering it was necessary to suppose that the scattering center was also very heavy compared with the alpha particle, and so Rutherford proposed that the whole mass of the atom (excluding the electrons) was contained within a *nucleus*. From the experimental results it appeared that this model correctly described the scattering even when the alpha particle and the nucleus approached within about 10^{-12} cm, and hence one could deduce that the nuclear radius was not greater than this amount. Rutherford's detailed calculation (Appendix V) gave a certain probability $dw(\phi)$ that an alpha particle should be scattered through an angle between ϕ and $\phi + d\phi$:

$$dw(\phi) \sim \frac{Z^2}{V^4} \cdot \frac{\cos(\phi/2)}{\sin^3(\phi/2)} \, d\phi$$

where V is the velocity of the alpha particle, and Z is the nuclear charge of the scattering atom. It is not relevant to the present discussion to go into all the details, but it may be mentioned that in 1913 Geiger and Marsden published a paper in which the dependence of scattering upon Z, V, and ϕ was shown to be in complete accord with Rutherford's formula, and could even be used to show that the value of Z for a given scatterer agreed with its chemical atomic number. For the purpose of introducing Bohr's theory, it is enough to realize that the nucleus is effectively a point charge $+Ze$ carrying mass AM_0, where A is the atomic weight and M_0 is a suitable unit, conventionally chosen to be not the mass of the hydrogen atom, but $\frac{1}{16}$ of the mass of an oxygen atom. (This apparently cumbersome choice of unit allows us often to approximate A by an exact integer, which is not possible if M_H is taken as a basis of measurement.)

5.4 BOHR'S THEORY OF THE HYDROGEN ATOM

Bohr in 1913 [*Phil. Mag.* **26**, 1, 476, 857, (1913)] took the step of uniting the concept of the nuclear atom with Planck's idea of the quantized nature of radiative processes. He supposed

1. That the electrons move in circular orbits about the atomic nucleus.
2. That, in contrast to all classical expectations, only certain discrete orbits are permitted. Furthermore, and again in complete opposition to classical ideas, the electrons radiate no energy while in these favored orbits.
3. That the radiation of a quantum of energy takes place when an electron jumps from one permitted orbit to another, the frequency of the radiation being defined by Planck's relation $E = h\nu$.

These postulates (particularly 2 and 3) were of an entirely arbitrary character, and could only be justified by the results. Their success was, however, beyond any doubt, and we shall outline the simple theory of the hydrogen atom in these terms.

1. Let us suppose that the mass of the nucleus of H (i.e. a proton) is infinitely great compared with the mass m of an electron. Since $m_p \approx 2000m$, this is not a very drastic approximation. For an

equilibrium orbit (Fig. 5.4), we must then have

$$\frac{e^2}{r^2} = \frac{mv^2}{r} \tag{5.1}$$

i.e., $\qquad T = KE$ of electron $= \tfrac{1}{2}mv^2 = \dfrac{e^2}{2r}$

The electrostatic potential energy of the electron in the field of the proton is given by

$$V = -\frac{e^2}{r}$$

Thus the total energy E is given by

$$E = T + V = -\frac{e^2}{2r} \tag{5.2}$$

The negative sign of the energy represents the fact that this is a "bound state"; work of the amount $e^2/2r$ must be done to pull the electron completely away from the proton.

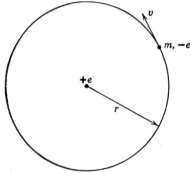

Figure 5.4. A circular orbit for an electron in a hydrogen atom, according to Bohr's original theory.

2. The above calculation is purely classical, but we now impose Bohr's condition for a permitted orbit. This is that the angular momentum of the electron is restricted to integral multiples of $h/2\pi$,* where h is Planck's constant. Now the angular momentum (otherwise called the moment of momentum) is given by mvr.

Thus $\qquad mvr = \dfrac{nh}{2\pi} \tag{5.3}$

and also $\qquad mv^2 r = e^2 \qquad$ (from 5.1)

Squaring the first of these equations, and dividing it into the second, we have

$$\frac{1}{mr} = \frac{4\pi^2 e^2}{h^2} \cdot \frac{1}{n^2}$$

---

* The combination $h/2\pi$ occurs very frequently in quantum physics and is given the special symbol $\hbar$. Its value for the purpose of rough calculations can be taken as 10^{-27} erg sec.

Quanta and Atoms

Thus the total energy E_n of this orbit is given, according to (5.2), by

$$E_n = -\frac{2\pi^2 e^4 m}{h^2} \cdot \frac{1}{n^2} \qquad (5.4)$$

3. The frequency of the light emitted when an electron falls from orbit m to orbit n $(m > n)$ is given by

$$\nu_{mn} = \frac{E_m - E_n}{h} = \frac{2\pi^2 e^4 m}{h^3}\left(\frac{1}{n^2} - \frac{1}{m^2}\right)$$

The wave number k_{mn} corresponding to this is simply ν_{mn}/c. Thus

$$k_{mn} = \frac{2\pi^2 e^4 m}{ch^3}\left(\frac{1}{n^2} - \frac{1}{m^2}\right) \qquad (5.5)$$

The resemblance of this to Rydberg's formula for the Balmer series is very plain. What is more, if Bohr's theory is correct, we are in a position to calculate the Rydberg constant:

$$R = \frac{2\pi^2 e^4 m}{ch^3}$$

With the most recent available values of e, m, c, h, we find $R = 109{,}737.3$ cm^{-1}, which is thus in extraordinarily close agreement with Rydberg's original value (109,720 cm^{-1}).

$$e = 4.8029 \cdot 10^{-10} \text{ esu}$$
$$m = 9.1085 \cdot 10^{-28} \text{ g}$$
$$c = 2.9979 \cdot 10^{10} \text{ cm/sec}$$
$$h = 6.6252 \cdot 10^{-27} \text{ erg sec}$$

The value of the energy E_n can be alternatively expressed as

$$E_n = -\frac{hcR}{n^2} \approx (2.18 \cdot 10^{-11})/n^2 \text{ erg}$$

5.5 HEAVIER ATOMS: ORBITAL ELECTRON STRUCTURE

The simple ideas of Bohr's theory for the hydrogen atom point to the existence of a number of possible orbits (an infinite number, in fact) having progressively greater radii as the quantum number n is increased. In the hydrogen atom, with its single electron, only one possible orbit can be occupied at any time. It is, however, a natural extension of the theory to apply it to an atom of nuclear charge Ze with Z orbital electrons. In the normal state of the atom, the electrons are distributed amongst the various orbits, working outward from the innermost orbit ($n = 1$) until the available number of electrons has been exhausted. The electrons will take up the states of lowest permitted total energy, which means (as we see from equation 5.4 of the previous section) the orbits of least n and least radius. There is, however, a definite limitation on the number of electrons that can be accommodated in a given orbit; the reasons for this will be deferred to Chapter 8 (Sections 1–4), because they involve a much more sophisticated picture of the atom than the one we are at the moment using. For the present we shall simply indicate the results.

Each type of orbit, characterized by a particular value of the quantum number n, is given a letter symbol, and any group of electrons having the same value of n—and hence, according to the Bohr theory, lying all at the same radial distance from the nucleus—constitutes what is called a "shell" that is labeled by its letter. We can think of a shell of electrons, on this view, as being all confined to a spherical surface, well separated from similar shells belonging to different n. In Table 5.1 we show how the first few shells are designated and occupied.

TABLE 5.1

Quantum Number, n	Shell Symbol	Electrons Accommodated
1	K	2
2	L	8
3	M	18
4	N	32

For hydrogen in its ground state we have just one electron in the K shell; in helium the K shell contains two electrons and is then filled. Thus in lithium, with three electrons, there are two electrons in the K shell and one in the L shell. The L shell can continue to accept electrons until we reach neon ($Z = 10$), after which the filling of the M shell begins. In order to provide places for all the electrons in the heaviest atoms known ($Z \approx 100$) it is necessary to go up to still higher shells, but our simple picture is so inadequate by this stage that it would be misleading to pursue it further. For the innermost one or two shells, however, it retains a rough validity in the face of all later refinements, and it will be particularly useful for the discussion of X-ray spectra [Section 11(3) of this chapter].

5.6 MOTION OF THE NUCLEUS

It must be remembered that in making our calculation of the wave number we should have taken account of the noninfinite mass of the nucleus. Strictly we should picture the nucleus and the electron as revolving about their common center of mass, just as the earth and the moon do (Fig. 5.5). The effect leads to a diminution of the

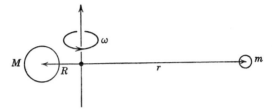

Figure 5.5. Illustrating the effect of noninfinite nuclear mass.

frequency with which the orbit is described. This time it is more convenient to introduce the *angular* velocity ω of the motion, since ω is the same for both particles. Then (a) By the definition of center of mass,

$$MR = mr, \qquad \therefore \quad R = \frac{m}{M} r$$

(b) The total angular momentum is given by

$$\frac{nh}{2\pi} = M\omega R^2 + m\omega r^2 = m\omega r^2 \left(1 + \frac{m}{M}\right)$$

(c) The equilibrium in the orbit is given by‡

$$m\omega^2 r = M\omega^2 R = \frac{e^2}{(R+r)^2} = \frac{e^2}{r^2}\left(\frac{M}{M+m}\right)^2$$

Thus from (b) and (c) we have

$$m\omega r^2 = \frac{M}{M+m} \cdot \frac{nh}{2\pi} \tag{5.6}$$

$$m\omega^2 r^3 = \left(\frac{M}{M+m}\right)^2 e^2 \tag{5.7}$$

‡ We are in fact limiting ourselves to a picture in which M includes the atomic nucleus and all but one of the electrons, and so has an effective charge of only $+e$. The student should work out for himself the consequences of associating a charge $+Ze$ with M.

Now the total energy E_n is given by

$$E_n = \tfrac{1}{2}m\omega^2 r^2 + \tfrac{1}{2}M\omega^2 R^2 - \frac{e^2}{(R+r)}$$

$$= \tfrac{1}{2}m\omega^2 r^2 \left(\frac{M+m}{M}\right) - \frac{e^2}{r}\left(\frac{M}{M+m}\right)$$

whence (by using 5.7)

$$E_n = -\frac{e^2}{2r}\left(\frac{M}{M+m}\right) \qquad (5.8)$$

But squaring (5.6) and dividing by (5.7) gives us

$$mr = \left(\frac{nh}{2\pi}\right)^2 \frac{1}{e^2} \qquad (5.9)$$

and so, combining (5.8) and (5.9),

$$E_n = -\left(\frac{M}{M+m}\right)\frac{2\pi^2 e^4 m}{h^2} \cdot \frac{1}{n^2}$$

We see that this is of the same form as for an infinite nucleus, except that a factor $M/(M+m)$ has entered. The Rydberg constant appropriate to a nucleus of mass M can thus be expressed as

$$R_M = \frac{M}{M+m} R_\infty$$

where R_∞ is the quantity we evaluated in Section 4.

This correction accounts, with complete success, for small discrepancies in the apparent value of Rydberg's constant as obtained from measurements of spectral series in different atoms. Indeed, the astonishing accuracy of spectroscopic wavelength measurements makes it possible to infer the ratio of electron and proton masses from such data. To take an example, the spectra of hydrogen and singly ionized helium are almost identical (after due allowance for the difference of nuclear charge) except for a small relative shift of corresponding lines. The results are described by putting

$$R_H = 109{,}677.58 \text{ cm}^{-1}$$

$$R_{He} = 109{,}722.27 \text{ cm}^{-1}$$

Now we know from atomic weight measurements that $M_H = 1.0081$, $M_{He} = 4.004$, on a scale in which the unit is $\tfrac{1}{16}$ the mass of the oxygen atom. Making use of these data, we find that $M_H/m = 1837.5$, which agrees almost perfectly with other determinations.

A more complete study shows that the Rydberg constants for atoms

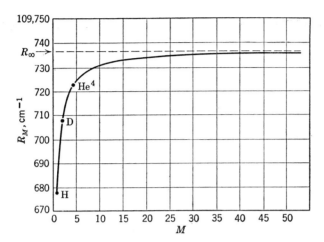

Figure 5.6. The effective Rydberg constant as a function of nuclear mass.

of all masses conform perfectly to the theoretical relationship, which is shown in Fig. 5.6.

5.7 THE CORRESPONDENCE PRINCIPLE

Although Bohr's postulates seem to be so completely at variance with classical ideas, we can establish a relation between the two if the quantum number n is large. This was pointed out by Bohr in 1923 and is called the correspondence principle. In essence it asserts that quantum and classical theories agree in the region of high quantum numbers (which for electron jumps within an atom means small wavenumbers and hence long wavelengths). We recognize this agreement for long wavelengths as being a feature of black-body radiation theory also (see Chapter 4, Section 9).

Let us consider an atom of effectively infinite mass, for simplicity. Then

$$E_n = -\frac{2\pi^2 e^4 m}{h^2} \cdot \frac{1}{n^2} \quad \text{(equation 5.4)}$$

Now let us find the energy change corresponding to a change of n by an amount Δn, where Δn is an integer. If n is sufficiently great, we may evaluate this energy by differentiation:

$$\Delta E \approx \frac{4\pi^2 e^4 m}{h^2} \cdot \frac{1}{n^3} \Delta n$$

The frequency of the light emitted in this transition is thus given by

$$\nu = \frac{\Delta E}{h} = \frac{4\pi^2 e^4 m}{n^3 h^3} \Delta n = \frac{e^4 m}{2\pi} \left(\frac{2\pi}{nh}\right)^3 \Delta n \qquad (5.10)$$

But we notice that $nh/2\pi$ is the angular momentum of the electron in its orbit; i.e., $m\omega r^2$. Thus we can put

$$\nu = \frac{e^4 m}{2\pi} \cdot \frac{1}{m^3 \omega^3 r^6} \Delta n \qquad (5.11)$$

Now, for equilibrium in the orbit, we have

$$m\omega^2 r = \frac{e^2}{r^2}$$

$$\therefore \frac{1}{r^3} = \frac{m\omega^2}{e^2}$$

Substituting this in equation (5.11), we find

$$\nu = \frac{e^4 m}{2\pi} \cdot \frac{m^2 \omega^4}{m^3 \omega^3 e^4} \Delta n$$

i.e.,
$$\nu = \frac{\omega}{2\pi} \Delta n \quad \text{simply}$$

Thus, for $\Delta n = 1$, the frequency of the light emitted is identical with the frequency of revolution in the orbit, which corresponds exactly to the classical requirements. If we allow Δn to take the higher values 2, 3, etc., we see that we generate all the harmonics of our fundamental frequency $\omega/2\pi$, and this is what Planck required of his harmonic oscillators in the thermal radiation problem. Thus Bohr's theory provides an interesting link between the quantum theory, in its first form, and the predictions of classical physics. The relationship is not satisfying or complete, but it represents an important step in the development of our ideas.

5.8 PHASE SPACE AND PHASE INTEGRALS

There exists an interesting and important way of representing the quantum conditions laid down by Planck and Bohr: namely, through

Quanta and Atoms

the concept of what is called "phase space." To approach this topic we consider a simple linear oscillator, consisting of a particle of mass m attracted to some fixed center by a force $-m\omega^2 x$, where x is its displacement from that center. If the total energy of the oscillation is E, we have

$$\tfrac{1}{2}m\dot{x}^2 + \tfrac{1}{2}m\omega^2 x^2 = E$$

We choose to rewrite this equation for the conservation of energy by introducing the momentum p ($= m\dot{x}$). Then we have

$$\frac{p^2}{2m} + \tfrac{1}{2}m\omega^2 x^2 = E$$

i.e.,
$$\frac{p^2}{2mE} + \frac{x^2}{2E/m\omega^2} = 1$$

Now this is the equation of an ellipse, and any instantaneous state of the motion of the particle in its oscillation is represented by some

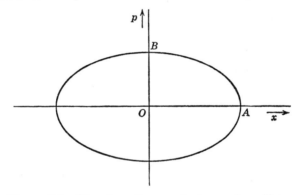

Figure 5.7. The phase diagram of a harmonic oscillator.

point (x, p) in a two-dimensional plot (Fig. 5.7). During one complete cycle of oscillation the representative point travels once round the ellipse. From the equation to the curve we see that the semiaxes OA and OB of the figure are given by

$$a = \frac{1}{\omega}\left(\frac{2E}{m}\right)^{1/2}$$

$$b = \sqrt{2mE}$$

The area of the ellipse is πab, and is also given by $\oint p\,dx$, and so we have

$$\oint p\,dx = \pi ab = \frac{2\pi E}{\omega} = \frac{E}{\nu}$$

where ν is the frequency of the motion. (By $\oint$ we understand integration round one complete circuit of the figure.) It is easy to see the relevance of this result to Planck's quantum conditions, for, if we suppose that $E = nh\nu$, then we have

$$\oint p\, dx = nh$$

Thus the permitted states of oscillation in Planck's quantum theory are represented by a set of ellipses, and the area enclosed between successive ellipses is always just h (Fig. 5.8). This particular space (of two dimensions in the present example) generated by taking a displacement and its associated momentum as co-ordinates is known as

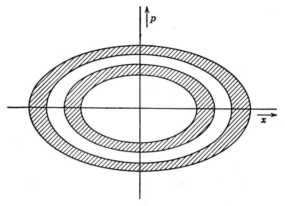

Figure 5.8. Quantization in phase space for a harmonic oscillator. The phase space is divided into elliptical annuli of equal area h. (The major and minor axes of successive ellipses are proportional to the square roots of the natural numbers.)

"phase space" and is of great importance in mechanics and theoretical physics. The quantity $\oint p\, dx$ is called the "phase integral" of the motion, and a feature that should be pointed out at once is that it represents the integral of the classical "action" over one oscillation. Thus Planck's quantum hypothesis may be regarded as a statement of the quantization of action, just as well as of energy.

The introduction of phase space and "action integrals" (i.e., phase integrals) brings out a close kinship between the forms of quantization adopted by Planck and by Bohr. We notice that the problem of atomic spectra is essentially one of rotational motions and of *angular momenta*. Let us give the angular momentum (mvr) the symbol p_ϕ, where ϕ is the angle between some fixed line and the radius vector to the moving electron (Fig. 5.9). Then $p_\phi\, d\phi$ has the dimensions of action, like $p\, dx$ in the linear oscillator problem. We can thus ask ourselves if the equation

$$\oint p_\phi \, d\phi = nh$$

represents an appropriate quantum condition, and it is at once evident that this is so. For in Bohr's circular orbits the angular momentum

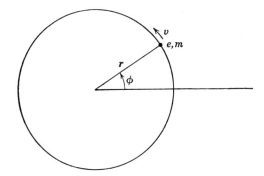

Figure 5.9. Angular momentum in terms of ϕ and p_ϕ.

mvr is independent of ϕ, and so we can put

$$p_\phi \oint d\phi = nh$$

i.e.,
$$p_\phi = \frac{nh}{2\pi}$$

which is precisely Bohr's condition for a stationary orbit.

5.9 ELLIPTIC ORBITS

It is known that the path of a particle under an inverse-square law of attraction is in general an ellipse (if the total energy is negative) with the center of force at one focus. The motion of the particle may then be described with the help of radial and angular co-ordinates (r, ϕ) which are both variables (Fig. 5.10). The kinetic energy of the motion is given by

$$T = \tfrac{1}{2}mv^2 = \tfrac{1}{2}m[(\dot{r})^2 + (r\dot{\phi})^2]$$

We can introduce the radial momentum $p_r \; (= m\dot{r})$ and the orbital angular momentum $p_\phi \; (= mr^2\dot{\phi})$, so that we may write

$$T = \frac{p_r^2}{2m} + \frac{p_\phi^2}{2mr^2}$$

The motion of an electron in an atom, if it is in an elliptic orbit, has then to be described by the co-ordinates (r, p_r) as well as (ϕ, p_ϕ); that is, we have two phase-space diagrams, and these are to some extent independent of each other. (The partial, as opposed to total, independence hinges on the fact that the energy of such an elliptic motion is

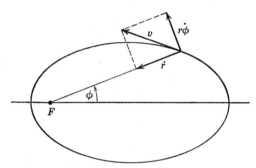

Figure 5.10. The radial and transverse components of velocity in an elliptic orbit.

the same for all ellipses having the same major axes.) It was suggested by Sommerfeld, who worked out the problem in 1916, that the motion was in this case doubly quantized, according to the equations

$$\oint p_r \, dr = nh$$
$$\oint p_\phi \, d\phi = kh$$

(n and k both being integers)

The quantum number n_0 that determines the total energy of the motion turns out to be equal to $(n + k)$, and, for $n = 0$ (and so $k = n_0$), we have one of Bohr's circular orbits. If we exclude the possibility $k = 0$, which represents an oscillation along a straight line passing through the center of force, we thus have just n_0 different orbits for a given value of n_0. An example is illustrated in Fig. 5.11. Although according to our simple ideas these distinct motions all have the same energy (a case of what is called "degeneracy" in mechanics), Sommerfeld showed that this is not exactly true. The more eccentric the orbit (i.e. the flatter the ellipse) the faster does the electron have to move at some stage of the motion, the highest speed being reached at the point of closest approach to the center of force. And this brings about a relativistic increase of the electron mass (Chapter 6, Section 16) which slows down the motion and lowers the total energy. We thus arrive at what is known as a *fine structure* in the energy levels of the atom, in consequence of which the number of possible quantum jumps for an electron within the atom is vastly increased. This goes a long way toward accounting for the multiplicity of lines in a typical spectrum, and for the occurrence of closely related spectral series. It is not the whole story, however, and to avoid obscuring the situation with additional facts and hypotheses we shall not develop this semiclassical

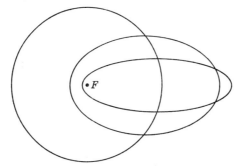

Figure 5.11. A sketch of the three orbits belonging to $n_0 = 3$ in the Sommerfeld theory.

atomic mechanics any further. We shall see later (Chapter 8) that the present-day picture of electrons in atoms is very different from the one that we have presented so far.

5.10 ATOMIC ENERGY LEVELS

It is of interest to know whether the existence of discrete stationary states for electrons in atoms, so clearly called for by the success of Bohr's theory, can be directly demonstrated. This question was answered within a year of the appearance of Bohr's papers, through an

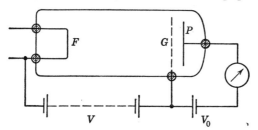

Figure 5.12. A schematic diagram of the Franck–Hertz experiment.

experiment carried out by Franck and G. Hertz (1914). A tube contains the vapor of an element at low pressure (Fig. 5.12). Electrons are accelerated through the vapor from a hot filament F towards a grid G and are subjected to a small retarding potential V_0 (about

$\frac{1}{2}$ volt) between G and a plate P. Thus electrons whose kinetic energy is less than eV_0 at G are unable to reach P. Now, if a free electron accelerated away from F collides with an electron bound in an atom of the vapor, it may give energy to it. In a head-on collision there is the possibility of a complete transfer of the kinetic energy of the incident electron to the struck electron. Such a transfer cannot take place, however, so long as the energy available is less than the energy needed

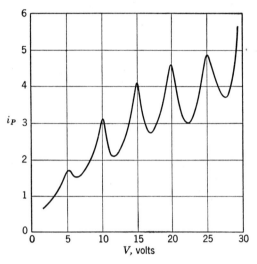

Figure 5.13. Variation of plate current with accelerating voltage in a Franck–Hertz experiment, showing characteristic excitation potentials.

to raise the bound electron from its orbit into the next higher available state. When this stage is reached, the incident electron may lose all its energy in a collision, and, if the collision takes place near G, there is then no chance for the electron to acquire enough energy subsequently to overcome the retarding potential V_0. Thus, as the accelerating potential V is raised from zero, the current arriving at P will increase until V reaches a characteristic excitation potential for the atoms of the vapor in the tube. Just beyond this point there is a sharp fall in plate current, followed by a recovery. Further falls will occur as V comes to exceed higher excitation potentials, if they exist, and also as electrons from F, after losing their energy in one inelastic collision between F and G, acquire sufficient energy to excite further atoms before reaching G (see Fig. 5.13).

The experiment is a rather crude one because the accelerating potential V required to stimulate an optical transition is only a few volts.

This may be seen if we assume a transition between two levels (orbits) corresponding to a spectral line of, say, $5 \cdot 10^{-5}$-cm wavelength. Then V is defined by

$$Ve = h\nu = \frac{hc}{\lambda}$$

$$\therefore V = \frac{hc}{e\lambda} = \frac{6.6 \cdot 10^{-27} \times 3 \cdot 10^{10}}{4.8 \cdot 10^{-10} \times 5 \cdot 10^{-5}} \text{ esu}$$

$$= 0.8 \cdot 10^{-2} \text{ esu}$$

$$= 2.4 \text{ volts} \quad \text{approximately}$$

Thus the energy Ve is by no means large compared with the spread in energies of thermoelectrons as they leave the filament (e.g., for a tungsten filament at 2400° K, about 10% of the electrons have energies greater than 0.5 ev§), and it follows that the characteristic width of one of the peaks in the experimental curve will at best be a few tenths of a volt, although the atomic excitation that it represents is quite perfectly sharp by such standards as these. Nevertheless, the transitions have been successfully identified with observed spectral lines in absorption (i.e. the equivalent of Fraunhofer lines). Thus, for mercury vapor, the following peaks have been found:

V, volts	Type		
4.9	a	Analysis shows	$a = 4.9$ volts
9.8	$2a$		$b = 6.7$ volts
13.5	$2b$		
14.7	$3a$	These correspond to known resonance absorption	
17.6	$a + 2b$	lines at 2537 A. ($= 2.537 \cdot 10^{-5}$ cm) and	
20.2	$3b$	1849 A.	

The picture has been completed through a beautiful experiment by Hertz. After exciting atoms in this way, he detected the spectral lines emitted as the electrons fell back into orbits of lower energy. By bombarding atoms with electrons of various energies, he was able to demonstrate that all lines requiring a specified amount of excitation energy were absent from the spectrum until the electron energy exceeded this threshold value. Figure 5.14 is a sketch of the spectra obtained from mercury vapor after bombardment with electrons of 8.7 and 9.7 ev, respectively. The arrows point to lines that are absent at the lower voltage.

§ 1 ev = 1 electron volt $\approx 1.6 \cdot 10^{-12}$ erg.

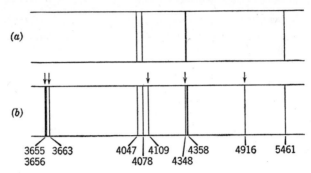

Figure 5.14. Sharp spectral lines following the excitation of mercury vapor by electron bombardment (a) at 8.7 volts and (b) at 9.7 volts. Arrows point to lines that are absent at the lower excitation. Wavelengths are given in angstroms (10^{-8} cm). [after G. Hertz, Z. *Physik*, **22**, 18 (1924)].

5.11 X RAYS

(1) Historical

The early study of X rays was one of the most exciting and important episodes in the history of modern physics. It was in 1895 that Roentgen observed the production of penetrating radiation when cathode rays (themselves not yet fully understood) were stopped in the walls of a discharge tube. It was some years before X rays were recognized as waves rather than particles; their apparent failure to be refracted by lenses or diffracted at pinholes created the same kind of doubts that Newton had had about the wave theory of visible light. But in 1899 it was shown (by Haga and Wind) that X rays could be diffracted by sufficiently narrow slits, and that their wavelengths must be about 10^{-8} cm (1 A): i.e., about 5000 times shorter than ordinary light. The conclusive proof that they were indeed transverse electromagnetic waves did not come until 1906, when Barkla demonstrated that they could be partially polarized in being scattered through an angle by matter, just as the light from the sky is partly polarized as a result of scattering by the air. We shall not dwell on these matters, but instead will study the production of X rays, and their subsequent interaction with matter, as providing an interesting and important link between classical and quantum physics.

(2) The X-ray Spectrum

X rays are customarily produced by accelerating electrons through a potential drop of about 10 to 100 kv in a high vacuum and then

Quanta and Atoms

stopping them abruptly in a "target" of some dense material. If the resulting radiation is analyzed with a suitable diffraction grating [Section 12(2) of this chapter], a spectrum of the type shown in Fig. 5.15 will often be obtained. The main features are: (a) The occurrence of intense, sharp lines (not always present), (b) A continuous background (sometimes called "white X rays" by analogy with white light), and (c) A sharp cutoff λ_0 on the short wavelength side.

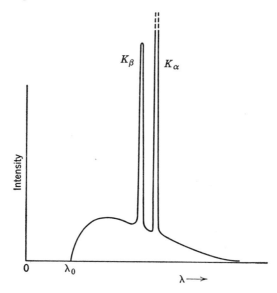

Figure 5.15. A continuous X-ray spectrum with two characteristic lines superimposed. Note the short wavelength cutoff.

Detailed observation shows that the sharp lines appear at well-defined wavelengths and are uniquely associated with the material of the target; they are therefore called "characteristic" X-ray lines. The continuous spectrum, on the other hand, has its most important dependence on the accelerating potential V applied to the electrons; as V is increased, the cut-off wavelength λ_0 becomes shorter in inverse proportion. Moreover, the value of λ_0 is the same for all target materials at a given value of V, and is correctly accounted for in absolute size by the equation

$$\lambda_0 = \frac{hc}{Ve} = \frac{1.24 \cdot 10^{-5}}{V} \text{ cm} \qquad (V \text{ in volts})$$

It is at once apparent that this feature, at least, is an exact inverse of the photoelectric effect (Section 1 of this chapter). The kinetic energy Ve of one electron can be entirely converted into the energy

$h\nu$ of one quantum. (There is no mention of the work function ϕ of the target material in the present discussion, because it is almost negligible compared with V in practice.) To this extent, therefore, the production of X rays is a clear case of a quantum phenomenon. The remainder of the continuous spectrum, which we shall next consider, is a more ambiguous problem, and is most easily approached from a classical point of view.

There can be no doubt about the fact that fast electrons, in entering a target, suffer a violent deceleration (e.g. the penetration by an electron of 40 kev into Al is about 10^{-3} cm, and this represents a mean deceleration of about $5 \cdot 10^{22}$ cm per sec^2). We can therefore refer to the theory of radiation by an accelerated charge [Chapter 3, Section 8(3)] from which one can estimate the rate of radiation of energy integrated over all directions. It is evidently to be expected that the intensity of radiation will be greatest at right angles to the electron's line of motion. The simplest classical theory of the spectral energy distribution is obtained by assuming that the electron continues with constant deceleration until it stops. The rate of radiation would then be constant during a limited time τ. This nonrepeated pulse could be analyzed, with the help of Fourier analysis, into a continuous distribution of frequencies, and would be found to have its most important contributions in the region of frequencies between zero and about $1/\tau$. In order to understand the same process from the point of view of a quantum hypothesis, we try to visualize in detail the progress of an electron through the target material. As it passes through the atoms, it will suffer a series of individual deflections in the Coulomb fields of the nuclei in the target. Each such deflection involves a brief acceleration process and hence a small burst of radiation, which according to quantum theory will appear as an individual quantum. In contrast, however, to the radiation by electrons bound in atomic orbits, there are no restrictions on the size of the quantum jump—except for the overriding one expressed by the short-wavelength limit.

The type of radiation exemplified by the continuous X-ray spectrum, and generated by the retardation of swiftly moving charged particles, is called *Bremsstrahlung* ("stopping radiation") and is an important contribution to the over-all process of energy loss for such particles.

The characteristic X-ray lines merit a separate discussion which we shall now give.

(3) X-ray Line Spectra. Moseley's Law

The bombardment of a target material by energetic electrons will result in collisions of the incident electrons with the orbital electrons of the target atoms, as well as with their nuclei. The nuclei are relatively so massive that they can be regarded as immovable, but an electron–electron collision (operating through the Coulomb repulsion) may well knock an orbital electron completely out of its atom. This is basically the same process as in the Franck–Hertz experiment (Section 10 of this chapter), and has basically the same consequences, namely that electrons falling back into the vacated positions will give rise to quanta and hence to sharp spectral lines.

Our particular interest for the consideration of X-ray spectra is that a sufficiently violent collision may drive one electron out of the K shell (see Section 5 of this chapter). We may then expect to observe a series of lines belonging to electron jumps terminating on the level $n = 1$. This is to all intents and purposes a generalization of the Lyman series for hydrogen (Section 2 of this chapter), and with the help of the Bohr theory we can calculate the frequencies or energies of the expected transitions.

If the Bohr theory is reformulated for an electron moving in the field of a central charge Q, we find that the energies of the various quantum states are given by

$$E_n = -\frac{2\pi^2 e^2 Q^2 m}{h^2} \cdot \frac{1}{n^2}$$

The question arises: What is the correct value of Q for transitions in an atom carrying a nuclear charge Ze? It is tempting to say that Q is equal to Ze, but this is not correct, because we have to recognize that the inner electrons of an atom provide what is called "screening," and effectively reduce the central charge. This is particularly clear-cut for the K electrons in the simple Bohr model, for we picture the two electrons in the K shell as circling much closer to the nucleus than any other electrons ($r_n \propto n^2$), so that by Gauss's theorem the electric field just outside the K shell is that due to a total positive charge $(Z - 2)e$. As we pass through the L shell, the effective charge Q falls to a still lower value. Q is therefore not constant for a given atom, but depends on the particular quantum state or transition considered. One can describe this by writing Q as equal to $(Z - s)e$, where s is the effective number of negative electronic charges providing screening in a given case.

Now if one electron has been ejected from the K shell, the effective nuclear charge as seen by an electron in the L shell will be $(Z - 1)e$. Hence the energy of a quantum jump from the L shell to the K shell will be given by

$$E_{12} = \frac{2\pi^2 e^4 (Z - 1)^2 m}{h^2}\left(1 - \frac{1}{2^2}\right)$$

If this same quantum jump is observed in a series of different atoms, we shall therefore expect to find that the frequency ν_{12} will vary systematically with Z, in such a way that a graph of $\sqrt{\nu_{12}}$ against Z will be a straight line. The same kind of linear relationship can be expected to hold even more generally when $\sqrt{\nu}$ is plotted against Z for a quantum jump between two specified orbits, no matter what the

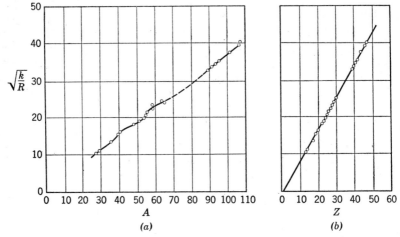

Figure 5.16. Graphs of $\sqrt{k/R}$ against (a) atomic weight A, (b) atomic number Z, for the characteristic K_α lines of the elements. The fact that Z, rather than A, is the significant quantity can be clearly recognized.

precise value of the screening constant s may be. It was H. G-J Moseley who, in 1913, discovered this relationship in the characteristic X-ray lines of the elements. What he did was to measure the wavelengths of what are called the K_α and K_β lines in the characteristic X-ray spectrum. These are a prominent pair of lines, always close together, with K_α having the longer wavelength. Moseley plotted the quantity $\sqrt{k/R}$ (where k is the wave number and R is Rydberg's constant—See Section 2 of this chapter) against the chemical atomic weight A or the chemical atomic number Z. Figure 5.16 shows the results of later, more extended measurements on the K_α lines of a series of elements. We can see the perfect linearity when Z is taken as abscissa, and Moseley found that the quantitative relationship was

$$k_\alpha = \tfrac{3}{4}R(Z - 1)^2$$

It may be verified that this is just what we expect according to Bohr's theory (Section 4 of this chapter) if the K_α line is identified as a jump from the L orbit ($n = 2$) to the K orbit ($n = 1$). The K_β lines are then found to represent a jump from the M orbit ($n = 3$) to the K orbit. It is worth remembering that Moseley's work was done at a time when Bohr's theory of the atom was only just being born, so that it represented a truly original contribution to the understanding of atomic structure. Moseley fully recognized the implications of his experimental results, and, in referring to the linearity of $\sqrt{k/R}$ as a function of Z (which is Moseley's law), he wrote: "We have here a proof that there is in the atom a fundamental quantity, which increases by regular steps as we pass from one atom to the next. This quantity can only be the charge on the atomic nucleus."

The proportionality of wave number k to $(Z - 1)^2$ makes the energy of the X-ray quanta rise rapidly with increasing Z. If we express the quantum energy in electron volts, we have for the K_α lines (by substituting values of the atomic constants in the theoretical formula):

$$E \approx 10(Z - 1)^2 \text{ ev} \quad \text{(a convenient "rule of thumb")}$$

The corresponding wavelength is given by

$$\lambda = \frac{hc}{E} \approx \frac{1200}{(Z - 1)^2} \text{ A}$$

(1 A = 1 angstrom = 10^{-8} cm.)

Thus, for example, for Cu ($Z = 29$), we should have $E \approx 7.8$ kev, $\lambda \approx 1.5$ A. These basically hydrogen-like spectra therefore appear in a region of wavelengths thousands of times shorter than the visible.

5.12 X-RAY SCATTERING AND DIFFRACTION

(1) Scattering by Single Atoms

When X rays pass through matter, they are scattered by the electrons in the atoms. If the electrons acted singly, each would (according to classical ideas) present an effective target area equal to the "Thomson cross section" (Chapter 3, Section 10):

$$\sigma_0 = \frac{8\pi}{3}\left(\frac{e^2}{mc^2}\right)^2 = 6.65 \cdot 10^{-25} \text{ cm}^2$$

This is what happens, to a good approximation, for scattering by the very lightest elements, but in general the electrons are more effective than this. The reason is that X-ray wavelengths (ranging from about 0.2 to 2 A) are not short compared to the distances between electrons within an atom. It is a curious fact that all atoms have about the same size, characterized by a radius equal to about 10^{-8} cm. (This is an experimental fact, which may be understood from Bohr's theory, since the outermost electrons of any atom "see" an effective nuclear charge that is only a small multiple of e, and it is this effective charge, reduced by screening, that determines the orbit radius, no matter how large Z itself may be.) Thus, as we go from lighter to heavier atoms, and from shorter to longer X-ray wavelengths, the more nearly do the Z electrons of a given atom behave as a kind of "superelectron" of charge $-Ze$. Since the size of the effective cross section presented to an electromagnetic wave by a charge Q is proportional to Q^2, the total scattering cross section presented by a heavy atom may approach the value $Z^2 \sigma_0$ rather than the much smaller value $Z \sigma_0$ that would apply if the electrons scattered independently. Expressed in other terms, this is a question of *coherence* in the scattering process. As the mean spacing of the electrons inside an atom becomes smaller compared to the wavelength of the incident radiation, the more nearly do the electrons scatter in phase, reinforcing each other to give an *amplitude* proportional to the number of electrons present, and hence an intensity proportional to the square of the number.

(2) Diffraction by Crystal Lattices

If we have an assemblage of separate atoms arranged in a regular structure, a new feature in X-ray scattering becomes apparent. (We now know that almost all solid materials exhibit some degree of regularity in the atomic arrangement.) Without concerning ourselves too much with details, let us consider a simple special case. The element copper has a density ρ of about 9 g per cm^3 and an atomic weight A of about 63. Thus the number of atoms per cubic centimeter is given by

$$n = \rho \frac{N}{A} \approx 10^{23} \text{ cm}^{-3}$$

The volume per atom is thus about 10^{-23} cm^3, and, if we imagine the atoms to be arranged in a regular cubic lattice (this is an oversimplification for copper), the distance between adjacent atoms is equal to the cube root of the volume per atom: i.e. about $2 \cdot 10^{-8}$ cm. This may be compared with a typical X-ray wavelength (10^{-8} cm); the ratio of wavelength to repetition distance is much the same as for the production of diffraction patterns by a diffraction grating with visible light.

If our picture is correct it is to be expected, therefore, that crystalline structures will behave like diffraction gratings for X rays.

In 1912 von Laue suggested (and Friedrich and Knipping carried out) the first experiment of this type. Its success confirmed both the wave nature of X rays and the regular internal structure of crystals, for it was found that a collimated beam of X rays, falling on a piece of crystal, gave strong emergent beams in certain directions only. The phenomenon is at first sight a formidable one to analyze, because we

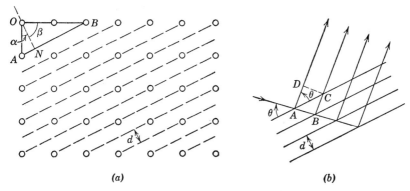

Figure 5.17. (a) A square lattice, with a set of effective atomic planes inclined to the principal directions. (b) Bragg reflections from the effective planes shown in (a).

have essentially a three-dimensional diffraction grating, composed not of ruled lines but of separate point-scattering centers (the individual atoms). But W. L. Bragg (1912) suggested a simple and ingenious way of treating the problem. He pointed out that a regular arrangement of atoms defines within itself a large variety of planes on which the atoms effectively lie. In Fig. 5.17a, we show a two-dimensional square array with a possible set of effective planes defined by a "knight's move" (one up and two along) from one atom to the next. If we draw a perpendicular ON to these planes from an atom at O, it makes angles α and β $(= 90° - \alpha)$ with the principal directions OA and OB. If the spacing of the square lattice is a, and if the perpendicular distance between successive planes is d, we have for our particular case

$$a \cos \alpha = 2d$$

$$2a \cos \beta = 2d$$

Hence
$$\cos \alpha = \frac{2d}{a}$$

$$\cos \beta = \sin \alpha = \frac{d}{a}$$

$$\therefore \cos^2 \alpha + \sin^2 \alpha = 1 = \frac{d^2}{a^2}(2^2 + 1^2)$$

and so
$$d = \frac{a}{(2^2 + 1^2)^{1/2}} = \frac{a}{\sqrt{5}}$$

More generally, for the three-dimensional cubic lattice, a possible spacing of reflecting planes is given by

$$d(hkl) = \frac{a}{(h^2 + k^2 + l^2)^{1/2}}$$

where h, k, l are any three integers having no common factor.

Now, if X rays of a certain wavelength λ are incident at an angle θ to our selected planes (Fig. 5.17b), the reflected rays from successive planes are in phase if the path differences are integral multiples of λ. But we have

$$\text{Path difference} = AB + BC - AD$$

$$= \frac{2d}{\sin \theta} - (2d \cot \theta) \cos \theta$$

$$= 2d \sin \theta$$

Hence the so-called Bragg equation for strong reflections is

$$2d \sin \theta = n\lambda$$

where n is an integer. The validity of this treatment is confirmed by a more rigorous analysis of the process, which, it must be remembered, is truly a scattering process and not a specular reflection. The atoms in a regular lattice simply behave *as if* they form reflecting planes.

If d is known, an X-ray spectrum can be analyzed by varying the angle of incidence θ and observing the intensity of the reflected X rays that are turned through an angle 2θ from the direction of an incident collimated beam. This is the basis of the Bragg crystal spectrometer. On the other hand, if monochromatic X rays with a known value of λ are used, it is possible to deduce the various effective values of d in a given crystal, and hence the basic atomic spacings a. This has, of

Quanta and Atoms   131

course, become an enormous and highly specialized field of study in its own right. It is possible to make absolute measurements of λ with the help of an ordinary ruled grating (of accurately known line spacing) at nearly grazing incidence. With this knowledge we can work backward to a determination of Avogadro's number, and such measurements provide the most accurate method of determining N.

5.13 THE COMPTON EFFECT

After the photoelectric effect, the most clear-cut evidence that radiant energy has the features that we associate with particles came in 1922, when A. H. Compton discovered the effect subsequently

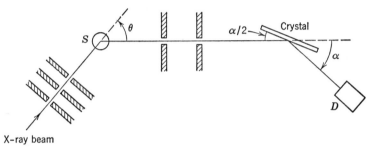

Figure 5.18. A schematic diagram of an experiment to observe the Compton effect.

named after him. It can even be said that this is the most direct manifestation we have of the corpuscular properties of quanta. The effect concerns the scattering of X rays by electrons. It had (as we have just seen) been clearly established that X rays, generated by the rapid deceleration of electrons in matter, were electromagnetic waves with all the properties of visible light. They showed the characteristic wave properties of diffraction and polarization, and had been used for many years in the analysis of crystal structures, which behaved like diffraction gratings for the X rays. Compton, however, made a careful study of the spectrum of an initially monochromatic beam of X rays after it had been scattered by matter. He found first of all that the scattered radiation was less penetrating than the primary incident radiation, and then showed that the scattered radiation in general contained radiation of two different wavelengths. One of these was always identical with the original wavelength, and the other was of a longer wavelength, of an amount varying with the angle of observation.

To carry out the experiment, X rays of a well-defined wavelength λ were collimated and allowed to fall on a scatterer S (Fig. 5.18). The X rays scattered in a given direction θ were allowed to fall on a crystal, which acted as a reflecting diffraction grating, so that a detector D showed a response only for particular angular settings α. Typical results are shown in Fig. 5.19 for four angles θ in the scattering of 0.7 A X rays by carbon. The longer wavelength line λ' moved away systematically from λ as the angle of scattering was increased, which is just what we might expect if the quantum is behaving like a particle with a certain concentrated amount of kinetic energy proportional to its frequency, and if the electron, initially supposed stationary, is struck by the quantum and set in motion. The line of unmodified wavelength is ascribed to scattering of the X rays by electrons that are so tightly bound to their atoms that they are not free to recoil; in this case a quantum is colliding effectively with a whole atom, and the energy transferred in a collision becomes negligibly small.

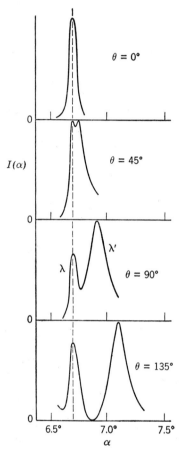

Figure 5.19. The spectrum of X rays scattered at four different angles, showing the Compton line and the unmodified line. [after A. H. Compton, *Phys. Rev.* **21**, 483 (1923)]

Let us treat the problem of Compton scattering in terms of what we have learned about radiation: viz., that a quantum of frequency ν has energy $h\nu$ and (because it is an electromagnetic disturbance) an associated momentum $h\nu/c$. We shall apply the principles of conservation of momentum and energy in ordinary Newtonian mechanics: i.e., disregarding any relativistic increase in mass of the recoiling electron. This last assumption will be justifiable so long as the momentum transferred to the electron is not too great. The collision is shown schematically in Fig. 5.20.

Quanta and Atoms

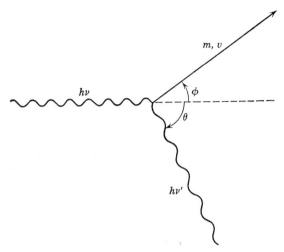

Figure 5.20. Diagram for the consideration of the balance of energy and momentum in the Compton effect.

The equations governing the collision are thus:

$$\frac{h\nu}{c} - \frac{h\nu'}{c}\cos\theta = mv\cos\phi$$

$$\frac{h\nu'}{c}\sin\theta = mv\sin\phi$$

$$h\nu - h\nu' = \tfrac{1}{2}mv^2$$

Put
$$\frac{h\nu}{mc^2} = \frac{h}{mc\lambda} = x, \qquad \frac{h\nu'}{mc^2} = \frac{h}{mc\lambda'} = y$$

Then our equations give

$$x - y\cos\theta = \frac{v}{c}\cos\phi$$

$$y\sin\theta = \frac{v}{c}\sin\phi$$

$$x - y = \frac{1}{2}\frac{v^2}{c^2}$$

Thus
$$\frac{v^2}{c^2} = x^2 - 2xy\cos\theta + y^2 = 2(x - y)$$

Let us put $y = x - \delta$, where δ is small (i.e., we assume the kinetic energy given to the electron to be much less than mc^2).

Then, approximately,

$$x^2 - 2x(x - \delta) \cos \theta + x^2 - 2x\delta = 2\delta$$

whence
$$\delta = \frac{x^2(1 - \cos \theta)}{1 + x(1 - \cos \theta)}$$

Thus
$$y \approx x - \frac{x^2(1 - \cos \theta)}{1 + x(1 - \cos \theta)} = \frac{x}{1 + x(1 - \cos \theta)}$$

$$\therefore \frac{x}{y} = \frac{\lambda'}{\lambda} \approx 1 + \frac{h}{mc\lambda}(1 - \cos \theta)$$

or
$$\lambda' - \lambda \approx \frac{h}{mc}(1 - \cos \theta)$$

This expression for the increase of wavelength, which appears in our calculation as only an approximate result, turns out to be *exactly* true when the problem is treated rigorously, taking account of relativity effects (see Problem 5, Chapter 6). But this nonrelativistic calculation, besides being simple and instructive, is also entirely justified for the X-ray energies with which Compton worked. In the typical experiment that we have quoted, for example, we have

$$x = \frac{h\nu}{mc^2} = \frac{h}{mc\lambda} \approx \frac{7 \cdot 10^{-27}}{10^{-27} \times 3 \cdot 10^{10} \times 7 \cdot 10^{-9}} \approx \frac{1}{30}$$

and so
$$\delta_{\max} = \frac{2x^2}{1 + 2x} \approx \frac{1}{500}$$

The observed variation of scattered X-ray wavelength with angle conformed perfectly to the theoretical formula, both as to the relative angular dependence and as to the absolute size of the shift. It may be noted that the constant h/mc appearing in the formula is a length equal to 0.024 A ($= 2.4 \cdot 10^{-10}$ cm). A few years after the Compton effect was discovered, Bothe and Geiger (1925) showed by the use of Geiger counters that the scattered X ray and the recoiling electron appeared simultaneously. It is thus quite certain that the collision of a quantum with an electron under these conditions may be treated as a problem in particle ballistics; we have traveled very far from the electromagnetic theory of light.

Quanta and Atoms 135

5.14 SOME COMMENTS

The main purpose of this chapter has been to collect together the evidence that radiant energy is absorbed, emitted, and scattered as though it has a corpuscular nature. Our development of this discussion has taken us into what is known as "the old quantum theory," which is essentially a treatment of atomic problems in terms of classical mechanics with some quite arbitrary quantum conditions superimposed. Although this is not logically satisfying, we are left in no doubt that the quantum aspects of radiation are inseparable from its wave nature, which leads us to ask if our conception of matter, and particles, is as clear-cut as the success of classical physics has led us to believe. This question was posed and quickly answered during the years 1924–1927, shortly after the experiments on the Compton effect had completed our picture of the quantum of radiation. The answer was, of course, that the duality of wave and corpuscle extends to particles also, and calls for a special type of mechanics, embodied in the "new quantum theory," to handle atomic problems. It is convenient to approach this subject through the results of relativity theory, and so we shall not begin our discussion of the problem until we have gone into the setting up of the relativity theory, and its consequences.

Problems

5.1. The surface of clean sodium is illuminated with monochromatic light of various wavelengths, and the retarding potentials required to stop the photoelectrons are observed as follows:

Wavelength, Å	2536	2830	3039	3302	3663	4358
Retarding potential, volts	2.60	2.11	1.81	1.47	1.10	0.57

Present these data graphically so as to verify the photoelectric equation. Deduce the value of h and the work function of the surface.

5.2. Calculate the series limit and the longest wavelength in the series for the cases $n = 1, 2, 3$ in the hydrogen emission spectrum according to equation (5.5), Section 4 of this chapter.

5.3. Calculate the radius of a Bohr orbit in hydrogen for which $n = 100$. Use the correspondence principle to calculate the wavelength of the emitted radiation for electrons falling into this orbit from orbits with $n = 101, 103$. Would such orbits have any physical significance?

5.4. Obtain the equation to an elliptic orbit for an electron in a hydrogen atom, ignoring the effects of quantization. (Make use of the treatment of Coulomb scattering, Appendix V, amended to the case of attractive forces and a negative total energy.) Show that all elliptic orbits having the same major axis have the same value of the energy.

5.5. In Section 10 some critical potentials for mercury vapor are given, and are ascribed to excitations within the mercury atom of 4.9 volts and 6.7 volts. Enumerate all the possible critical potentials that we might expect to observe in a Franck–Hertz experiment with mercury vapor, for accelerating potentials up to 25 volts, on the basis of the stimulation of these two transitions. Sketch the form of collector current against potential that you might expect to observe. [Then refer to Einsporn, *Z. Physik.*, **5**, 208 (1921).]

5.6. In Section 11 of this chapter, we discuss the emission of X-ray quanta as atoms return to their ground states after electron bombardment. The reverse of this is a photoelectric effect, with a sharp rise in X-ray absorption when a critical potential of the inner electron shells is reached. Calculate the energies (in kev) and the wavelengths (in A) required to remove a K electron completely from the following atoms: Li, Mg, Ca, Cr, Mo, Ba, W, U.

5.7. A two-dimensional square lattice of the type shown in Fig. 5.17 is studied with K_α X rays from tungsten, of mean wavelength 0.21 A. Bragg reflections are obtained at the following angles θ: 5.2°, 8.2°, 10.4°, 11.7°, 13.3°, 15.2°, 15.7°, 16.6°, 18.5°. Identify the indices (h, k) and the orders (n) characterizing these reflections, and find the lattice spacing. [The notation is that of Section 12(2) of this chapter.]

6 Relativity

6.1 DESCRIPTION OF MOTION IN MECHANICS

To appreciate the content of relativity, it is essential to have a clear understanding of the description of motion in ordinary Newtonian mechanics, and we shall begin with a very simple problem. Let us consider the motion of a body in one dimension, and plot its displace-

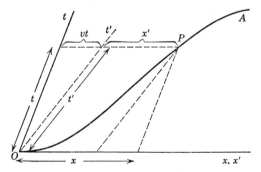

Figure 6.1. A motion OPA referred to two different co-ordinate systems in Newtonian mechanics. The system $S'(x', t')$ has a velocity v relative to $S(x, t)$.

ment against the time t (Fig. 6.1). The "life history" of the body is then represented by some line OPA, and the point P represents some particular stage of the motion. Let us now consider referring the motion to an origin which is itself moving with velocity v along the x axis. The motion of the particle is now described by (x', t') instead

137

of by (x, t), and we have the following relationship:

$$x' = x - vt$$
$$t' = t$$

The equation $x' = 0$ defines the t' axis, which is thus represented as a sloping line $x = vt$ on the original graph. The equation $t' = 0$ defines the x' axis, which is thus coincident with the x axis (by the condition $t = t' = 0$). We can evidently define an infinite number of possible co-ordinate systems of this type, in which the axis of x always has the same direction and the corresponding axis of t is obliquely inclined to it. The motion of a moving body can thus be represented by a point in some system of oblique co-ordinates, which we shall call a "frame of reference."

Now, for any pair of such possible co-ordinate systems, we can put

$$x' = x - vt, \qquad t' = t$$

and so
$$\frac{dx'}{dt'} = \frac{dx}{dt} - v$$
$$\frac{d^2x'}{dt'^2} = \frac{d^2x}{dt^2}$$
if $v = \text{const}$

Thus Newton's laws of motion are true in all such frames of reference, and a body moving under no forces in a straight line is equally well described in any system. We know that this ceases to be true in accelerated systems—e.g., in the comparison of (a) a fixed frame of reference and (b) a rotating frame in which centrifugal forces make their appearance. Thus we cannot tell if our local frame of reference is in steady motion, but we *can* tell if it is being accelerated. This is the real basis of Newtonian mechanics. It allows us to prefer an *inertial frame* (i.e., one in which a body under no forces can remain at rest) to any other as labeling absolute space in which events take place. Phenomena such as the bulging of the earth at the equator, or Foucault's pendulum, point to the rotation of the earth in absolute space, and could not (on this view) take place if the earth were at rest with the universe rotating around it.

6.2 THE ATTEMPT TO DISCOVER ABSOLUTE REST

We next ask ourselves if it is possible to discover, from physical measurements, whether we are in motion or at rest in the "absolute space" of Newtonian mechanics. To approach this problem let us consider the propagation of sound in air. Let the sound wave velocity in still air be c. Suppose we have two observers, A and B, a distance l apart (Fig. 6.2). Let a wind be blowing with velocity v from A to B.

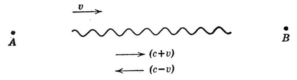

Figure 6.2. Transmission of a sound signal between two points when a wind is blowing.

Sound then has a velocity $c + v$ in going from A to B, and $c - v$ in going from B to A. Thus we have different times of transit for sending a signal each way:

$$t_1 = \frac{l}{c+v}, \qquad t_2 = \frac{l}{c-v}$$

Thus, if A and B have synchronized watches, they can establish the existence of the wind. Suppose they are 1000 ft apart, and that $v = 100$ ft per sec ($c = 1000$ ft/sec). Let A send off a signal at $t = 0$. B receives it at $t_1 = \frac{1000}{1100} = \frac{10}{11}$ sec $= 0.91$ sec. Let B send off a signal the moment he receives A's: e.g., by simply reflecting it. A receives this at $t = t_1 + t_2 = t_1 + \frac{1000}{900}$

$$= \tfrac{10}{11} + \tfrac{10}{9} = \tfrac{200}{99} \text{ sec}$$

$$\therefore \quad t = 2.02 \text{ sec}$$

If there were no wind, we should have $t_1 + t_2 = 2.00$ sec. Thus, it needs very accurate measurements to establish the existence of the wind from the total time t, but, if A and B compare notes about the separate times t_1 and t_2, the effect of the wind is easily noticed.

We say that the effect of v on t_1 or t_2 (or $t_2 - t_1$) is of the first order; whereas the effect of v on $t_1 + t_2$ is of the second order.

In either case, however, we should be able by accurate measurement to establish the existence of the wind experimentally.

Now suppose that A and B are on a ship moving with velocity v in still air (Fig. 6.3). The signal does not move any faster or slower as a result of any motion of the sources of sound, but the distance it has to travel is affected. Let A and B be moving to the left, so that t_1 is again the shorter time. B moves toward the sound signal, and we have

$$ct_1 = l - vt_1$$

i.e., $$t_1 = \frac{l}{c+v}$$

and similarly $$t_2 = \frac{l}{c-v}$$

This is exactly what we had when A and B were at rest and a wind was blowing.

We therefore have a relativity principle already. We can establish whether A and B are moving or at rest relative to the air, but we cannot

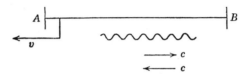

Figure 6.3. Transmission of a sound signal between two points that are fixed on an object which is itself moving through still air.

say that AB are moving and that the air is at rest, or vice versa. We can, however, assert that, if we were on a ship in still air, in a fog, it would always be possible to find out how we were moving, and whether we were at rest.

6.3 ABSOLUTE MOTION IN SPACE

In 1879 Maxwell proposed a method for detecting motion of the solar system through the ether. This is based on the observation of eclipses of Jupiter's moons, a phenomenon used long before by Römer to find the velocity of light. Jupiter has a period of twelve years, and so in six months, while the earth moves from A to B (Fig. 6.4) it does not travel very far in its orbit. Thus by observing the apparent times of eclipses with the earth successively at A and at B, we can infer the time taken for light to travel a distance equal to the diameter of the earth's orbit. If this time is measured when Jupiter is first at A', and then, six years or so later, at B', we can hope to discover whether the whole solar system is moving through the luminiferous ether with some speed v. For, if the diameter of the earth's orbit is l, we have

$$t_1 = \frac{l}{c+v}, \qquad t_2 = \frac{l}{c-v}$$

Relativity 141

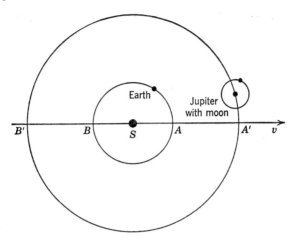

Figure 6.4. The attempt to measure the velocity of the solar system through space, by observations on the eclipses of Jupiter's moons.

using the same notation as for the acoustic problem we have just discussed.

Thus

$$\Delta t = t_2 - t_1 \approx \frac{2lv}{c^2} = 2t_0 \cdot \frac{v}{c}$$

where $t_0 = 16$ min approximately. If we could detect $\Delta t = 1$ sec, this would then correspond to

$$v = \frac{1 \times 3 \cdot 10^{10}}{2 \times 16 \times 60} = 1.5 \cdot 10^7 \text{ cm/sec} = 150 \text{ km/sec}$$

This is rather high compared with the known velocities of stars relative to the solar system (20 km/sec is a typical figure), but it is not excessive. The experimental result is that no delay as small as 1 sec has been detected, though the difficulty of discovering such a difference by measurements made six years apart is obvious.

6.4 MOTION OF A SOURCE OF LIGHT

It is well established that the velocity of light is independent of the velocity of the source, as indeed it is bound to be on any undulatory theory. The behavior of light as quanta might lead us to question this fact, and so it is fortunate that we can assert it with confidence. The proof comes from a study of double stars, which revolve about each other and reveal their motions through the Doppler effect on the spectra they emit. If the velocity of light depended in the least way on the speed of the source, the operation of this difference of velocity over the enormous distances of interstellar space would completely distort the apparent motions and eclipses of such double stars, but no such effects are observed.

6.5 ABERRATION OF LIGHT IN A VACUUM

In 1725 the astronomer Bradley observed a variation of apparent direction of the stars at different times of the year. This could be ascribed to the earth's finite velocity. The effect could not be detected if the earth moved always in a straight line, but it seems to show that the earth does move in the fixed ether, for, if the ether moved bodily with the earth, we would not expect such an aberration to occur. (It is quite distinct from a parallax effect of the type used in triangulation and surveying, which depends on the object being at a finite distance.) But it must be noted that only the *changes* in such motion can be detected; any uniform translation superimposed on the earth's motion would have no effect.

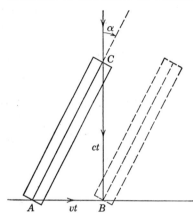

Figure 6.5. The principle of stellar aberration.

If we consider the light coming from a star in a direction perpendicular to the plane of the earth's orbit (Fig. 6.5), and imagine that this light enters the center of the object lens C of a telescope, then it will emerge through the center of the eyepiece A if the telescope is tilted through an angle α, such that in the same time t the light travels through CB and the telescope through AB. Hence

$$\alpha \approx \tan \alpha = \frac{v}{c}$$

For v we put the earth's speed in its orbit, viz. 30 km per sec, and this gives

$$\alpha = 20.5'' \text{ of arc}$$

This figure was accurately confirmed in observations on a large number of stars. In general a given star appears to describe an ellipse, of semimajor axis 20.5'', during the course of one year.

6.6 LIGHT IN A MOVING MEDIUM

It was suggested that Bradley's aberration experiment could be modified so as to reveal the absolute velocity of the earth through the ether. The method was to observe the apparent direction of a star (a) with a normal telescope, and (b) with the same telescope filled with water. The principle was that the light would be slowed down in passing through the water, so that in (b) the telescope would have to be pointed in some different direction from that used in (a), if the image of the star were to be centered on the cross-wires of the instrument in each case. But the outcome of the experiment was totally negative, and to explain this fact it was necessary to suppose that the water dragged the light sideways to some extent.

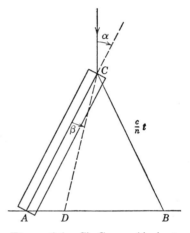

Figure 6.6. Sir George Airy's attempt to find the earth's velocity through the ether by studying the stellar aberration with a water-filled telescope.

We suppose that the incident light makes an angle α with the axis AC of the telescope (Fig. 6.6), and would be refracted along CD were it not for the fact that the drag brings it to B. Let the velocity of the earth through the ether be v, and the drag velocity be v'. Then, if light takes a time t to pass through the water-filled telescope, we have (remembering that α, β, etc. are very small angles)

$$AD + DB = AB$$

i.e., $$\frac{c}{n} t \times \beta + v't = vt \qquad (n = \text{refractive index of water})$$

or $$v' = v - \frac{c}{n} \beta$$

Now by the law of refraction, $\beta = \dfrac{1}{n}\alpha$

$$\therefore\ v' = v - \frac{c}{n^2}\alpha$$

But here we recall that the telescope has to be pointed in the direction α for experiment (a): i.e.,

$$\alpha = v/c \quad \text{or} \quad c\alpha = v$$

Hence

$$v' = v\left(1 - \frac{1}{n^2}\right)$$

Thus the negative result seems to require that the water drags the light with a certain fraction of its own velocity; this fraction is called the "drag coefficient" f:

$$f = 1 - \frac{1}{n^2}$$

Fresnel had in fact arrived at this idea of a drag coefficient as long ago as 1818, as a result of a negative experiment of a similar kind. This was that, when an observer looked through a prism at two stars in succession, one in the direction of the earth's motion in its orbit, and the other in an opposite direction, the deviation of the image by the prism was found to be exactly the same in both cases. But the velocity of light relative to the prism would be expected to take the two different values $c \pm v$, and so lead to a difference of deviations. Its absence could be explained if the glass dragged the light with the velocity fv.

6.7 EXPERIMENTS WITH A MOVING MEDIUM

If a medium drags the light waves, it should be possible to demonstrate the effect by passing light through flowing water. This was done by Fizeau (1851) and later by Michelson and Morley (1885).

Water was made to flow through a system of tubes as shown in Fig. 6.7, and, by means of a half-silvered plate P and a set of mirrors M_1, M_2, M_3, light from a monochromatic source S was split into two beams that traversed the flowing water in opposite directions. One beam (full line) traveled always against the water flow, and the other beam (broken line) always with it. The arrangement was thus a modified form of Michelson interferometer and could be set up to give straight fringes, observed with a telescope T. The existence of a drag coefficient brings about a time difference between the two beams in one complete circuit of the apparatus:

$$\Delta t = 2l \left(\frac{1}{\frac{c}{n} - fv} - \frac{1}{\frac{c}{n} + fv} \right)$$

$$\approx \frac{2ln^2}{c^2} 2fv$$

The time difference can also be expressed in terms of the distance $c\,\Delta t$

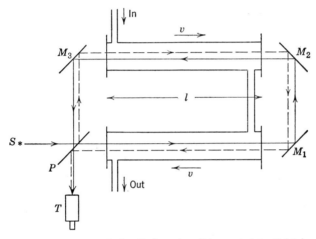

Figure 6.7. Measurement of the "ether drag" imparted to light in a moving medium. The apparatus is a modified form of the Michelson interferometer.

by which two initially coincident wave crests are separated by the time they reach the telescope T, and this lag in distance is most conveniently described as a certain fringe shift, $c\,\Delta t/\lambda$, by which the system of interference fringes is displaced when the water is made to flow. By making the water flow first in one direction, and then in the other, the effect can be doubled. Hence we have

$$\text{Total fringe shift} = \frac{2c\,\Delta t}{\lambda} = 8n^2 \left(\frac{v/c}{\lambda/l} \right) f$$

Clearly the dimensionless ratios v/c and λ/l are both small. In the experiment by Michelson and Morley, v was about 8 m per sec, and l was about 6 m. This gives $v/c \approx 3 \cdot 10^{-8}$, $\lambda/l \approx 10^{-7}$ (with $\lambda \approx$

$6 \cdot 10^{-5}$ cm), and so a fringe shift of the order of unity. A definite fringe shift was observed, and gave the result $f = 0.44 \pm 0.02$. This is in complete agreement with the theoretical value given by $(1 - 1/n^2)$ with $n = \frac{4}{3}$.

The essential point here is that motion of the medium *relative to the observer* produces a positive effect. But motion of medium and observer together through the ether is found to give a null result.

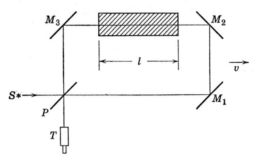

Figure 6.8. The attempt to observe an ether drag in a medium that is moving with the interferometer through space.

This was studied by Hoek (1868) in a very similar arrangement (shown in Fig. 6.8). The limb M_2M_3 of an interferometer contained a length l of material of refractive index n, whereas the other limb PM_1 was empty. No shift of interference fringes was observed as the whole apparatus was turned through 180°, which ought to reverse any effects due to motion through the ether. We deduce from this that light takes the same time to go around the apparatus clockwise or anticlockwise, which implies that

$$\frac{l}{c-v} + \frac{l}{(c/n+v) - fv} = \frac{l}{c+v} + \frac{l}{(c/n-v) + fv}$$

$$\therefore \quad \frac{1}{c-v} - \frac{1}{c+v} = \frac{1}{c/n - (1-f)v} - \frac{1}{c/n + (1-f)v}$$

$$\therefore \quad \frac{2v}{c^2} \approx \frac{2(1-f)v}{(c/n)^2}$$

i.e., $\quad f = 1 - \dfrac{1}{n^2} \quad$ once again

Of course, there would also be a null result if $v = 0$. It is as if nature did not intend us to detect our motion through the ether, and the drag coefficient f is of just such a size as to frustrate our attempts.

6.8 THE MICHELSON–MORLEY EXPERIMENT

In 1880 Maxwell wrote to Michelson, commenting that in terrestrial studies of the velocity of light the light is made to retrace its path, so that effects due to motion through the ether are only of the second order, and probably unobservable. But Michelson decided that such effects ought to be measurable with his interferometer, and went on to

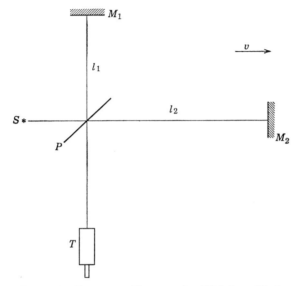

Figure 6.9. Schematic diagram to illustrate the Michelson–Morley experiment.

investigate them. The exact measurement, which has become famous as the Michelson–Morley experiment, was carried out in 1887.

Let us suppose that the earth is moving through the ether with speed v in a direction parallel to one limb PM_2 of the interferometer (Fig. 6.9). Then, to make a ray of light go along PM_1, we must head it into the "ether wind" so that the resultant of c and v lies along PM_1. The resultant velocity of the light in these circumstances is $(c^2 - v^2)^{1/2}$. Thus, for the passage of light from P to M_1 and back, we have

$$t_1 = \frac{2l_1}{(c^2 - v^2)^{1/2}} \approx \frac{2l_1}{c}\left(1 + \frac{v^2}{2c^2}\right)$$

For light traveling to M_2 and back, we have simply

$$t_2 = \frac{l_2}{c - v} + \frac{l_2}{c + v} \approx \frac{2l_2}{c}\left(1 + \frac{v^2}{c^2}\right)$$

$$\therefore \quad \Delta = t_1 - t_2 = \frac{2(l_1 - l_2)}{c} + \frac{l_1 v^2}{c^3} - \frac{2 l_2 v^2}{c^3}$$

Let the apparatus be turned through 90°, thereby interchanging the roles of l_1 and l_2. We now have

$$\Delta' = t_1' - t_2' = \frac{2(l_1 - l_2)}{c} + \frac{2 l_1 v^2}{c^3} - \frac{l_2 v^2}{c^3}$$

Thus

$$\Delta' - \Delta = \frac{(l_1 + l_2) v^2}{c^3}$$

The corresponding fringe shift is

$$c(\Delta' - \Delta)/\lambda$$

In the experiment, the distance $l_1 + l_2$ was 22 m, and $\lambda = 5.9 \cdot 10^{-5}$ cm. If we take for v the orbital velocity of the earth (30 km/sec), we should then have a fringe shift δ given by

$$\delta = \frac{(v/c)^2}{\lambda/l} = \frac{(10^{-4})^2}{2.7 \cdot 10^{-8}} = 0.37$$

Michelson estimated that a shift one hundred times smaller than this should be detectable, though perhaps not measurable, but the result of the experiment was totally negative, although it was repeated at several times of the year. This brought the question of our motion through space to a crucial stage, for it would appear to be an inescapable conclusion that if, by chance, the earth were at rest in space when the experiment was first performed, it could not possibly be again at rest six months later, when the direction of its orbital velocity had been reversed.

Michelson took his null result to mean that the earth dragged some of the surrounding ether along with it, although it was hard to square this idea with the picture of the ether as an all-pervasive, frictionless medium. It became even more difficult to hold to this interpretation after Lodge (1892) measured the speed of light near rapidly rotating bodies, and concluded that not more than 1/200 of the velocity of the bodies was communicated to the ether.

6.9 THE CONTRACTION HYPOTHESIS

We can express the results of all the experiments we have described by saying that

Relativity

1. Relative motions of source and observer, or of (source + observer) and material medium, give rise to observable effects.

2. Motion of source, material medium, and observer together relative to a luminiferous ether gives rise to no observable effects.

Now there is an alternative explanation of the null result of the Michelson–Morley experiment (the question of the drag coefficient for moving material media must be put aside for the moment). It was pointed out by Fitzgerald (1892), and independently by Lorentz, that a contraction of bodies along the direction of their motion through the ether would modify the calculations that we have made hitherto. In the arrangement that we have already discussed, we had

$$t_1 = \frac{2l_1}{(c^2 - v^2)^{1/2}}$$

$$t_2 = \frac{2l_2 c}{c^2 - v^2}$$

If now we suppose that the arm l_2 of the apparatus (Fig. 6.9) is shortened to l_2', such that

$$l_2' = l_2 \left(1 - \frac{v^2}{c^2}\right)^{1/2}$$

we then have

$$\Delta = t_1 - t_2 = \frac{2(l_1 - l_2)}{(c^2 - v^2)^{1/2}}$$

and the magnitude of this time difference is completely unaffected by rotation of the apparatus through 90°.

A contraction of this sort was obtained by Lorentz in his theory of electromagnetism, which was based on the idea that all physical properties of matter reduce to the properties of electrons. He found that the field equations of electron theory are left unchanged if a contraction by the factor $(1 - \beta^2)^{1/2}$ takes place ($\beta = v/c$), provided also that a new measure of time is used in a uniformly moving system. (He called this "local time.") At this period the belief that all matter and energy were electrical in origin was becoming prevalent.

6.10 EINSTEIN'S SPECIAL PRINCIPLE OF RELATIVITY

The outcome of the Lorentz theory is that an observer perceives the same phenomena in his system, no matter whether it is at rest in the ether or moving with constant velocity. Different observers are thus equally unable to say whether they are moving or at rest in the ether, so that for optical phenomena, just as for Newtonian mechanics, we have no means of identifying absolute rest. It has become the scientific tradition not to postulate things that are by their nature unobservable, and so it is tempting to suggest that there is no such concept as motion through the ether. This is tantamount to saying that the ether does not exist, but this, nevertheless, was the bold step that Einstein took in 1905.

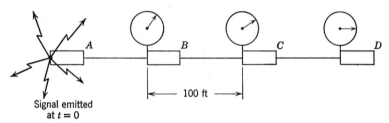

Figure 6.10. To illustrate the principle of synchronization of clocks by a sound signal.

To prepare the way for the relativity theory, it is perhaps helpful to restate the dilemma of physics in 1900 in the following contradictory statements:

1. According to classical mechanics the velocity of any motion has different values for observers moving relative to each other.

2. Experiment tells us that the apparent velocity of light relative to ourselves is not affected by our motion through the ether.

The second statement is based directly on observation; the first is an extrapolation from mechanical experience. Einstein decided that the conflict between them must be due to an imperfection of the classical ideas about measuring space and time, and he first attacked our preconceived ideas about simultaneity. To see the nature of his argument, consider the following situation:

A tug with a series of barges (call it a system S) is at sea in a fog. The distances between barges are equal, say 100' (see Fig. 6.10). It is agreed that clocks on the different barges will be synchronized with the help of a sound signal from the tug. The velocity of sound in air being 1000 ft per sec, there is a correction of $\frac{1}{10}$ sec to be made for each successive barge. Thus, if a signal is sent out from A at $t = 0$, the clock on B must be set at $t = 0.1$, that on C at $t = 0.2$, etc., when the signal is received. Now, if the vessels happen to be moving, unknown to them, the clocks will be wrong, because the time of transit of sound does depend on the motion. In the optical analog of this situation, however, our procedure is entirely justifiable (and in fact the only one) for defining simultaneity.

Now suppose a second tug and chain of barges, S'. It too can define simultaneity by the time of arrival of a sound signal, corrected for the known velocity of sound. But, if S' is moving relative to S, then there will be a progressive difference of readings between the clocks A–A', B–B', etc. Each would say that the other was wrong. In the acoustic problem only one system would in fact be correct, namely that which was at rest relative to the air. But, in the exchange of electromagnetic signals, experiment fails to reveal any preferred frame of reference. Thus, though observers moving relative to each other disagree in their findings, each is entitled to define simultaneity in his own system. But the inescapable inference is that there is no such thing as *absolute* simultaneity.

6.11 EINSTEIN'S DESCRIPTION OF MOTION

We now go back to the description of motion in co-ordinates (x, t). Let us suppose that three observers A, B, C, are equally spaced along

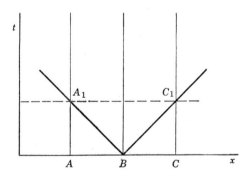

Figure 6.11. Simultaneity as established by signals in a stationary system.

the x axis and that they are at rest in a system of reference S. This means that their "life histories" are straight lines parallel to the t axis as shown in Fig. 6.11. Suppose that a light signal is sent out from B at $t = 0$. It travels at the same speed c forward and backward along the x axis, so that its motion is described by two sloping lines $x = \pm ct$ (referred to B as origin). The arrival of the light at the positions of A and C is thus given by the intersections A_1, C_1, and simultaneity at the positions of A and C is defined by the line $A_1 C_1$ parallel to the x axis (t = constant).

Now consider that A, B, C are at rest in a system S' which is moving with respect to S at a speed v (Fig. 6.12). A light signal sent from B at $t = 0$ is again described (in S) by the lines $x = \pm ct$. But the life

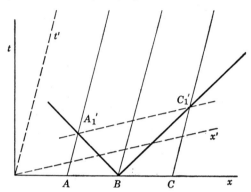

Figure 6.12. Apparent simultaneity as established by signals in a moving system.

histories of A, B, C as seen by S are now described by straight lines of the type $x = x_0 + vt$ (parallel to the axis of t'). Thus the light signals now arrive at A and C when they are at positions and times denoted by A_1', C_1'. These are clearly not simultaneous for S, for the line $A_1'C_1'$ is manifestly not parallel to the x axis. But we *demand* that they should be simultaneous so far as S' is concerned. The only way of securing this is to draw our x' axis parallel to $A_1'C_1'$, and in doing this we must remember that there is nothing special or preferred about a system of reference in which the axes of x and t are perpendicular to each other. But whereas, in classical relativity, the transformation from one system of reference to another merely entails a tilting of the t axis, we see that in Einstein's relativity both the x and t axes have to be adjusted, as indicated in Fig. 6.13.

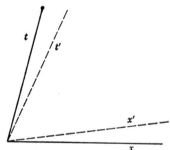

Figure 6.13. Modification of both x and t axes according to special relativity.

6.12 THE EINSTEIN–LORENTZ TRANSFORMATIONS

Our next step is to put these ideas into quantitative form. We consider first the fact that the t' axis of S' is defined by $x' = 0$, and also (from the point of view of S) by $x - vt = 0$. If these are to denote the same thing we can put

$$\alpha x' = x - vt,$$

where α is some scaling factor to be determined. We now reverse the roles played by the two systems of reference, and so, with equal justification, put

$$\alpha x = x' + vt'$$

The sign of the relative velocity between S and S' has to be reversed as we change our point of view, but otherwise there is complete symmetry in the problem. Making use of these two equations, we find

$$t' = \frac{\alpha x - x'}{v} = \frac{\alpha x - (x - vt)/\alpha}{v} = \frac{t - x(1 - \alpha^2)/v}{\alpha}$$

A motion in S, defined by $x = ut$, is represented in S' by $x' = u't'$, where

$$u' = \frac{x'}{t'} = \frac{x - vt}{t - x(1 - \alpha^2)/v}$$

i.e.,
$$u' = \frac{u - v}{1 - u(1 - \alpha^2)/v}$$

We now add the principle that the velocity of light has the same value c for all systems of reference; i.e., in this formula we put $u = u' = c$. We then find

$$1 - \alpha^2 = \frac{v^2}{c^2} = \beta^2 \quad \text{say}$$

and so
$$\alpha = (1 - \beta^2)^{1/2}$$

From this result we at once obtain the Einstein–Lorentz transformations:

$$x' = \frac{x - vt}{(1 - \beta^2)^{1/2}}, \qquad t' = \frac{t - vx/c^2}{(1 - \beta^2)^{1/2}}$$

For v/c small and $x \ll ct$ these reduce to $x' \approx x - vt$, $t' \approx t$, which we recognize as the so-called Galilean transformations of Newtonian mechanics. For a three-dimensional system we have the additional conditions

$$y' = y; \qquad z' = z \qquad \text{(assuming v to be along the x direction)}$$

6.13 GRAPHICAL REPRESENTATIONS

Let us introduce two variables ξ, η defined by

$$\xi = x + ct$$
$$\eta = x - ct$$

Then the history of a light signal traveling in the direction $\pm x$ can be written as $\eta = x - ct = 0$, defining an axis of ξ in the (x, t) diagram (Fig. 6.14) and $\xi = x + ct = 0$, defining an axis of η. Let the ξ and η axes be drawn as perpendicular lines; this can be done if we draw our unit of t to be c times as large as our unit of x. Now, when we change from frame $S(x, t)$ to frame $S'(x', t')$, we have

$$x' - ct' = \frac{1}{(1 - \beta^2)^{1/2}} \times \left[x\left(1 + \frac{v}{c}\right) - ct\left(1 + \frac{v}{c}\right) \right]$$

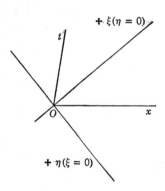

Figure 6.14. Co-ordinate system (ξ, η) for relating all co-ordinate systems connected to each other through a Lorentz transformation.

i.e., $$x' - ct' = \left(\frac{1 + \beta}{1 - \beta}\right)^{1/2} (x - ct)$$

Similarly $$x' + ct' = \left(\frac{1 - \beta}{1 + \beta}\right)^{1/2} (x + ct)$$

Multiplying these together, we have

$$x'^2 - c^2 t'^2 = x^2 - c^2 t^2 = G \quad \text{say}$$

This expresses the constancy of the velocity of light as measured in different frames of reference. But we also have $\xi \eta = G = \xi' \eta'$, so that a rectangular hyperbola with the (ξ, η) axes as asymptotes defines situations of constant G in all frames of reference.

Now if we put $G = 1$, $t = 0$, we see that this defines for us a line of unit length in system S. And if we put $G = -1$, $x = 0$, this defines a unit of duration (equal to the time for light to travel unit distance) again as measured in S. We therefore draw the hyperbolas defined by $G = \pm 1$ (Fig. 6.15). It may then be verified (see Appendix VI) that, if an x axis Ox is drawn anywhere between the $+\xi$ and $+\eta$ directions, the slope of the tangent at the intersection A between Ox and the hyperbola $G = +1$ is the direction of the time axis Ot belonging to the space axis Ox. Conversely, the tangent at the intersection B, between Ot and the hyperbola $G = -1$, is parallel to the x axis. Thus, once we have drawn the hyperbolas $G = \pm 1$, which are called "calibration curves," it is possible to construct all possible reference systems that differ only by a uniform velocity of relative motion along the x direction.

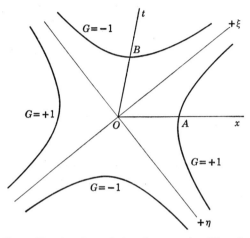

Figure 6.15. The calibration hyperbolas $G = \pm 1$ for Einstein–Lorentz transformations.

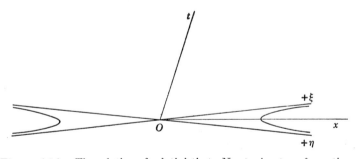

Figure 6.16. The relation of relativistic to Newtonian transformations.

It is to be noted that a possible x axis may lie anywhere between the lines $+\xi$ and $+\eta$, which appears to give a great deal of latitude. The impression is misleading, however, because our diagrams have been magnified by a factor c in the t direction. If we modify the picture to a more conventional one, in which 1 cm and 1 sec are represented as units of equal length, the diagram is squashed by a factor c in the t direction (Fig. 6.16). This allows very little play in the x axis, and corresponds more nearly to the Newtonian transformations with which we began this chapter.

6.14 LENGTH CONTRACTION AND TIME DILATION

The application of the special principle of relativity at once gives rise to the contraction which, as Fitzgerald and Lorentz recognized, would account for the negative results of the Michelson–Morley experiments. It must be understood, however, that Einstein's version dispenses with the question; "Does the contraction *really* take place?" The whole emphasis is on defining what we mean when we speak of measuring the length of some body that is in motion relative to ourselves. And Einstein's answer is that we determine the position of both ends of the body at one and the same time, as judged by us. Let us see how this is described with the help of our graphical treatment of the Einstein–Lorentz transformations. (This essentially geometrical presentation of special relativity was elegantly developed, about 1908, by H. Minkowski.)

(1) Length Contraction

We draw the calibration curve $G = +1$, and an x axis Ox. Then, as we have seen, OA is a line of unit length in the system S. Suppose a body of unit length is at rest in S, with its ends at O and A at $t = 0$ (Fig. 6.17a). The "life history" of its ends is then given by the lines OO', AA', both parallel to the tangent at A. Now, if we wish to measure the length of the body from some other system of reference S', we must do so by noting the positions of its ends at the same time. Let us choose $t' = 0$. This means that we have to find the intersections of the lines OO', AA' with the line $t' = 0$—which is the x' axis of our new system. The figure shows that these intersections are the points O and C. But we then note that the distance OD represents unit length in S', because it is the intersection of the line Ox' with the calibration curve $G = +1$. It is clear that OC is always less than OD; i.e., the apparent length of a body is reduced when it is in motion relative to an observer.

It is important to recognize that this is a truly reciprocal relationship. It is equally true that a body of unit length OD at rest in S' (Fig. 6.17b) appears to be shorter than unity as seen from S, since by an exactly similar construction we see that $OE < OA$.

For an analytical treatment of the problem we can simply consider

Relativity

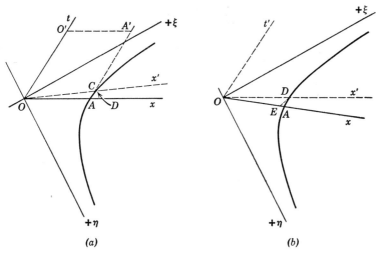

Figure 6.17. (a) Graphical treatment of the Lorentz contraction of length. (b) Illustrating the symmetrical nature of the contraction—each system sees the other's lengths as being shortened.

two points of a body, at the co-ordinates x_1, x_2 in system S. Then

$$x_1 = \frac{x_1' + vt_1'}{(1 - \beta^2)^{1/2}}, \qquad x_2 = \frac{x_2' + vt_2'}{(1 - \beta^2)^{1/2}}$$

The apparent distance between them as measured in S' is obtained by putting $t_1' = t_2'$. Then clearly

$$x_2 - x_1 = l_0 = \frac{l'}{(1 - \beta^2)^{1/2}}$$

i.e., $$l' = l_0(1 - \beta^2)^{1/2}$$

(2) Time Dilation

Let $x = x' = 0$ at $t = t' = 0$. A clock at rest in S (and at $x = 0$) ticks off 1 sec represented by OA in Fig. 6.18. We are interested in knowing when this appears to occur for an observer at rest in some other system S'. Now a line through A parallel to the x' axis represents a series of events that are simultaneous in S', since such a line has the equation $t' = $ constant. This line intersects the axis of t' ($x' = 0$) at C. Thus OC corresponds to unit time in S, as measured by a clock at rest in S' (at $x' = 0$). But OB is a measure of unit time for S'. We see that $OC > OB$, so that a duration of 1 sec in S looks like more than 1 sec to S'. The converse is, of course, also true.

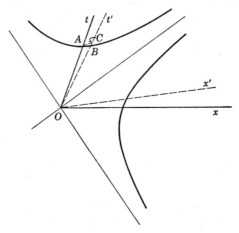

Figure 6.18. Graphical construction to illustrate time dilation in special relativity.

From the Lorentz transformations, we can describe the problem by putting

$$t_1' = \frac{t_1 - vx_1/c^2}{(1 - \beta^2)^{1/2}}, \qquad t_2' = \frac{t_2 - vx_2/c^2}{(1 - \beta^2)^{1/2}} \quad \text{with} \quad x_1 = x_2$$

Then

$$t_2' - t_1' = t' = \frac{t_0}{(1 - \beta^2)^{1/2}}$$

(Note: The analytic geometry of the calibration curves and the Lorentz transformations is given in Appendix VI.)

6.15 ADDITION OF VELOCITIES

Suppose that a particle is moving with a uniform velocity u in the system S':

$$x' = ut'$$

Let us view this motion from the system S, in which the origin of S' is moving with velocity v. Then we have

$$x = \frac{x' + vt'}{(1 - \beta^2)^{1/2}} = \frac{t'(u + v)}{(1 - \beta^2)^{1/2}}$$

$$t = \frac{t' + vx'/c^2}{(1 - \beta^2)^{1/2}} = \frac{t'(1 + uv/c^2)}{(1 - \beta^2)^{1/2}}$$

The particle therefore appears to have a velocity w, given by

$$w = \frac{x}{t} = \frac{u + v}{1 + uv/c^2}$$

The maximum value of w is given (for fixed u) by the condition $dw/dv = 0$.

$$\therefore \quad 0 = (1 + uv/c^2) - (u + v)u/c^2$$

$$\therefore \quad u = c$$

Under these conditions,

$$w_{\max} = \frac{c + v}{1 + v/c} = c$$

Thus the velocity of light represents a maximum observable velocity. Note, however, that if two bodies are moving in S with speeds of $+0.9c$ and $-0.9c$, respectively, S is entitled to assert that their relative velocity is $1.8c$. But an observer fixed on one of the bodies would not see things in this light. For him the velocity of the other body would be given by our relativistic formula for addition of velocities, with $u = 0.9c$, $v = 0.9c$.

i.e.,
$$w = \frac{1.8c}{1 + 0.81} = 0.995c$$

6.16 MASS AND ENERGY

By considering a collision between two bodies, we can infer that the mass of a body appears to vary with its velocity. Suppose that we have two experimenters, one at rest in a system S, and the other at rest in S', which has a speed v relative to S in the x direction. Let these experimenters have identical perfectly elastic spheres, and suppose that, as they pass, each of them projects his sphere with a speed u (as judged by himself) in a direction perpendicular to x (again as judged by himself), so that a collision takes place. Figure 6.19 depicts the situation as observed in the two systems. We shall suppose $u \ll v$.

Now, from the point of view of S, the y component of the velocity of ball B is different from u. For we have

$$y = y', \qquad t = \frac{t' + vx'/c^2}{(1 - \beta^2)^{1/2}}$$

$$\therefore \quad dy = dy', \qquad dt = \frac{dt'}{(1 - \beta^2)^{1/2}}$$

(x' being constant for B as viewed from S').

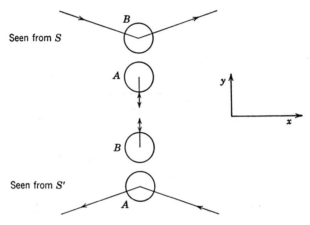

Figure 6.19. The elastic collision of two identical spheres as viewed from systems in relative motion.

The velocity as judged by S is therefore w, where

$$w = \frac{dy}{dt} = \frac{dy'}{dt'}(1 - \beta^2)^{1/2} = u(1 - \beta^2)^{1/2}$$

If energy is not to be lost in the collision, the individual velocities of the spheres along their line of centers are simply reversed. Thus, from the point of view of S, momentum can only be conserved if we put

$$m_0 u - mw = mw - m_0 u$$

where m_0 is the mass of a sphere A at rest relative to S (except for an arbitrarily small velocity u), and m is the apparent mass of an identical sphere B which is passing S with velocity v.

It then follows that

$$m(v) = \frac{m_0}{(1 - \beta^2)^{1/2}}$$

We can expand this result if β ($= v/c$) is small:

$$m(v) = m_0 \left(1 - \frac{v^2}{c^2}\right)^{-1/2} \approx m_0 \left(1 + \frac{1}{2}\frac{v^2}{c^2}\right)$$

$$\therefore \quad mc^2 \approx m_0 c^2 + \tfrac{1}{2} m_0 v^2$$

But we see that $\tfrac{1}{2} m_0 v^2$ is the classical kinetic energy T of a body, and so we can put

$$T = c^2 \times \Delta m$$

This equivalence between energy and mass proves to be universal, so that, if a body is radiating energy, it is at the same time losing mass

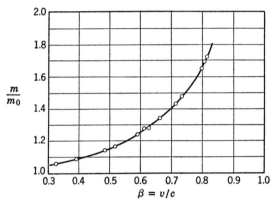

Figure 6.20. The variation of electron mass with velocity.

thereby. But since, as we see from the equation, it takes $9 \cdot 10^{20}$ ergs to make up 1 g, the effect is in general slight. As we are all aware, it does become alarmingly important in nuclear physics, where the transformation of large quantities of one kind of element into another can be made to liberate prodigious amounts of energy.

From 1905 to 1909 there was no experimental test of the conclusions of relativity theory. But in 1909 Bucherer, by subjecting fast electrons (beta particles) from radium to the combined effects of electric and magnetic fields, was able in effect to measure e/m as a function of velocity. (The basic equations are the same ones that J. J. Thomson used in designing his experiment to find e/m, although the experimental arrangement was quite different.) The experiment was repeated with greater accuracy by Zahn (1937) and a number of other workers. The size of the effect can be judged from Table 6.1 and from Fig. 6.20.

TABLE 6.1

v/c	m/m_0	v/c	m/m_0
0.01	1.000	0.9	2.294
0.1	1.005	0.95	3.203
0.5	1.155	0.98	5.025
0.75	1.538	0.99	7.089
0.8	1.667	0.998	15.82

6.17 NEWTON'S SECOND LAW IN SPECIAL RELATIVITY

Let us consider a force F applied in the x direction to a body initially at rest and of mass m_0. We set the force equal to the rate of change of momentum:

$$F = \frac{dp}{dt} = \frac{d}{dt}(mv) = \frac{d}{dt}\left[\frac{m_0 v}{(1-\beta^2)^{1/2}}\right]$$

The work done on the body in an element of distance dx is given by

$$dW = F\,dx = Fv\,dt$$

$$\therefore \quad dW = m_0 v\, d\left(\frac{v}{(1-v^2/c^2)^{1/2}}\right)$$

Integrating this from zero to a final velocity v, we have

$$W = m_0 \int_0^v v\, d\left(\frac{v}{(1-v^2/c^2)^{1/2}}\right)$$

$$= \left[\frac{m_0 v^2}{(1-v^2/c^2)^{1/2}}\right]_0^v - m_0 \int_0^v \frac{v\,dv}{(1-v^2/c^2)^{1/2}}$$

$$= \frac{m_0 v^2}{(1-v^2/c^2)^{1/2}} + m_0 c^2\left[\left(1-\frac{v^2}{c^2}\right)^{1/2}\right]_0^v$$

i.e., $$W = \frac{m_0 c^2}{(1-\beta^2)^{1/2}} - m_0 c^2 = c^2[m(v) - m_0]$$

Since the work done on the body must represent its gain of kinetic energy, we see here a more convincing demonstration of the mass-energy equivalence than was offered in the previous section.

6.18 RELATIVITY THEORY OF OPTICAL EFFECTS

(1) Doppler Effect

The equation to a plane wave of light of frequency ν, transmitted in the x direction in a medium of refractive index n, is

$$\psi = \cos 2\pi\nu \left(t - \frac{nx}{c}\right)$$

ψ represents the wave amplitude as a function of x and t. Since a crest (or other specific displacement) is recognized as such by observers moving with different speeds, we can put

$$\nu'\left(t' - \frac{n'x'}{c}\right) = \nu\left(t - \frac{nx}{c}\right)$$

The quantity $\nu(t - nx/c)$ is called an *invariant*. Considering light in free space we have, by substituting for t' and x' in the Lorentz transformations,

$$\frac{\nu'}{(1-\beta^2)^{1/2}}\left[\left(t - \frac{vx}{c^2}\right) - \frac{1}{c}(x - vt)\right] = \nu\left(t - \frac{x}{c}\right)$$

$$\therefore \quad \nu'\left(t - \frac{x}{c}\right)\frac{(1+\beta)}{(1-\beta^2)^{1/2}} = \nu\left(t - \frac{x}{c}\right)$$

i.e.,
$$\nu' = \nu\left(\frac{1-\beta}{1+\beta}\right)^{1/2}$$

This may be compared with the classical theory of the Doppler effect:
(a) Source moving away from observer; observer at rest in medium:

$$\nu' = \frac{\nu}{1+\beta}$$

(b) Observer moving away from source; source at rest in medium:

$$\nu' = \nu(1 - \beta)$$

We see that the relativity result embraces both these formulas so long as β may be taken as small.

(2) The Drag Coefficient

It is satisfying to discover that the apparent drag exerted by a material medium on a light wave comes as a natural consequence of special relativity. We consider light propagating in a medium, of refractive index n, that is at rest in a system S. Its velocity as measured in S is c/n. If now the apparent velocity w of the light is measured by an observer in S', which has as usual a velocity v relative to S, then w is given by the velocity addition formula:

$$w = \frac{c/n + v}{1 + (vc/n)/c^2}$$

$$= \frac{c}{n}\left(1 + \frac{nv}{c}\right)\left(1 + \frac{v}{nc}\right)^{-1}$$

$$\approx \frac{c}{n}\left(1 + \frac{nv}{c}\right)\left(1 - \frac{v}{nc}\right)$$

$$\approx \frac{c}{n}\left(1 + \frac{nv}{c} - \frac{v}{nc}\right)$$

i.e.,
$$w \approx \frac{c}{n} + v\left(1 - \frac{1}{n^2}\right)$$

Thus the "ether drag" explanation is correct so long as one ignores terms higher than the first order in v. This same formula for w can be obtained by considering the invariance of the phase of a wave, as in our treatment of the Doppler effect.

(3) Aberration

This is essentially a two-dimensional problem, and we shall consider a plane wave traveling in some direction s, inclined to the x and y axes as shown in Fig. 6.21. Then our wave is described by the equation

$$\psi = \sin 2\pi\nu\left(t - \frac{s}{c}\right)$$

i.e., the quantity $\nu(t - s/c)$ is the invariant quantity that is relevant here. Let us suppose that this wave appears to be traveling in the direction α with respect to the y axis of a system S, but appears to be traveling exactly parallel to the y' axis of system S'. Then we can put

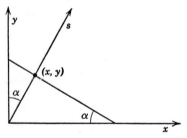

Figure 6.21. To illustrate the problem of stellar aberration.

$$\nu\left(t - \frac{s}{c}\right) \equiv \nu'\left(t' - \frac{y'}{c}\right)$$

$$= \nu'\left(\frac{t - vx/c^2}{(1 - \beta^2)^{1/2}} - \frac{y}{c}\right)$$

i.e., $$\nu\left(t - \frac{s}{c}\right) \equiv \frac{\nu'}{(1 - \beta^2)^{1/2}}\left\{t - \frac{1}{c}[\beta x - (1 - \beta^2)^{1/2}y]\right\}$$

But, if we express s in terms of x and y, we have

$$s = x \sin \alpha + y \cos \alpha$$

Comparison of the two sides of our equation then tells us that

$$\tan \alpha \; (\approx \alpha) = \frac{\beta}{(1 - \beta^2)^{1/2}} \approx \frac{v}{c}$$

6.19 LIFETIME OF MESONS

We are familiar today with the existence of a variety of particles intermediate in mass between the electron and the proton. Such particles are given the generic name "meson," although the precise nature of the relationships between them is not fully understood. All of them are unstable and undergo radioactive decay of one sort or another. One of these particles, the μ meson, was discovered long before the others (in 1937), and its lifetime τ has been carefully measured and found to be equal to about $2.15 \cdot 10^{-6}$ sec. By "lifetime," in this context, we mean the *average* time that a meson lasts before decaying; the decay process is found to be governed by the laws of chance, such that, if n_0 mesons are present at $t = 0$, the number left at time t is given by

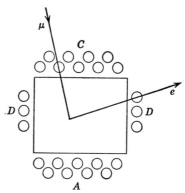

Figure 6.22. Schematic diagram of apparatus used to measure the lifetime of μ mesons stopped in matter.

$$n(t) = n_0 \exp(-t/\tau)$$

A direct method of finding τ consists in setting up a block of material and surrounding it with Geiger counters as shown in Fig. 6.22. Mesons (from the cosmic radiation) can enter the block and may, in doing so, trigger off one of the counters C. The fact that they stop inside the material, rather than passing right through, is established by the absence of any count from the row of counters A. Now the decay of the μ meson produces a fast electron, and this may escape from the block and cause a count in one of the counters D. It is possible to record the frequency of cases in which (a) C gives a count, (b) A gives no count, and (c) D gives a count at a time t later than (a). This phenomenon is called a "delayed coincidence" between C and D, and a

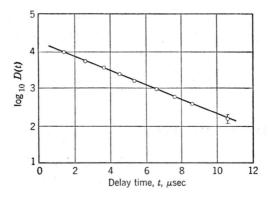

Figure 6.23. The decay curve for μ mesons, expressed as a semilogarithmic plot of the delayed μ—e coincidences against the time [after Bell and Hincks, *Phys. Rev.* **84**, 1243 (1951)].

semilogarithmic plot of the delayed coincidence rate $D(t)$ against t yields a straight line whose slope is $-1/\tau$ (Fig. 6.23).

The theory of the method is simple. One is effectively measuring the rate of decay of mesons at a chosen time t after they arrive within the block (and stop there).

$$\therefore \quad D(t) = -\frac{dn(t)}{dt} = \frac{n_0}{\tau} \exp\left(-\frac{t}{\tau}\right) = D_0 \exp\left(-\frac{t}{\tau}\right)$$

$$\therefore \quad \log D(t) = \log D_0 - \frac{1}{\tau} t$$

Now there is another (and earlier) way of finding the lifetime. Mesons are less abundant at the earth's surface than at high altitude, because they are created high in the earth's atmosphere by incoming primary cosmic rays (largely protons) and many are lost before reaching sea level. The loss on the way down is due partly to scattering and absorption, partly to decay in flight. If we let the mesons pass through a thickness of solid material (e.g., lead or graphite) equivalent in absorbing power to the atmosphere, the attenuation is less, because the time available for decay in traversing the solid material is almost negligible in this case. To be specific, the whole atmosphere has a weight of about 1000 g per cm^2 (this is the atmospheric pressure at sea level), which is equivalent to a height of about 10 km at the sea-level value of the atmospheric density. This same weight of material represents only about 1 m of lead. Now the mesons are mostly traveling with speeds comparable to the speed of light, so the time taken to traverse a distance l is roughly l/c. Thus, if the attenuation of mesons due to absorption and scattering alone is f, we have the following relationships:

In the solid material,

$$\frac{n}{n_0} = f$$

In the equivalent atmosphere,

$$\frac{n'}{n_0} = f \exp\left(-\frac{l}{c\tau'}\right)$$

The effective lifetime τ' deduced from such measurements is very much greater—by a factor of about 15—than the lifetime τ obtained from the delayed coincidence experiment. This discrepancy is explained in terms of the relativity transformations in a simple and direct way:

(a) From the point of view of the meson, as it were, the height of the atmosphere is not l, but its Lorentz-contracted value $l(1 - \beta^2)^{1/2}$, where βc is the meson velocity.

(b) From our point of view, the height of the atmosphere is definitely l, but time τ measured by the meson appears to us as $\tau/(1 - \beta^2)^{1/2}$.

These descriptions are strictly equivalent and are equally valid. From the experimental results we deduce that $1/(1 - \beta^2)^{1/2}$ is about 15, so that the mean velocity of μ mesons in cosmic rays is about $0.998c$.

6.20 LIGHT IN A GRAVITATIONAL FIELD

The equivalence of energy and mass, demanded by special relativity, leads naturally to the question whether a quantum of energy $h\nu$ behaves like a mass $h\nu/c^2$ under gravitational forces. Let us explore this question, but with a clear recognition that our procedure here is tentative and arbitrary, because it is not based on any understanding of gravitation or on any prior knowledge of how light behaves in a gravitational field. This problem cannot, in fact, be properly treated without the help of the general theory of relativity, to which we shall refer shortly.

(1) Gravitational Red Shift of Spectral Lines

Suppose that a quantum of frequency ν is emitted by an atom at the surface of a star of mass M and radius R (Fig. 6.24). The effective

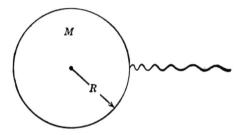

Figure 6.24. To illustrate the gravitational red shift of spectral lines.

mass of the quantum is given by

$$m = \frac{h\nu}{c^2}$$

This mass has a gravitational potential energy V, given by $V = -GMm/R$, where G is the constant of universal gravitation. Thus the total energy of the quantum is given by

$$E = h\nu + V = h\nu - \frac{GM}{R} \cdot \frac{h\nu}{c^2}$$

Now when the quantum reaches the earth, it has escaped altogether from the gravitational pull of the star, and its total energy E can now be expressed as $h\nu'$, where ν' is the frequency of the quantum as measured by us. We thus find

$$\nu' = \nu\left(1 - \frac{GM}{c^2 R}\right)$$

or

$$\lambda' \approx \lambda\left(1 + \frac{GM}{c^2 R}\right)$$

The apparent wavelength is thus shifted toward the red, compared with a quantum produced in the same electronic transition in an atom on the earth. An effect of this sort has been detected in the spectra of the remarkable type of stars known as "white dwarfs"; they possess abnormally large values of M/R through having quite enormous densities—about 10,000 times as great as ordinary solid matter.

(2) Deflection of Light by the Sun

We consider a quantum, regarded as a point particle P, grazing the edge of the sun (Fig. 6.25). The force F acting on it at any instant is

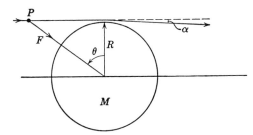

Figure 6.25. The deflection of light by the gravitational field of the sun.

given by

$$F = \frac{GMm}{R^2 \sec^2 \theta} \qquad \left(m = \frac{h\nu}{c^2} \right)$$

The component of F transverse to the direction of motion is

$$F \cos \theta = \frac{GMm \cos \theta}{R^2 \sec^2 \theta}$$

This force acts for a time dt, defined by

$$dt = \frac{1}{c} d(R \tan \theta) = \frac{R \sec^2 \theta \, d\theta}{c}$$

Thus the quantum acquires a certain transverse momentum, given by

$$F \cos \theta \, dt = \frac{GMm}{cR} \cos \theta \, d\theta$$

The total transverse momentum p is thus given by

$$p = \frac{GMm}{cR} \int_{\theta=-\pi/2}^{\pi/2} \cos \theta \, d\theta = \frac{2GMm}{cR}$$

But the longitudinal momentum is mc. Thus the direction of the quantum is changed by the angle α, where

$$\alpha = \frac{2GM}{c^2 R} \text{ radians}$$

Putting in the values $G = 6.66 \cdot 10^{-8}$ cgs, $M = 1.97 \cdot 10^{33}$ g, $R = 6.96 \cdot 10^{10}$ cm, we find $\alpha = 0.87''$ of arc. An effect of this sort has indeed been observed, by measuring the apparent positions of stars lying just outside the sun's disk. Such observations are possible only during total eclipses of the sun, and even then the magnitude of the deflection is only just outside the limits of experimental error. It appears to be well established, however, that the observed deflection is about twice the value calculated above.

6.21 THE GENERAL RELATIVITY PRINCIPLE

We have seen that the special principle of relativity asserts the equivalence of all inertial frames, in which a body under no forces moves in a straight line. This still allows us, as we said, to prefer an inertial frame to an accelerated frame in which extra forces, such as centrifugal forces, make their appearance. It is, however, difficult to justify this choice. The most challenging problem is presented by gravity, which appears as an acceleration that is the same for all bodies at a given place. In the language of mechanics, the gravitational force acting on a body is strictly proportional to its inertial mass (i.e. the mass that enters into Newton's law $F = ma$); the force would seem to disappear if all measurements were referred to a frame of reference accelerating with the local value of g. Should gravity then be regarded as a real force, or as a fictitious one that arises from a natural but unfortunate choice of reference frame on our part? In 1916 Einstein sought to answer this question by putting forward his general principle of relativity, according to which all frames of reference in *uniformly accelerated* motion relative to each other are equivalent, in the sense that no preferred system of this type exists. The detailed working out of this principle led to several observable consequences, and one of these—the so-called "precession of the perihelion" of the planet Mercury—does not seem to be explicable in any other way. The effect is that the major axis of Mercury's orbit rotates at the rate of about $43''$ of arc per century. General relativity gives precisely this result; special relativity does predict a precession, as in Sommerfeld's theory of fine structure of spectra (Chapter 5, Section 9), but the value is six times too small.

Relativity 171

When the problem of light in a gravitational field is treated by general relativity, it is found that the formula for the gravitational red shift is exactly the same as in our simple calculation, but that the theoretical deflection of light by the sun is just twice as great (1.75″) as we found it, and is therefore in full accord with observation. Thus, although special relativity gives a useful qualitative insight into the problems, it cannot be relied upon when gravitational accelerations are involved.

6.22 RELATIVITY OF INERTIA

It would be a pity to leave the subject of relativity without some mention of a very simple but profound idea that was first put forward by Ernst Mach (1893). The starting point was a famous experiment carried out by Newton with a bucket of water. Newton hung a bucket of water at the end of a twisted rope and then let it go. He noticed that the bucket began to spin rapidly while the water remained almost stationary and with a flat surface. Later, however, viscous drag made the water spin with the bucket, and its surface became concave owing to the rotation. Newton then stopped the bucket suddenly, but the water continued to rotate and its surface remained curved. Newton concluded from this that it was not rotation of the water relative to the bucket that was important, since this relative rotation was associated initially with a flat water surface, but finally with a curved surface; and he took the experimental results to imply that one could speak of rotation in absolute space. The bulging of the earth at the equator, and the behavior of Foucault's pendulum, could likewise be taken as evidence that the earth had an absolute rotation.

It was Mach who subjected these conceptions to a logical test. In effect he asked the question: "How do we measure the inertial mass of a body?" The conventional answer is that we measure the ratio of an applied force to the acceleration that it produces. But the measurement of absolute acceleration requires measurement of absolute displacement, whereas all that we can do physically is to measure displacement relative to other bodies. Mach therefore suggested that it is only by virtue of the presence of other bodies that a given body can be said to have inertial mass at all. Furthermore, Newton's bucket experiment and Foucault's pendulum seem to show that large masses at great distance are more important than small masses nearby in defining our inertial frame.

Now the universe of our experience is nearly but not quite isotropic.

In particular our own galaxy (the Milky Way) seems to be more or less disk-shaped, and, although it represents only a tiny fraction of all the matter visible to us, it would have a slight disturbing influence. The inertial mass of a body, inferred from its acceleration under a given force, would change by about 1 part in 10^7 if measured first in the plane of the Milky Way and then perpendicular to the plane. Attempts have been made to discover this effect by studying the frequency of oscillation of a mass on a spring, but so far the experimental errors have been just big enough to mask the expected variation. This principle of the relativity of inertia is, however, very much in keeping with the modern approach to physics, and it is to be hoped that it will in time be given a decisive test.

Problems

6.1. By what amount would the earth be shortened along its diameter (according to the Lorentz contraction hypothesis) as a result of its orbital motion round the sun?

6.2. The earth receives energy from the sun at the rate of $1.35 \cdot 10^3$ watts/m². Deduce from this the rate at which the sun is losing mass in the form of radiant energy.

6.3. Prove the identity

$$E^2 = p^2c^2 + (m_0c^2)^2$$

where p and E are the relativistic momentum and total energy, respectively, of a body of rest mass m_0.

6.4. Calculate the relativistic change of mass (in %) for an electron moving in the first Bohr orbit of (a) hydrogen ($Z = 1$), and (b) uranium ($Z = 92$).

6.5. Obtain the formula connecting wavelength and scattering angle in the Compton effect:

$$\lambda' - \lambda = \frac{h}{m_0 c}(1 - \cos\theta)$$

by treating the problem relativistically. This means that the conservation of energy must be stated in terms of total energy. [The electron initially at rest has energy m_0c^2; its final energy is $m_0c^2/(1 - \beta^2)^{1/2}$, and its final momentum is $m_0\beta c/(1 - \beta^2)^{1/2}$.]

6.6. Protons of 6 Bev (i.e., of a kinetic energy equivalent to that of one electronic charge accelerated through $6 \cdot 10^9$ volts) are produced in a high energy accelerator. Find the fractional difference between the velocity of the protons and the velocity of light: i.e. the value of $1 - \beta$. Show in general that, if a particle is highly relativistic (i.e., β approaching 1) the fractional difference between c and v is given approximately by $\frac{1}{2}(m_0c^2/E)^2$.

6.7. Obtain the formula for the drag coefficient [Section 18(2) of this chapter] by considering the invariance of phase of a wave in a medium of refractive index n, as judged by two observers—one at rest relative to the medium, the other moving with a relative velocity v.

6.8. A stationary electron is struck by a highly energetic photon of energy $E_\gamma \gg m_0c^2$. Show that the maximum possible kinetic energy of recoil of the electron is approximately equal to $E_\gamma - \frac{1}{2}m_0c^2$.

6.9. A particle has momentum p, kinetic energy T, and rest mass m_0. Prove that

$$\frac{p\beta c}{T} = \frac{(T/m_0c^2) + 2}{(T/m_0c^2) + 1}$$

Point out the significance of the extreme cases $T \ll m_0c^2$ and $T \gg m_0c^2$.

6.10. A beam of protons of kinetic energy 900 Mev is stopped in a target. The average current represents $5 \cdot 10^{13}$ protons per sec striking the target. Find (a) the power in watts dissipated in the target, and (b) the force exerted on the target by the incident beam. (1 Mev = $1.6 \cdot 10^{-6}$ erg; proton mass = $1.66 \cdot 10^{-24}$ g.)

6.11. The dark companion of the star Sirius is a dense star that causes an observable gravitational red shift of its emitted light. The amount of the shift is equal in size to the Doppler shift that would be observed for a source receding at a speed of about 19 km/sec. The mass of the star is $1.68 \cdot 10^{33}$ g. Find its density. (The constant of universal gravitation is $6.66 \cdot 10^{-8}$ dynes cm^2 g^{-2}.)

7 Wave Mechanics

7.1 MECHANICS AND GEOMETRICAL OPTICS

In 1650 Fermat enunciated a principle by which the known laws of geometrical optics could all be derived. This was the so-called "principle of least time," and it stated that, of all the possible paths that a ray of light might take between two points, the actual path was that for which the time taken was a minimum. If light is passing from A to B through a medium whose refractive index n varies from one place to another, then Fermat's principle can be expressed mathematically in the form

$$\delta \int_A^B \frac{n\,ds}{c} = 0$$

The operation of δ is to vary the path subject to the condition that the end points A and B remain fixed. When expressed in this form the principle is one of *stationary* time, i.e., either maximum or minimum, and so is more general than Fermat's original statement. Examples of both possibilities can be found in the solution of optical problems: e.g., in the reflection or refraction of light at a plane boundary a condition of minimum time applies, but, in determining the focusing properties of a concave mirror, the appropriate condition may be one of maximum time.

We have already mentioned (Chapter 4, Section 12) the principle of least action put forward by Maupertuis (1740) and given mathematical form by Euler and Lagrange. This, like the principle of least time, can be enlarged into a stationary condition:

$$\delta \int_A^B mv\, ds = 0, \quad \text{i.e.} \quad \delta \int_A^B [2m(E - V)]^{1/2}\, ds = 0$$

where E is the total energy of a particle moving in a conservative field of force in which its potential energy V is a function of position. There is thus a formal similarity between the laws governing the path of a ray of light in a refracting medium and the laws governing the path of a particle in a potential field.

In 1924 Louis de Broglie examined the possibility that this analogy was more than a formal one. The starting point was the knowledge that light is transmitted in discrete quanta, although all the laws governing its propagation, refraction, and interference are characteristic of wave motion. It is of very great interest to read how de Broglie summed up his appraisal of the situation:*

I shall take it that there is reason to suppose the existence, in a wave, of points where energy is concentrated, of very small corpuscles whose motion is so intimately connected with the displacement of the wave that a knowledge of the laws regulating one of these motions is equivalent to a knowledge of the laws governing the other. Conversely, I shall suppose that there is reason to associate wave propagation with the motion of all the kinds of corpuscles whose existence has been revealed to us by experiment. · · · I shall take the laws of wave propagation as fundamental, and seek to deduce from them, as consequences which are valid in certain cases only, the laws of dynamics of a particle.

It is not difficult to recognize this as a remarkably bold and imaginative step, and we must next see how these ideas were given definite form.

7.2 THE FORMAL BASIS OF WAVE MECHANICS

Let us consider a particle of mass m_0. For such a particle de Broglie assumed that there was some characteristic vibration of a frequency ν_0 given by Einstein's mass-energy relation:

$$h\nu_0 = m_0 c^2$$

It is thus possible to write the equation of the vibration associated with

* L. de Broglie, *J. Phys. Rad.* **7**, 1 (1926): "Sur le parallélisme entre la dynamique du point matériel et l'optique géométrique." (Quoted in this translated form by kind permission of M. Louis de Broglie and the publishers of *Le Journal de Physique*.)

a particle:
$$\psi = \sin 2\pi \nu_0 t_0$$

This clearly represents some stationary vibration attached to the particle. Let us now consider what this vibration looks like when viewed from some fixed origin, relative to which the particle itself is moving with a velocity v, say in the x direction. Then, according to the transformations of special relativity (Chapter 6, Section 12) we have

$$t_0 \rightarrow \frac{t - vx/c^2}{(1 - \beta^2)^{1/2}}$$

and so
$$\psi = \sin 2\pi \nu_0 t_0 \equiv \sin 2\pi \nu \left(t - \frac{vx}{c^2}\right)$$

with
$$\nu = \frac{\nu_0}{(1 - \beta^2)^{1/2}}$$

The vibration attached to the particle thus appears to us as a progressive wave. We compare our equation for ψ with the equation to a wave of velocity w:

$$\psi = \sin 2\pi \nu \left(t - \frac{x}{w}\right)$$

Evidently we must put
$$w = \frac{c^2}{v}$$

The fact that the wave velocity is inversely proportional to the particle velocity is in accord with what one would expect from a comparison of the principles of Fermat and Maupertuis.

We can associate a wavelength λ (the de Broglie wavelength) with the vibration, by putting

$$\lambda = \frac{w}{\nu} = \frac{c^2}{v} \cdot \frac{(1 - \beta^2)^{1/2}}{\nu_0} = \frac{h(1 - \beta^2)^{1/2}}{m_0 v} = \frac{h}{mv}$$

In the last step of the calculation we have used the formula expressing the relativistic variation of mass with velocity (Chapter 6, Section 16). We thus have a very important result:

$$\lambda = \frac{h}{p}$$

where p = momentum of particle. Finally, we can evaluate the

Wave Mechanics 177

energy $h\nu$ of the vibration as observed by us. We have

$$h\nu = \frac{h\nu_0}{(1-\beta^2)^{1/2}} = \frac{m_0 c^2}{(1-\beta^2)^{1/2}} = mc^2$$

Thus, if we write the total energy of the particle as E, we have the condition

$$E = h\nu$$

in all frames of reference: i.e., the familiar Einstein relation between energy and frequency.

7.3 WAVE AND GROUP VELOCITIES

We have shown that the postulation of a progressive wave associated with a moving particle is consistent with special relativity. But does it have any physical meaning? If the particle velocity v is less than c, as it must be, then the wave velocity w is greater than c. It would therefore appear that the particle and its associated wave are bound to part company, even if they are initially together. To draw this inference would, however, be to overlook a very important feature of classical wave motion—the distinction between what are called *wave* and *group* velocities. If a wave is described by the equation

$$\psi = \sin 2\pi(\nu t - kx)$$

we have the relations:

$$\text{Wave velocity} \quad w = \frac{\nu}{k}$$

$$\text{Group velocity} \quad g = \frac{d\nu}{dk}$$

Now, for the wave that we have supposed to belong to a particle, we have

$$\nu = \frac{\nu_0}{(1-\beta^2)^{1/2}}, \quad \therefore \frac{d\nu}{d\beta} = \frac{\beta\nu_0}{(1-\beta^2)^{3/2}}$$

$$k = \frac{\beta\nu_0}{c(1-\beta^2)^{1/2}}, \quad \therefore \frac{dk}{d\beta} = \frac{\nu_0}{c(1-\beta^2)^{3/2}}$$

Hence
$$g = \beta c = v$$

Thus we have the very satisfactory result that the velocity with which energy is transmitted in a small group or packet of waves with frequencies in the neighborhood of ν is just the velocity of the particle itself.

[**Derivation of Group Velocity Formula**

Suppose two waves of equal amplitude but slightly differing frequency and wavelength:

$$y_1 = a \sin 2\pi(\nu t - kx)$$

$$y_2 = a \sin 2\pi[(\nu + \delta\nu)t - (k + \delta k)x]$$

Each of these represents a pure sinusoidal wave extending to infinity along the x axis. Together they give a resultant disturbance y, where

$$y = y_1 + y_2 \approx 2a \cos 2\pi \left(\frac{\delta\nu}{2} t - \frac{\delta k}{2} x \right) \sin 2\pi(\nu t - kx)$$

This represents an oscillation of the original frequency ν, but with a modulated amplitude as shown in Fig. 7.1. A given segment of the

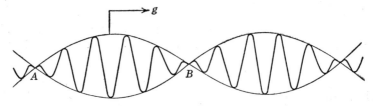

Figure 7.1. To illustrate the concept of group velocity in the superposition of two waves.

wave system, such as AB, can be regarded as a wave group and moves with a certain velocity g. Since $\delta k \ll k$, this segment contains a large number of oscillations of the primary wave, and the crests of this primary wave travel with a velocity w that is in general different from g. By inspection of the equations, we see that we must have

$$w = \frac{\nu}{k}, \qquad g = \frac{\delta\nu/2}{\delta k/2} \rightarrow \frac{d\nu}{dk}$$

Of course the modulation of the wave is indefinitely repeated in this case. More generally, however, the technique of Fourier analysis can be used to demonstrate that any isolated packet of oscillatory disturbance of frequency ν, e.g. a single quantum of light, can be described in terms of a combination of infinite trains of frequencies

Wave Mechanics

distributed around ν, and the velocity of the packet is given by the formula for the group velocity at ν as we have just derived it.

Of course, it is conceivable that g and w are equal. If so, we have

$$\frac{d\nu}{dk} = \frac{\nu}{k}$$

$$\therefore \log \nu = \log k + \log \text{const}$$

i.e.,
$$\frac{\nu}{k} = \nu\lambda = \text{const}$$

Thus the velocity of the pure sinusoidal waves must be independent of frequency (or wavelength). This is true, for example, with light in vacuum. More usually, however, we have the phenomenon of dispersion: i.e., a dependence of wave velocity on wavelength, and in such cases g and w are different, and may be radically so.]

7.4 EXPERIMENTAL CONFIRMATION OF MATTER WAVES

De Broglie's treatment of this problem stood for three years before being tested, and then in the years 1927–1928 the diffraction of electrons was experimentally demonstrated by Davisson and Germer (reflection from a crystal face) and by G. P. Thomson (transmission through thin foils). To consider the orders of magnitude involved, let us calculate the wavelength of an electron that has been accelerated through a potential difference V. To simplify the calculation we shall suppose that the kinetic energy acquired (i.e., Ve) is much less than the rest mass energy, so that we can use nonrelativistic formulas.

Then
$$mv = (2meV)^{1/2}$$

$$\therefore \lambda = \frac{h}{(2meV)^{1/2}}$$

If V is expressed in volts, we have

$$\lambda = \frac{6.625 \cdot 10^{-27}}{(2 \times 9.11 \cdot 10^{-28} \times 4.80 \cdot 10^{-10} \times \frac{1}{300})^{1/2}} \frac{1}{\sqrt{V}} \text{ cm}$$

i.e.,
$$\lambda = \left(\frac{150}{V}\right)^{1/2} \cdot 10^{-8} \text{ cm}$$

Thus for V equal to about 10 to 100 volts, the electron wavelength λ is comparable with interatomic distances in crystals.

The original experiments of Davisson and Germer were done with a nickel cubic crystal cut perpendicular to a cube diagonal (Fig. 7.2a). The nickel lattice is a face-centered cubic structure, and this cut gives a simple diffraction grating structure, with a line spacing d of 2.15 A ($= 2.15 \cdot 10^{-8}$ cm). The electrons were directed normally onto this face, and a reflected maximum was observed at 50° to the normal when the accelerating potential was 54 volts (Fig. 7.2b). Of course, there

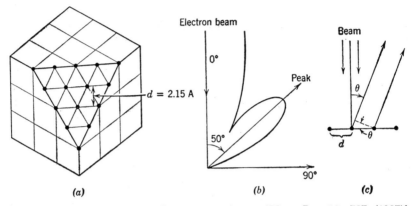

Figure 7.2. The Davisson–Germer experiment [*Phys. Rev.* **30**, 707 (1927)]. (a) Showing the face-centered Ni crystal cut to reveal a simple grating of spacing 2.15 A. (b) An experimental curve obtained for electrons accelerated through 54 volts, showing an intense back reflection and a secondary peak at 50°. (c) To show the interference condition. The low-energy electrons used in this experiment do not penetrate the crystal appreciably; so the interference depends primarily on scattering by atoms in the two-dimensional surface layer, as for an optical diffraction grating.

was also an intense reflection back along the path of the original electron beam. By strict analogy with the behavior of an ordinary optical grating, we expect a first-order maximum of intensity at an angle given by

$$d \sin \theta = \lambda$$

as may be seen from Fig. 7.2c. Thus we infer $\lambda = 2.15 \sin 50° = 1.65$ A. On the other hand, the calculated wavelength is given by

$$\lambda = (\tfrac{150}{54})^{1/2} = 1.67 \text{ A}$$

The agreement is as nearly perfect as the experimental uncertainties allow.

The experiments by G. P. Thomson were made with much faster electrons, accelerated through some tens of kilovolts. A beam of such electrons was allowed to pass through a thin foil (about 1000 A thick)

Wave Mechanics

of some polycrystalline material: i.e. one in which large numbers of microscopic crystals were oriented at random. The technique was similar to that used in the powder method for X rays, the principle being that rays of a given wavelength will pick out those crystals whose planes are suitably oriented so as to give strong reflections according to the Bragg equation [Chapter 5, Section 12(2)]:

$$2d \sin \theta = n\lambda$$

where n is an integer, and d is the spacing between planes of atoms (Fig. 7.3a). If all possible crystal orientations are present, the diffracted

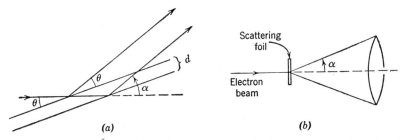

Figure 7.3. Diffraction of fast electrons [after G. P. Thomson, *Proc. Roy. Soc. A,* **117**, 600 (1928)]. (a) Illustrating the Bragg condition for reflection from successive planes within the structure. (b) A cone of scattered electrons from similar planes, all oriented at an angle $\alpha/2$ to the incident beam, in a polycrystalline aggregate.

rays (or electrons) corresponding to some particular angle of scattering $\alpha(=2\theta)$ will generate a cone whose axis lies along the direction of the incident beam (Fig. 7.3b). Thus the electrons received on a screen some distance behind the scatterer will fall on a series of concentric circles. Evidently this method is facilitated if the wavelength λ is made much less than the atomic spacings d, so that the angles α are small. For example at 60 kev the electron wavelength is $5 \cdot 10^{-10}$ cm; so with $d = 2 \cdot 10^{-8}$ cm we should have

$$\alpha \approx n \frac{5 \cdot 10^{-10}}{2 \cdot 10^{-8}} = 0.025n \text{ radians}$$

i.e., $\quad \alpha \approx 1.5n$ degrees

Under these conditions, therefore, a screen or photographic plate set up say 50 cm behind the scattering foil would show a ring pattern a few centimeters across.

This pioneer work left no doubt that electrons, so confidently regarded as particles, exhibited wave properties just as certainly as

X rays were known to do. The duality of wave and corpuscle that had been forced on physicists in order to describe the properties of radiation must be extended to matter also, exactly as de Broglie had proposed. It was clearly of interest, however, to verify such behavior for ordinary atoms also, and in 1930 Estermann and Stern succeeded in doing this. The experiment was performed with neutral molecules at ordinary thermal energies. In this case the energy is small—the mean energy of one molecule at 300° K is equivalent to only $\frac{1}{40}$ ev—but the mass is of course much greater than for electrons, and the result is that the de Broglie wavelength again comes out to be comparable with lattice spacings in crystals. To take an example, for helium atoms at 400° K, we have

$$\tfrac{1}{2}mv^2 = \tfrac{3}{2}kT$$

$$\therefore \ mv = (3mkT)^{1/2}$$

$$= (3 \times 6.7 \cdot 10^{-24} \times 1.4 \cdot 10^{-16} \times 400)^{1/2}$$

$$\approx 10^{-18} \text{ g cm/sec}$$

$$\therefore \ \lambda \approx \frac{6.6 \cdot 10^{-27}}{10^{-18}} \approx 0.6 \text{ A}$$

Estermann and Stern scattered atoms of this and similar energies at the face of a LiF crystal. The scattered atoms in a specified direction were detected with an extremely sensitive pressure gage (able to respond to changes of 10^{-8} mm Hg), and clearly showed, in addition to a normal reflection, two "wings" of just the same type as Davisson and Germer had found for electrons (see Fig. 7.4).

In recent years, long after the establishment of wave mechanics, the wave properties of material particles have been widely exploited through the use of neutrons. With the help of slow neutrons of a controlled speed and wavelength, many diffraction studies have been carried out. They can be particularly useful in revealing the presence and position of hydrogen atoms in crystal structures, since hydrogen atoms are very effective in scattering neutrons, whereas they are almost undetectable in ordinary X-ray diffraction photographs. Neutrons as used in this way have wavelengths of the order of 1 A (10^{-8} cm), but it is also possible to confirm their wave properties for very much smaller wavelengths. For example, neutrons with energies of about 10 Mev are rather readily produced in nuclear reactions. Such neutrons have wavelengths of about 10^{-12} cm, a distance that is comparable with the diameters of most nuclei; and when these fast neutrons are scattered by nuclei, the result is a characteristic diffraction pattern, of just the sort we would get for plane waves of sound scattered by a solid sphere, or for light diffracted by a tiny obstacle [see Chapter 9, Section 18(2)]. In short, the wave behavior of matter is a universal property, and whenever it has been tested quantitatively the result has been to demonstrate the correctness of de Broglie's theory.

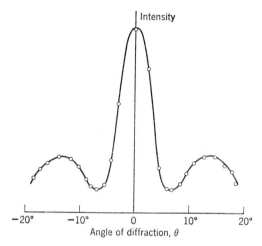

Figure 7.4. The diffraction of neutral atoms at the surface of a crystal [after Estermann and Stern, *Z. Physik*, **61**, 95 (1930)]. θ is the angle of departure from a regular (specular) reflection.

7.5 WAVE MECHANICS AND THE ATOM

We know that geometrical optics, i.e. the treatment of optical problems in terms of rays, breaks down when the width of the light beam or of obstacles in its path becomes comparable with the wavelength. In the same way we may expect ordinary mechanics to break down when we are considering systems whose size is of the order of magnitude of the wavelength of a particle. It is plain that this will not happen even in the smallest ordinary physical systems in our experience (Compare, for example, the de Broglie wavelength to the diameter for a dust particle of radius 10^{-5} cm moving with a speed of 1 cm per sec.) But let us consider an electron in a hydrogen atom. The ionization potential of H is about 13 volts, and the kinetic energy of the electron in its orbit is also equal to about 13 ev, as the solution of the problem shows (Chapter 5, Section 4). Thus we have

$$T = \tfrac{1}{2}mv^2 = Ve \qquad \text{with } V = 13 \text{ volts}$$

This is identical in form with the expression for the energy of electrons accelerated through the potential V, so we can use our formula:

$$\lambda = \left(\frac{150}{V}\right)^{1/2} 10^{-8} \text{ cm} = 3.4 \cdot 10^{-8} \text{ cm}$$

Also, from Bohr's theory of the hydrogen atom, we have the radius of the orbit given by

$$r = \frac{e^2}{2T} = \frac{2.3 \cdot 10^{-19}}{2 \times 13 \times 1.6 \cdot 10^{-12}}$$

$$= 0.53 \cdot 10^{-8} \text{ cm}.$$

The general correctness of this estimate of r is confirmed by the known sizes of atoms (cf. Chapter 1). Thus we have $r < \lambda$, so that we are in the region where wave properties predominate, and the question now arises: Can we use this result to predict special features of the situation?

7.6 STATIONARY STATES

We have seen that, when he introduced the quantum theory of spectra and the atom in 1913, Bohr had to state, without justification, that an electron in an atom could not be in any arbitrary orbit, but only in orbits for which one could write

$$mv \times 2\pi r = nh$$

where n is an integer. We have seen further (Chapter 5, Section 8) how this represents a quantization of the "action" as described by the phase integral. In Bohr's theory there is no way of explaining why the electron should fail to radiate so long as it remains in such a quantum state, since it is still, on Bohr's view, an accelerated point charge. The wave-mechanical description really does away with this difficulty, however, since we now think of a wave whose wavelength is $\lambda = h/mv$. Putting $mv/h = 1/\lambda$, Bohr's quantum condition becomes

$$\frac{2\pi r}{\lambda} = n$$

i.e., $$2\pi r = n\lambda$$

Thus the allowed orbits are just those that will exactly accommodate the de Broglie waves, allowing a perfect join when the circuit is complete. In such a picture there is no suggestion of progressive motion;

Wave Mechanics 185

we have instead the kind of situation presented by a stretched string with both ends fixed. The only permitted solutions in this latter case are those having nodes at the ends, and this at once defines for us the wavelengths, and hence the frequencies, of the possible stationary vibrations of the system. Any one such stationary vibration (or normal mode, as it is sometimes called) goes on indefinitely once it has been excited, subject only to the effects of damping, so that wave mechanics has given us an entirely natural description of stationary states in atoms.

7.7 SCHRÖDINGER'S EQUATION

In 1925, soon after de Broglie had put forward his ideas, Schrödinger used them to set up a wave equation to describe this new mechanics of particles, so that it could be applied to various atomic problems.

We have already supposed a wave of amplitude ψ associated with a particle, with the equation

$$\psi = \sin 2\pi\nu \left(t - \frac{x}{w}\right)$$

ψ is called the *wave function*. Let us differentiate ψ twice with respect to x.

Then
$$\frac{\partial^2 \psi}{\partial x^2} = -\frac{4\pi^2 \nu^2}{w^2}\psi = -\frac{4\pi^2}{\lambda^2}\psi \tag{7.1}$$

We can substitute

$$\frac{1}{\lambda^2} = \frac{m^2 v^2}{h^2} = \frac{2m(\tfrac{1}{2}mv^2)}{h^2}$$

For particles that are not moving too fast (i.e., $v \ll c$), we can write

$$m = \text{rest mass} = \text{const}$$

$$\tfrac{1}{2}mv^2 = E - V(x)$$

where E is the total energy, $V(x)$ the potential energy of the particle at x. For this case, therefore, equation (7.1) becomes

$$\frac{\partial^2 \psi}{\partial x^2} + \frac{8\pi^2 m}{h^2}[E - V(x)]\psi = 0 \tag{7.2}$$

This is Schrödinger's equation (or one form of it) and can be regarded as the equation governing the propagation of a wave in a medium whose refractive index varies from one place to another. We must remember that the wave velocity is inversely proportional to the particle velocity, so that acceleration of the particle from one region of space to another corresponds to moving it into a region of higher refractive index and hence of *lower* wave velocity. If such an acceleration takes place across a plane boundary, the ray direction (normal to the wave front) is therefore bent toward the normal, and we have the

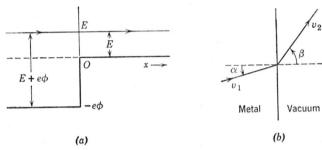

Figure 7.5. (a) The potential energy diagram for electrons crossing the potential step at the surface of a metal. (b) Refraction of an electron beam at the boundary.

same law of refraction whether we treat the problem as one of particle motion or of wave motion. It is to be noted that in this case it is the wave velocity, not the group velocity, that determines the behavior of the wave when refracted; the same is true for light entering a dispersive medium. To make the situation quite explicit, consider the problem (illustrated in Fig. 7.5a) of an electron crossing a potential "step," as it would do, for example, in emerging from a metal into a vacuum and so losing an amount of kinetic energy corresponding to the work function ϕ. Outside the metal the potential energy is zero, so we have

$$v_2 = \left(\frac{2E}{m}\right)^{1/2}$$

where E is the total energy of the electron, excluding the rest mass energy mc^2. Inside the metal the potential energy is $-e\phi$, so the kinetic energy is $E + e\phi$, and hence

$$v_1 = \left[\frac{2(E + e\phi)}{m}\right]^{1/2}$$

Since the electron suffers no tangential force in crossing the boundary, the tangential component of velocity must be unchanged. Hence

$$\frac{\sin \beta}{\sin \alpha} = \frac{v_1}{v_2} = \left(\frac{E + e\phi}{E}\right)^{1/2} \quad \text{(See Fig. 7.5}b\text{).}$$

This represents the corpuscular description. But, if we wish to describe such a refraction process in terms of waves, we make use of Snell's law:

$$\frac{\sin \beta}{\sin \alpha} = \frac{w_2}{w_1}$$

and, since w is proportional to $1/v$, we at once see that this is identical with the other result.

Schrödinger's equation in the form (7.2) can be used for all problems in which one has either a stationary state (as typified by the electron in a Bohr orbit) or a steady state (as in determining the path of a continuous beam of electrons in a fixed potential field of some kind). There are other types of problem, however, which are essentially nonstationary (e.g., a transition between two states of an atom resulting in the production of a quantum, or the motion of a single particle, described by a wave packet, from one point to another) and for such cases as these another form of Schrödinger's equation may be set up. We have so far always described the wave function ψ by the equation $\psi = \sin 2\pi\nu(t - x/w)$. What we are primarily concerned with, however, is the wave nature of the motion, and it would seem that we could just as well have used a cosine function, or an arbitrary combination of sine and cosine, to take account of all possible phases of the wave. Equation (7.2) would be valid for all such cases. This apparent freedom of choice is, however, illusory; we must take account of the connection between energy and frequency ($E = h\nu$), together with the relationship (nonrelativistic) between energy and momentum:

$$E = \frac{p^2}{2m} + V(x) \tag{7.3}$$

When these conditions are imposed, we find that ψ must be essentially a single exponential; it is described, not by real sine or cosine functions, but by a complex expression of the form

$$\psi = \exp[\pm i \cdot 2\pi\nu(t - x/w)] = \exp[\pm i \cdot 2\pi(\nu t - kx)]$$

It is customary to take the negative value of the exponent, in which case we have

$$\frac{\partial \psi}{\partial t} = -i \cdot 2\pi\nu\,\psi = -i\,\frac{2\pi E}{h}\,\psi$$

i.e.,
$$E\psi = \frac{ih}{2\pi}\frac{\partial \psi}{\partial t}$$

Substituting this value of $E\psi$ in equation (7.2), we arrive at the so-called *time-dependent* Schrödinger equation:

$$\frac{ih}{2\pi}\frac{\partial \psi}{\partial t} = -\frac{h^2}{8\pi^2 m}\frac{\partial^2 \psi}{\partial x^2} + V(x)\psi \qquad (7.4)$$

It may be noted that equation (7.4) corresponds term by term to the energy conservation expressed by equation (7.3), since we also have the identity

$$-\frac{h^2}{8\pi^2 m}\frac{\partial^2 \psi}{\partial x^2} = \frac{h^2}{2m}k^2\psi = \frac{(h/\lambda)^2}{2m}\psi = \frac{(mv)^2}{2m}\psi$$

7.8 PROPERTIES OF THE WAVE FUNCTION

Let us fix attention on the steady-state form of the wave equation:

$$\frac{\partial^2 \psi}{\partial x^2} + \frac{8\pi^2 m}{h^2}[E - V(x)]\psi = 0$$

In looking for solutions of this equation, we recognize two distinct situations:

(a) $E > V$

In this case we may put

$$\frac{8\pi^2 m}{h^2}(E - V) = 4\pi^2 k^2$$

where k is the wave number, i.e. the reciprocal wavelength. The solution to the equation becomes

$$\psi_1 = \exp[\pm\, i \cdot 2\pi(\nu t - kx)],$$

as we have already found, and describes a progressive motion.

(b) $E < V$

This represents a situation that is inconceivable in the classical mechanics of particles, since it would imply a negative kinetic energy. There is nothing to rule it out in wave mechanics, however, and we put

$$\frac{8\pi^2 m}{h^2} (V - E) = \alpha^2$$

so that the wave function assumes the form

$$\psi_2 = \exp(\pm i \cdot 2\pi \nu t) \exp(\pm \alpha x)$$

This is no longer a progressive wave, but a stationary vibration whose amplitude varies exponentially with distance. Here we have the basis of the "tunnel effect," which allows particles to cross barriers that are quite impenetrable classically. It is important to realize, however, that this behavior need come as no surprise if we take a wave description seriously, for we have precisely this type of phenomenon in optics. Consider a ray of light (I) traveling through a block of glass (Fig. 7.6) and striking the boundary between glass and air at A. Then if the angle of incidence θ is greater than the critical angle, we have total internal reflection; i.e., all the incident energy is carried away in a reflected ray R, and in a ray description the light never emerges into the air. The detailed solution of Maxwell's equations shows, however, that there exists an oscillating electromagnetic field just outside the glass surface; its form is that of ψ_2: i.e., a stationary vibration without transport of energy. If now a second block of glass is brought close to the first one, this oscillating field is picked up at B and transformed once more into a progressive wave T. The effect was observed experimentally by Newton (*Opticks*, 4th ed, 1730, Book 3, Query 29).

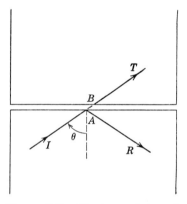

Figure 7.6. The transmission of a ray of light across an air gap for an angle of incidence θ exceeding the critical angle.

The fact that ψ is essentially complex means that it has no direct physical interpretation, but it was proposed by Born (1926) that the quantity $|\psi|^2$ should be understood to represent a probability density.

For example, in the case of a single particle constrained to move only along the x axis, the quantity $|\psi(x)|^2\, dx$ is the probability that the particle should be found between x and $x + dx$. This interpretation has been generally accepted, and it carries with it the concept of the "normalization" of wave functions, by which we mean that the amplitude of ψ is adjusted so that, if we are dealing with a single particle,

$$\int_{\text{all space}} |\psi|^2\, d\tau = \int_{\text{all space}} \psi^*\psi\, d\tau = 1$$

(The function ψ^* is called the complex conjugate of ψ, and is obtained from ψ by replacing $+i$ with $-i$ throughout.) When the wave function is normalized in this way, the total probability that the particle it describes should be *somewhere* is unity, in accordance with the conventions of probability theory.

The interpretation of $\psi^*\psi$ as a probability density can be understood if we make use of the time-dependent Schrödinger equation. Let us consider a one-dimensional problem. We have

$$\psi \sim \exp[-i \cdot 2\pi(\nu t - kx)]$$
$$\psi^* \sim \exp[+i \cdot 2\pi(\nu t - kx)]$$

and the corresponding equations governing ψ and ψ^* are

$$+\frac{ih}{2\pi}\frac{\partial \psi}{\partial t} = -\frac{h^2}{8\pi^2 m}\frac{\partial^2 \psi}{\partial x^2} + V(x)\psi$$

$$-\frac{ih}{2\pi}\frac{\partial \psi^*}{\partial t} = -\frac{h^2}{8\pi^2 m}\frac{\partial^2 \psi^*}{\partial x^2} + V(x)\psi^*$$

Multiply the first of these by ψ^* throughout, the second by ψ, and subtract: Then

$$\frac{ih}{2\pi}\left(\psi^*\frac{\partial \psi}{\partial t} + \psi\frac{\partial \psi^*}{\partial t}\right) = -\frac{h^2}{8\pi^2 m}\left(\psi^*\frac{\partial^2 \psi}{\partial x^2} - \psi\frac{\partial^2 \psi^*}{\partial x^2}\right)$$

i.e.,
$$\frac{\partial}{\partial t}(\psi^*\psi) = \frac{ih}{4\pi m}\left(\psi^*\frac{\partial^2 \psi}{\partial x^2} - \psi\frac{\partial^2 \psi^*}{\partial x^2}\right)$$

or
$$\frac{\partial}{\partial t}(\psi^*\psi) = \frac{ih}{4\pi m}\frac{\partial}{\partial x}\left(\psi^*\frac{\partial \psi}{\partial x} - \psi\frac{\partial \psi^*}{\partial x}\right)$$

Let us now integrate both sides of this last equation between definite limits x_1 and x_2. Then

$$\frac{\partial}{\partial t}\int_{x_1}^{x_2} \psi^*\psi\, dx = \frac{ih}{4\pi m}\left[\psi^*\frac{\partial \psi}{\partial x} - \psi\frac{\partial \psi^*}{\partial x}\right]_{x_1}^{x_2}$$

But, with the forms assumed for ψ, ψ^*, we have

$$\frac{\partial \psi}{\partial x} = +i \cdot 2\pi k \psi, \qquad \frac{\partial \psi^*}{\partial x} = -i \cdot 2\pi k \psi^*$$

Hence

$$\frac{\partial}{\partial t} \int_{x_1}^{x_2} \psi^* \psi \, dx = \left[-\frac{hk}{m} \psi^* \psi \right]_{x_1}^{x_2}$$

But by the de Broglie relationship,

$$\frac{hk}{m} = \frac{h}{m\lambda} = v \quad \text{simply}$$

Thus finally,

$$\frac{\partial}{\partial t} \int_{x_1}^{x_2} \psi^* \psi \, dx = (v\psi^* \psi)_{x_1} - (v\psi^* \psi)_{x_2}$$

If we compare this equation with the equation for the flow of a fluid or an electric current, we see that $\psi^* \psi$ plays the role of a density ρ. The left-hand side is the change with time of the total amount of a quantity contained between planes at x_1 and x_2 (Fig. 7.7); the right-hand side is the net rate at which the quantity flows into the region, as given by the difference between the currents entering and leaving it:

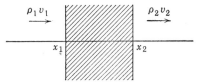

Figure 7.7. Illustrating the probability current in wave mechanics.

$$\frac{\partial}{\partial t} \int_{x_1}^{x_2} \rho \, dx = \rho_1 v_1 - \rho_2 v_2$$

The normalization of the wave function implies that, when it describes one particle or a limited number of particles, the value of ψ (and so $\partial \psi / \partial x$) must tend to zero at $\pm \infty$. This is an important condition in determining the possible stationary states of a system. For example, if the mathematical solution of Schrödinger's equation is of the form $e^{\pm \alpha x}$ (α real and positive) in a region of positive x extending to infinity, then only the solution with a negative exponent is physically admissible in this region. We shall consider actual examples of this shortly.

7.9 THE UNCERTAINTY PRINCIPLE

The description of material particles in terms of waves, and the setting up of a connection between energy and momentum, on one side, and frequency and wavelength, on the other, lead to an indefiniteness that had not been previously suspected in the mechanics of particles. This indefiniteness has become famous as the "uncertainty principle," put forward by Heisenberg in 1927. We have seen how the only type of disturbance that can be described by a pure frequency in wave motion is an infinitely extended wave train. If we wish to construct a disturbance that exists over a limited range of distance, we must combine a number of pure waves of neighboring frequencies and wavelengths. The essential features of the problem are shown by the

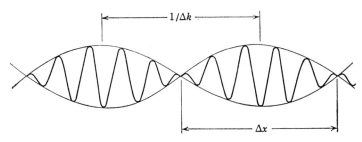

Figure 7.8. The relation between the length of a wave group (or wave packet) and the range of wave numbers contained in it.

simple combination of two sine waves which we discussed in connection with group velocities. For waves differing in wave number by Δk, the distance between maxima of the modulated disturbance (Fig. 7.8) is $1/\Delta k$. If, then, we denote the length of one segment, between nodes, as Δx, we have the simple relation

$$\Delta x \cdot \Delta k = 1$$

If we do not restrict ourselves to a combination of two waves, but draw on a continuous spectrum of frequencies within the range Δk, then a more detailed analysis (see Problem 3 at the end of this chapter) shows that the length of the corresponding wave packet can be reduced, such that

$$\Delta x \cdot \Delta k \approx \frac{1}{2\pi}$$

We have to use an approximate equality here, because the details of the appropriate spectrum will depend on the exact profile (i.e. shape) of the single wave group to be described.

Now, according to wave mechanics, we put

$$k = \frac{mv}{h} = \frac{p}{h}$$

Wave Mechanics 193

Hence we have $\Delta p \cdot \Delta x \approx \dfrac{h}{2\pi}$, and so it is impossible to define both the momentum and the position of a particle to an arbitrarily high degree of precision. This is one statement of the uncertainty principle. But if a particle of velocity v may be located anywhere within a distance Δx, its description in time is uncertain by the amount $\Delta t = \Delta x/v$. Also, if its kinetic energy is E, we have

$$\Delta E = mv \, \Delta v = v \, \Delta p$$

We can therefore state the uncertainty principle in the alternative form

$$\Delta E \cdot \Delta t \approx \frac{h}{2\pi}$$

The energy of a particle cannot be measured perfectly unless we take an infinite time to do it. We can relate this result, also, directly to ordinary wave theory if we consider a short pulse of oscillation lasting a time Δt. This cannot be described by a pure frequency but must be built up (or analyzed) in terms of a band of frequencies extending over a range $\Delta \nu$. (Hence, for example, the need for wide-band amplifiers to handle the short "pips" contained in radar signals.) The relation $\Delta \nu \cdot \Delta t \approx 1/2\pi$, together with the Einstein equation $E = h\nu$, then leads at once to this second form of the uncertainty principle.

One of the most interesting features of the uncertainty principle is that it calls for a review of our ideas about the process of *measuring* such quantities as momentum and energy. Despite the purely formal conclusions that we have reached, it might seem possible to devise an experiment to measure the precise position of a particle of known momentum. But here we run into a situation that is peculiar to quantum physics. The fact of making an observation disturbs any physical system, and nothing can be done to avoid or allow for the disturbance; its size is such as to be an expression of the uncertainty principle. This can be illustrated by a famous example suggested by Bohr (1928). Suppose that we wish to look at an electron through a microscope (Fig. 7.9). Then we must provide illumination of some sort, but this means (because of the Compton effect) that the electron recoils somewhat. To reduce this disturbance as far as possible, we use a very weak source of light and suppose that the presence of the electron can be detected if just one scattered photon enters the microscope objective. The direction of the photon may, however, lie anywhere within the angle θ subtended at the electron by the objective lens. Thus the electron is given a recoil momentum that is uncertain by $\Delta p = (h\nu/c) \times \theta$ approximately: i.e., $h\theta/\lambda$. By increasing λ we can reduce this undetermined momentum indefinitely, but another consideration comes in—the imperfect resolution of the microscope. If the diameter of the objective is d, diffraction makes the direction of light passing through it uncertain by an angle λ/d approximately. At distance l below the objective this represents an uncertainty in position given by $\Delta x = l\lambda/d$ approximately. Hence $\Delta p \cdot \Delta x \approx (h\theta/\lambda) \cdot (\lambda/\theta) = h$. Much ingenuity has been expended in attempts to cheat the uncertainty principle by suitable hypothetical experiments, but always without success.

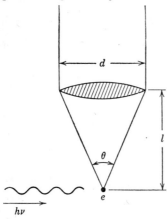

Figure 7.9. The hypothetical "microscope experiment" to show the limitations imposed by Heisenberg's uncertainty principle.

Another matter that is related to the uncertainty principle is the impossibility of observing, at one and the same time, both the corpuscular and the wave aspects of either material particles or quanta. Let us consider a beam of electrons or X rays of a single energy or wavelength, falling on two parallel slits, as in Young's experiments to show the interference of light (Fig. 7.10). The small width of the slits leads to spreading of the waves by diffraction, and in the shaded region (see figure) the two diffracted beams overlap, so that a screen placed in this region will exhibit the evenly spaced interference bands characteristic of a two-slit system. The whole situation can be described in terms of a pure wave theory, with a division of the wave front so that part of the wave passes through A, and part through B. Yet we must remember that the apparently continuous distribution constituting the interference pattern is really built up by the arrival of individual electrons or quanta, as a short-term exposure of a photographic plate placed at the position of the screen will show, and it would seem to be incontrovertible that a single entity of this kind could pass through one slit or the other, but certainly not through both at once. The paradox is resolved when we examine the problem more closely. We *can* determine whether a given electron or photon passes through a particular slit by placing some suitable detector at that position—but the process of observation disturbs the state of affairs in just such a way as to destroy the interference pattern on the screen. We can observe either the wave or the particle aspects of a physical system,

Wave Mechanics 195

but we are precluded, through the uncertainty principle, from observing both together; the alternative descriptions are complementary to each other.

In the two-slit interference experiment we see a physical illustration of the connection between probability and the wave function. The relative amplitude and phase of ψ at the two slits defines the interference pattern on the screen: i.e., the distribution along the screen of the intensity $\psi^*\psi$. But, as we have said, the same intensity distribution

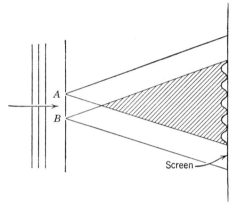

Figure 7.10. Double-slit interference for light or particles.

can be built up by counting the numbers of particles arriving in a fixed time at different parts of the screen, and the ratio of the number arriving at some given place to the total number recorded measures the *probability* that any single particle will arrive at that specified place. It will always be true that the point of arrival of any single particle can never be predicted in advance, but the statistical distribution in position for a very large number of particles must conform to the definite interference pattern required by wave theory.

7.10 EIGENFUNCTIONS

We have seen how the introduction of wave mechanics provides a natural account of the permitted states of an electron in a hydrogen atom. For a stationary state in a circular orbit we had

$$2\pi r = n\lambda$$

This at once defines particular values of the orbital radii and the corresponding energies of the system. We have here one expression of the application of wave mechanics to physical problems. More generally the possible stationary states of a system are given by solutions to the time-independent form of Schrödinger's equation, and it is at once found that, as a rule, the energy E of the system is limited in the values that it can assume. Moreover, to a permitted value of E there will correspond only a limited number of forms (perhaps only one) for the wave function ψ. The possible wave functions for a system are called "proper functions," or more usually "eigenfunctions" (from the German), and the form that they take is determined by the boundary

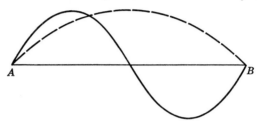

Figure 7.11. The two lowest modes for a stretched string.

conditions in any particular case. One can go so far as to say that it is the imposition of boundary conditions which is responsible for the appearance of quantized states and quantized energy levels. (There is no restriction, for example, imposed by quantum mechanics on the energy of a freely moving particle, except that it shall be positive.)

The nature of the problem is illustrated by considering a stretched string that is fixed at its ends A and B (Fig. 7.11). Let the length of the string be l. The differential equation governing transverse waves on the string is

$$\frac{\partial^2 \psi}{\partial x^2} = \frac{m}{T} \frac{\partial^2 \psi}{\partial t^2}$$

where T is the tension in the string and m is the mass of the string per unit length. Let us look for solutions of the type

$$\psi = A \exp[i(2\pi \nu t - Kx)]$$

Then
$$\frac{\partial^2 \psi}{\partial t^2} = -4\pi^2 \nu^2 \psi$$

so that we can obtain a time-independent equation

$$\frac{\partial^2 \psi}{\partial x^2} + \frac{4\pi^2 \nu^2 m}{T} \psi = 0$$

i.e.,
$$\frac{\partial^2 \psi}{\partial x^2} + K^2 \psi = 0$$

(Note that K may be identified with $2\pi k$ in making comparison with Schrödinger's equation.)

In the absence of any boundary conditions, this would have an acceptable solution for any value of K, but we must meet the requirement that ψ falls to zero at all times for $x = 0$ and $x = l$. The general solution is

$$\psi(x) = A \sin Kx + B \cos Kx$$

From the condition
$$\psi(0) = 0$$

we have $B = 0$ for all K. From the condition

$$\psi(l) = 0$$

we have either $A = 0$ (which is uninteresting since ψ then vanishes altogether), or else the nontrivial condition

$$\sin Kl = 0$$

whence
$$Kl = n\pi$$

where n is an integer. The only possible solutions are therefore of the form

$$\psi_n(x, t) = A_n \exp(i \cdot 2\pi\nu t) \sin\left(\frac{n\pi x}{l}\right)$$

These are the eigenfunctions for the stretched string, and they define a discrete set of permitted frequencies for the string under these conditions. For we have

$$K_n = \frac{n\pi}{l} = 2\pi\nu_n \left(\frac{m}{T}\right)^{1/2}$$

i.e.,
$$\nu_n = \frac{n}{2l}\left(\frac{T}{m}\right)^{1/2}$$

the familiar result for the set of harmonics of, say, a piano string. It may be verified that for any two distinct eigenfunctions we have the condition

$$\int_0^l \psi_n(x)\psi_m(x)\, dx = 0 \qquad (m \neq n)$$

One says that the different eigenfunctions are *orthogonal* to each other, since mathematically this result is rather similar to the condition for orthogonality of two vectors **a**, **b**:

$$\sum_i a_i b_i = 0$$

where i labels the component of a vector along a given axis.

We can recognize a far-reaching similarity between the solution of problems in wave mechanics and the classical solution for wave motions on strings. There is, however, at least one important distinction. When we solve the vibrating string problem we find no restriction on the amplitude of the vibration. But in wave mechanics we have the condition of normalization by which the total probability is set equal to unity in a one-particle problem, and the two conditions governing the set of eigenfunctions for the wave-mechanical case are:

$$\int_{\text{all space}} \psi_n^* \psi_n \, d\tau = 1 \quad \text{(normalization)}$$

$$\int_{\text{all space}} \psi_m^* \psi_n \, d\tau = 0 \quad \text{(orthogonality)}$$

It is to be noted that our time-independent equation for the vibrating string is quite identical in form with the time-independent Schrödinger equation. In the simple problem that we have just treated, K was a constant; but, by considering a string whose mass per unit length was not uniform, we could define a problem in which K is a function of x. In this case we can simulate completely the problem of a particle moving through a region in which its potential energy changes with position, for we can set up the identity

$$\frac{2m}{h^2}[E - V(x)] = [k(x)]^2 = \frac{1}{4\pi^2}[K(x)]^2$$

7.11 PARTICLE IN A BOX

Suppose that we have a particle constrained to move along the x axis, and let the potential energy it experiences be zero in the range $0 \leqslant x \leqslant l$, and equal to a constant V for all $x < 0$ and all $x > l$. Let the total energy of the particle be E (setting aside the rest mass), and suppose $V > E$. Then we can distinguish three different regions, separated by these discontinuous changes of potential (Fig. 7.12). In the first and third regions we have

$$\frac{\partial^2 \psi}{\partial x^2} - \frac{8\pi^2 m}{h^2}(V - E)\psi = 0$$

In the central region,

$$\frac{\partial^2 \psi}{\partial x^2} + \frac{8\pi^2 m}{h^2} E\psi = 0$$

These can be rewritten in the abbreviated forms used before:

$$\frac{\partial^2 \psi}{\partial x^2} = \alpha^2 \psi \qquad (x < 0, x > l)$$

$$\frac{\partial^2 \psi}{\partial x^2} = -K^2 \psi \qquad (0 \leqslant x \leqslant l)$$

(α and K both real and positive).

From the first equation we infer

$$\psi_1(x) = A e^{+\alpha x} \quad (x < 0), \qquad \psi_3(x) = B e^{-\alpha x} \quad (x > l)$$

The negative exponent in region 1 and the positive exponent in region 3 are ruled out by the condition that ψ shall not become infinite anywhere. From the second equation we have

$$\psi_2(x) = C e^{+iKx} + D e^{-iKx}$$

Now the condition that ψ remain everywhere finite also imposes conditions at the boundaries between the different regions, because from the Schrödinger equation itself we see that $\partial^2\psi/\partial x^2$ must in turn be finite everywhere, and this can be so only if $\partial\psi/\partial x$ has no discontinuity at any boundary. Furthermore, the existence of $\partial\psi/\partial x$ as a continuous function implies that ψ too is continuous across a boundary. Thus we always have two boundary conditions, one on ψ and the other on $\partial\psi/\partial x$, which can be applied in determining acceptable solutions to Schrödinger's equation.

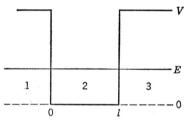

Figure 7.12. A one-dimensional rectangular potential well, with a bound particle of energy E ($<V$).

In considering our problem further, let us suppose that V becomes arbitrarily large. Then $\alpha \to \infty$, and so ψ_1 and ψ_3 fall effectively to zero at points indefinitely close to $x = 0$ and $x = l$, respectively. It follows then that we can come arbitrarily close to the boundary conditions

$$\psi_2(0) = 0, \qquad \psi_2(l) = 0$$

We therefore require, in the expression for $\psi_2(x)$,

$$C + D = 0$$

$$C e^{iKl} + D e^{-iKl} = 0$$

$$\therefore \quad D = -C \quad \text{and so} \quad \sin Kl = 0$$

Hence $Kl = n\pi$, and mathematically the problem has become identical with that of the vibrating string fixed at both ends. When account is taken of the normalization of ψ, the complete solution to the wave-mechanical problem is

$$\psi_2(x) = \left(\frac{2}{l}\right)^{1/2} \sin\left(\frac{n\pi x}{l}\right)$$

The problem in this form is described as that of a particle in a box, because it corresponds to the classical situation of a particle constrained to move between completely rigid boundaries, but otherwise

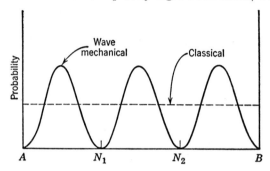

Figure 7.13. The probability distribution for finding a particle of given energy at points between the walls of a rigid box, according to classical and wave mechanics.

free, so that inside the box it moves with constant speed. We see, however, that the wave-mechanical treatment brings out two features that could never appear in classical particle mechanics:

(a) The energy of the particle is quantized, for we have

$$K = \frac{n\pi}{l}, \qquad K^2 = \frac{8\pi^2 m}{h^2} E$$

Hence

$$E_n = \frac{h^2}{8\pi^2 m} \cdot \frac{n^2 \pi^2}{l^2} = n^2 \frac{h^2}{8ml^2}$$

(b) Whereas, classically, the particle spends the same proportion of its time within any element dx of distance, no matter where that element may be situated in the box, the conclusion drawn from wave mechanics is that the probability of finding the particle at a given point x is an undulating function of x: viz.,

$$|\psi(x)|^2 \, dx = \frac{2}{l} \sin^2\left(\frac{n\pi x}{l}\right) dx$$

Wave Mechanics

There even exist nodes (e.g., N_1, N_2 in the curve of Fig. 7.13, which is drawn for $n = 3$) where the particle is never found at all. We must simply accept this as a necessary consequence of the wave-mechanical description.

If we now go back to the case where the potential V is not indefinitely large, it is easy to see that the wave functions are modified (Fig. 7.14) in such a way that there are no longer any nodes at $x = 0$ and $x = l$.

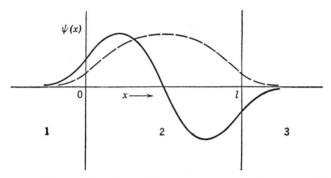

Figure 7.14. The wave functions of the two lowest states for a particle in a one-dimensional box with walls of finite height (comparable to a stretched string connected between yielding supports).

This means that, to secure a smooth fit in the wave functions at the potential discontinuities, the characteristic wavelengths and hence energies of the permitted states are shifted somewhat in the direction of longer wavelength and hence lower energy with respect to those for a box with infinitely high walls. The possible energies are still quantized, however. Figure 7.14 shows the forms of solution corresponding to $n = 1$ and $n = 2$. (See also Appendix VIII for the relevance of this problem to nuclear energy states.)

7.12 POTENTIAL STEPS AND BARRIERS

As another example of a typical wave-mechanical problem we may take the question of a particle traversing a "potential step"; i.e. a region in which its potential energy changes suddenly. This could be realized physically, for example, with an electron passing out of one grounded metal cylinder into another cylinder held at potential $-\phi$ so that $V = e\phi$ (Fig. 7.15). We shall suppose $E > V$. This means classically that the particle is certain to pass over the potential step, suffering merely a drop in speed in the process. We have in fact

$$v_1 = \left(\frac{2E}{m}\right)^{1/2}, \qquad v_2 = \left[\frac{2(E - V)}{m}\right]^{1/2}$$

According to wave mechanics, however, there is a finite chance that the particle will be reflected, in the same way that a beam of light falling normally on the boundary between two transparent media (e.g. air and glass) will be partially reflected and partially transmitted.

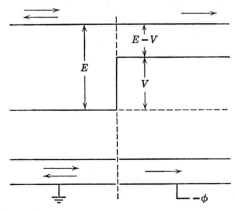

Figure 7.15. The partial reflection and transmission of electrons approaching a potential step from the left.

Strictly speaking, we should treat this situation as one involving the time-dependent Schrödinger equation, with a particle (represented by a wave packet) moving along the x axis. But it is also possible to solve the problem as a time-independent one, by imagining that, instead of a single particle, we have a continuous stream of particles moving toward the step, some being reflected and some passing straight on. Let us suppose that particles are initially incident from the left, as shown in Fig. 7.16. Then our Schrödinger equations for regions 1 and 2 may be assumed to have solutions of the following forms:

$$\psi_1(x) = Ae^{iK_1 x} + Be^{-iK_1 x}$$
$$\psi_2(x) = Ce^{iK_2 x}$$

By putting $\psi_1(0) = \psi_2(0)$, $(\partial \psi_1/\partial x)_0 = (\partial \psi_2/\partial x)_0$, we can solve for the amplitudes B and C in terms of A. It is not difficult to verify (Problem 6 at the end of this chapter) that the reflection coefficient R is given by

$$R = \frac{|B|^2}{|A|^2} = \left(\frac{K_1 - K_2}{K_1 + K_2}\right)^2$$

Wave Mechanics

It may also be verified that

$$v_1|A|^2 = v_1|B|^2 + v_2|C|^2$$

We recognize in this the role of $\psi^*\psi$ as a probability density ρ such that the total current (given by ρv) is conserved in the steady state. What is perhaps surprising is that, if we had considered the particles coming from the right instead of from the left, so that they experienced a sudden acceleration in crossing the step, there would again be a partial

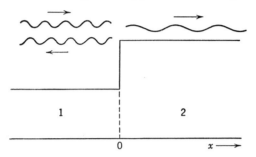

Figure 7.16. The potential step as viewed in wave mechanics.

reflection—described by the *same* value of R, in fact, as we have just obtained.

Finally, we may refer briefly to the basis of the tunnel effect, which we have already mentioned as being important in the processes of nuclear disintegration, and which will be discussed further in Chapter 9, Section 15. We suppose a beam of particles, of energy E, approaching from the left a potential barrier of height $V(>E)$ extending from $x = 0$ to $x = l$. We recognize three distinct regions of space (Fig. 7.17) for the purpose of solving Schrödinger's equation (again in the time-independent form). Let us put

$$\frac{8\pi^2 m}{h^2} E = K^2 \quad \text{(for regions 1 and 3)}$$

$$\frac{8\pi^2 m}{h^2} (V - E) = \alpha^2 \quad \text{(for region 2)}$$

Then

$$\psi_1(x) = Ae^{iKx} + Be^{-iKx}$$
$$\psi_2(x) = Ce^{\alpha x} + De^{-\alpha x}$$
$$\psi_3(x) = Ge^{iKx}$$

It is to be noted that in the barrier region itself we must admit the possibility of both increasing and decreasing exponentials. The

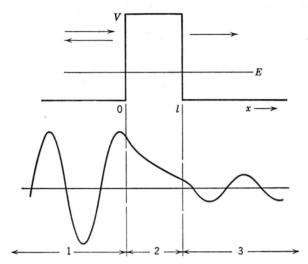

Figure 7.17. Partial penetration and partial reflection at a potential barrier, with exponential attenuation through the barrier from left to right.

existence of two boundary conditions at $x = 0$ and two more at $x = l$ allows us to solve for B, C, D, and G in terms of A. The ratio $|G|^2/|A|^2$ is then what is called the "penetrability" of the barrier, and represents the finite probability that a particle incident on the barrier from one side may appear subsequently on the other side—an effect that is inconceivable in classical physics but is a familiar fact of experience to the nuclear physicist.

Problems

7.1. Calculate the de Broglie wavelengths of (a) An atom of mercury ($A \approx 200$) at 500° C, (b) A molecule of hydrogen at 20° K, (c) An atom of helium at 1° K, (d) An electron accelerated through 1 Mv, (e) A proton of kinetic energy 500 Mev. [Note that (d) and (e) involve relativistic energies for the particles concerned.]

7.2. Many solid substances present an attractive potential of about 10 volts to incoming electrons. Consider an electron, accelerated through 50 volts, entering such a substance at an angle of incidence of 70°. Find the angle of refraction, and the wavelength of the electron inside the material.

7.3. To illustrate the relation between a spread of distance and a spread of wave number for a wave packet, consider the disturbance

$$f(x) = e^{-\alpha x^2} \cos 2\pi k_0 x$$

Obtain the Fourier analysis of this disturbance by evaluating the amplitude $g(k)$ associated with wave number k, and check this against the discussion in Section 9 of this chapter. [See Problem 2, Chapter 4. In the present problem, express $\cos 2\pi k_0 x$ in exponential form, and complete the square in the exponent of the integrand of $g(k)$. The definite integral so obtained is essentially the error integral mentioned in Appendix I.]

7.4. Consider in detail the wave functions and energy states for a particle in a one-dimensional potential well of the type shown in Fig. 7.12. Take the top of the well as the zero of potential energy, and consider the states of negative total energy. Construct a graphical method for solving for the eigenvalues of the energy. Find the two lowest states for the particular case in which the particle is an electron and the potential well is of depth 10 ev and of width 10^{-7} cm. Draw the corresponding normalized wave functions. [For a neat treatment of this type of problem see Pitkanen, *Am. J. Phys.* **23**, 111 (1955).]

7.5. One of the physical systems of basic importance is a particle moving under an elastic restoring force proportional to the displacement, i.e. in a "harmonic oscillator" potential defined by

$$V(x) = \tfrac{1}{2}\beta x^2 \quad \text{(for all } x\text{)}$$

Write down the Schrödinger equation for this potential, and look for solutions of the type

$$\psi(x) = e^{-\alpha x^2} f(x)$$

where $f(x)$ is a simple series in positive powers of x. [Start with the lowest possibility, $f(x) =$ constant, and then proceed to try out the linear and quadratic forms.] Solve for the eigenvalues of the energy, taking $V(0)$ as the zero of energy.

7.6. Verify that the reflection coefficient for a beam of particles approaching a potential step is as given in Section 12 of this chapter, for either direction of incidence. A beam of electrons traveling in a field-free region encounters a sharply defined potential step equivalent to a retarding potential of 0.1 volt; the initial electron energy is 1 ev. Calculate the reflection coefficient, and the ratio of amplitudes of the incident and transmitted waves. Consider qualitatively (and by analogy with classical wave theory) what would be the effect of spreading the potential change over a range of distance several times greater than the wavelength.

7.7. Solve in detail the problem of partial reflection and transmission at a rectangular potential barrier (Section 12 of this chapter). Find the probability of transmission for protons of energy 1 Mev through a barrier of height 4 Mev and of thickness 10^{-12} cm.

7.8. Treat (algebraically and numerically) a modified form of the previous problem, in which the potential barrier is replaced by a negative rectangular potential of the same numerical size and width.

8 Some Applications of Quantum Mechanics

8.1 THE HYDROGEN ATOM

In our development of wave mechanics we have confined ourselves to a treatment of one-dimensional problems; this simplifies the mathematics without destroying any essential features of the subject. But we must now set up the Schrödinger equation for three dimensions so as to solve the problem of an electron moving under a central force. This is readily done if we compare the classical energy equation with the time-independent Schrödinger equation:

$$E = \frac{p_x{}^2}{2m} + V(x)$$

$$E\psi = -\frac{h^2}{8\pi^2 m}\frac{\partial^2 \psi}{\partial x^2} + V(x)\psi$$

We see that the following correspondence holds:

$$p_x{}^2 \equiv -\frac{h^2}{4\pi^2}\frac{\partial^2}{\partial x^2}$$

Thus, for motion in three dimensions, such that a particle has component momenta p_x, p_y, p_z, the energy equation gives rise to the following form of the wave equation:

$$[E - V(x, y, z)]\psi = -\frac{h^2}{8\pi^2 m}\left(\frac{\partial^2}{\partial x^2} + \frac{\partial^2}{\partial y^2} + \frac{\partial^2}{\partial z^2}\right)\psi = -\frac{h^2}{8\pi^2 m}\nabla^2\psi$$

Hence the equation to be solved is

$$\nabla^2\psi + \frac{8\pi^2 m}{h^2}[E - V(\mathbf{r})]\psi = 0 \quad \text{where} \quad \mathbf{r} = (x, y, z)$$

For a hydrogen atom, or more generally for what one may call a "hydrogen-like" atom, we put

$$V(\mathbf{r}) = -\frac{Ze^2}{r}$$

The fact that the potential energy depends only on the radial distance r and not on its direction (or, what amounts to the same thing, on its separate components x, y, z) suggests that we should attempt to solve for ψ in spherical polar co-ordinates (r, θ, ϕ) and, furthermore, that we should seek a solution which factorizes into separate radial and angular parts. To this end we put

$$\psi(r, \theta, \phi) = R(r)\,\Theta(\theta)\,\Phi(\phi)$$

where R, Θ, Φ are functions to be determined. We also express the differential operator ∇^2 in spherical polar form:

$$\nabla^2\psi = \frac{1}{r^2}\frac{\partial}{\partial r}(r^2\psi) + \frac{1}{r^2\sin\theta}\frac{\partial}{\partial\theta}\left(\sin\theta\,\frac{\partial\psi}{\partial\theta}\right) + \frac{1}{r^2\sin^2\theta}\frac{\partial^2\psi}{\partial\phi^2}$$

To avoid obscuring the argument with mathematical details, we shall simply assert that the factorization of ψ is indeed possible, and that its first result concerns the function Φ:

$$\frac{1}{\Phi}\frac{d^2\Phi}{d\phi^2} = \text{a function independent of } \phi = -m^2 \quad \text{say}$$

Hence $\quad\Phi(\phi) = A e^{im\phi}$

and, since any acceptable Φ must be a single-valued function of position, we require

$$e^{im\phi} = e^{im(\phi + 2\pi)}$$

from which we deduce that m must be an integer, and is a quantum number for the system.

The elimination of Φ and ϕ from the equation next makes possible an equation in θ alone and an equation in r alone. The condition that the solutions for Θ and R should also be well-behaved brings to light two more quantum numbers. The first of these is an integer l, which appears in the differential equation for the radial part of ψ:

$$\frac{1}{r^2}\frac{d}{dr}\left(r^2\frac{dR}{dr}\right) + \frac{8\pi^2 m}{h^2}\left[E + \frac{Ze^2}{r} - \frac{l(l+1)h^2}{8\pi^2 mr^2}\right]R = 0$$

It is instructive to make the substitution

$$R = \frac{S}{r}$$

because then we obtain a much simpler differential equation: viz.,

$$\frac{d^2 S}{dr^2} + \frac{8\pi^2 m}{h^2}\left[E + \frac{Ze^2}{r} - \frac{l(l+1)h^2}{8\pi^2 mr^2}\right]S = 0$$

It emerges that this equation has an acceptable solution for any *positive* value of E, but that E may only take up discrete *negative* values given by

$$E_n = -\frac{2\pi^2 me^4 Z^2}{h^2} \cdot \frac{1}{n^2}$$

where n is integral. This, it may be noted, is exactly the same expression for the quantized energy states as that given by Bohr's original theory (Chapter 5, Section 4). In Appendix VII we discuss the simple particular case corresponding to $l = 0$, $n = 1$.

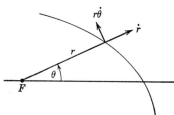

Figure 8.1. A two-dimensional motion in polar co-ordinates.

The purely mathematical properties of the solutions to the equations for Φ, Θ and R (or S) show that (a) n may be any integer, and defines the total energy of the system, (b) l may take any integral value between 0 and $n - 1$, (c) m may take any integral value between $-l$ and $+l$ for a given value of l.

Now l is closely related to the angular momentum of the electron in its orbit. We can recognize this if we compare our differential equation for $S(r)$ with a classical equation governing motion of a particle under a central force. For a motion in two dimensions, described by co-ordinates r, θ, as illustrated in Fig. 8.1, we have

$$E = \tfrac{1}{2}m\dot{r}^2 + \tfrac{1}{2}m(r\dot\theta)^2 + V(r)$$

But, since we are dealing with a central force, the angular momentum

Some Applications of Quantum Mechanics

is a constant (L):
$$L = mr^2\dot{\theta}$$

Hence
$$E = \frac{p_r^2}{2m} + \frac{L^2}{2mr^2} + V(r)$$

where p_r = radial momentum of particle.

The above equation is formally identical with that for motion in one dimension, and so, it may be noted, is the differential equation for S. But in each equation an effective addition to the potential energy has appeared. We know that in the classical problem it describes the effects of angular momentum, so by comparison of the two equations we propose the identity

$$\frac{L^2}{2mr^2} \equiv \frac{l(l+1)h^2}{8\pi^2 mr^2}$$

i.e., Angular momentum = $L = [l(l+1)]^{\frac{1}{2}} \dfrac{h}{2\pi} = [l(l+1)]^{\frac{1}{2}}\hbar$

We next recognize that angular momentum, as a vector quantity, is not properly specified unless its direction as well as its magnitude is known, and we see that it is possible to regard the quantum number m ($-l \leqslant m \leqslant +l$) as defining a limited number of orientations ($2l+1$ in fact) that **L** may take up with respect to a specified axis. This is shown in Fig. 8.2. We have argued these results merely on the basis of plausible analogies between classical and wave-mechanical systems, but a more thorough-going treatment does confirm them completely. It is interesting to notice that the vector angular momentum **L** can never lie entirely along a chosen axis, but only at a minimum inclination given by $\arccos [l/(l+1)]^{\frac{1}{2}}$, which of course tends to zero in the limit of large l.

The wave function amplitude for an electron in a hydrogen-like atom is given as a result of this analysis by

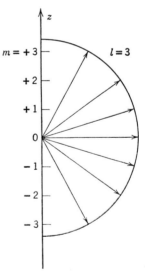

Figure 8.2. The space quantization of orbital angular momentum.

$$\psi_{nlm}(r, \theta, \phi) = R_{nl}(r)\,\Theta_{lm}(\theta)\,\Phi_m(\phi)$$

This is a continuous function, and the quantity $e\psi^*\psi$ gives a measure of the charge density at any given point. The radial variation of charge density passes through a series of maxima and zeros, and the distance between the nucleus and the centroid of the wave-mechanical charge (or probability) distribution bears a close relationship to the radius of the corresponding orbit in Bohr's picture of the atom. For example, in the lowest state of the H atom ($n = 1, l = 0$) the peak of the graph of $r^2 R_{10}{}^2(r)$ occurs exactly at a value of r equal to the radius a_0 of the first Bohr orbit ($a_0 = h^2/4\pi^2 m e^2 = 0.53 \cdot 10^{-8}$ cm) (cf. Appendix VII).

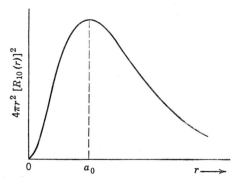

Figure 8.3. The radial probability distribution for an electron in the first Bohr orbit ($n = 1, l = 0$).

The factor r^2 multiplying $R^2(r)$ is due to the fact that the probability of finding the electron between r and $r + dr$ is proportional to the volume $4\pi r^2\, dr$ of the spherical shell contained between these limits. The shape of this particular charge distribution is shown in Fig. 8.3.

8.2 DOUBLET FINE STRUCTURE OF ATOMIC ENERGY LEVELS

Even before the development of wave mechanics it was realized that Bohr's theory was not adequate to describe all the features of spectra. Although the Bohr–Sommerfeld theory, taking into consideration relativistic effects, was able to account for much of the fine structure of spectra, one feature remained a mystery. This was that in many of the characteristic series of spectral lines, e.g., in hydrogen and the alkali metals, each supposedly single line was in fact a close doublet. A good example is provided by the D lines of sodium, with wavelengths of 5890 and 5896 A. This consistent appearance of a doubling of the

Some Applications of Quantum Mechanics 211

lines led Uhlenbeck and Goudsmit (1925) to postulate that the electron had a certain intrinsic "spin," so that it could be thought of as rotating about its own axis at the same time as it revolved in an orbit about a nucleus—rather like the earth in its motion around the sun. To describe the spin a certain characteristic angular momentum for the electron was introduced. We associate with this a quantum number s that is to be analogous to the orbital quantum number l. But we have seen that the angular momentum vector described by l has $2l + 1$ distinct orientations. If we postulate the same behavior for the spin momentum, and if we take the fact of doublet structure to mean that only *two* orientations of the spin are possible, then we require

$$2s + 1 = 2, \quad \text{i.e.,} \quad s = \tfrac{1}{2}$$

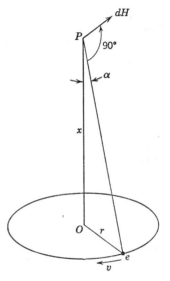

Figure 8.4. The magnetic field of an orbital electron.

It still remains, of course, to account for the existence and the magnitude of the splitting between the two components of a doublet. The explanation is to be found in the fact that a moving charge generates a magnetic field; what we do is to assume (*a*) that an electron, by virtue of its spin, has a magnetic moment, and (*b*) that by virtue of its orbital motion it finds itself in a magnetic field which interacts with the magnetic moment to cause a change in the energy of the system.

Let us consider an electron moving in a circular orbit of radius r with speed v, as shown in Fig. 8.4. It is equivalent to a current i'. (We revert to the original Bohr picture for simplicity.) We will evaluate the magnetic field at some point P on the axis, distance x from the center of the circle. The field dH due to current i' (emu) in an arc of length dl is given by

$$dH = \frac{i' \, dl}{x^2 + r^2} \approx \frac{i' \, dl}{x^2} \quad \text{for} \quad x \gg r$$

Evidently the component of dH perpendicular to OP will be canceled out by other contributions, so that the resultant field H is along OP, and we have

$$H = \int dH \sin \alpha \approx \int \frac{i'\,dl}{x^2} \cdot \frac{r}{x}$$

i.e., $$H \approx \frac{2\pi r^2 i'}{x^3}$$

By comparing this to the expression for the magnetic field of a small magnetic dipole we see that the electron has an effective magnetic moment μ given by

$$\mu = \pi r^2 i'$$

Now, since i' is the charge in emu passing a given point per second, we have

$$i' = \frac{e}{c} \cdot \frac{v}{2\pi r}$$

Hence $$\mu = \frac{evr}{2c} = \frac{e}{2mc} mvr$$

But mvr is the orbital angular momentum L of the electron, and so we finally obtain the result

$$\mu_l = [l(l+1)]^{1/2} \frac{eh}{4\pi mc}$$

The quantity $eh/4\pi mc$ represents a natural unit for the measurement of magnetic moments in atoms. It is called the *Bohr magneton*, and is

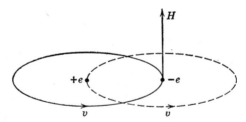

Figure 8.5. The magnetic spin-orbit interaction.

numerically equal to $0.927 \cdot 10^{-20}$ emu. It is reasonable to suppose that the intrinsic magnetic moment of a spinning electron must be comparable with one magneton, and this is found to be so. (We shall return to this question in more detail later—Section 9 of this chapter.)

Now consider this magnetic electron in a circular orbit around a proton. From the point of view of the electron, so to speak, it is the

Some Applications of Quantum Mechanics 213

proton that is describing the orbit; i.e., the electron is exposed to a magnetic field due to a charge $+e$ circulating around it (Fig. 8.5). Now, for a current i' flowing in a circle, we have

$$H = \frac{2\pi i'}{r}$$

Hence $$H = \frac{2\pi}{r} \cdot \frac{e}{c} \cdot \frac{v}{2\pi r} = \frac{e}{mc} \cdot \frac{mvr}{r^3} = \frac{e}{mc} \cdot \frac{L}{r^3}$$

or $$H = [l(l+1)]^{1/2} \cdot \frac{eh}{2\pi mc} \cdot \frac{1}{r^3}$$

Thus the order of magnitude of H is represented by a magnetic dipole equal to one Bohr magneton acting at a distance equal to one electron orbital radius. For r let us substitute $a_0 \approx 0.5 \cdot 10^{-8}$ cm (the first Bohr orbit radius).

Then $$H \approx \frac{10^{-20} \times 8}{10^{-24}} \approx 10^5 \text{ gausses}$$

An electron in an atom may thus find itself in an immensely strong magnetic field, and this field then provides a physically meaningful axis along which the electron spin momentum component must take up the value $\pm h/4\pi$.

We are now in a position to make a crude estimate of the doublet splitting for an electron level. The magnetic potential energy of a dipole μ in a field $\mathbf{H}$ is given by

$$\Delta W = -\mu \cdot \mathbf{H}$$

Thus, with respect to a situation in which electron spin is ignored, the magnetic interaction leads to the appearance of two energy levels, one

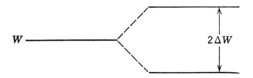

Figure 8.6. The doublet fine structure splitting.

higher and one lower than the original level of the Bohr theory, as shown in Fig. 8.6. Putting $\mu = 1$ magneton, $H = 10^5$ gausses we have

$$\Delta W \approx 10^{-20} \times 10^5 \text{ erg} = 10^{-15} \text{ erg} \approx 10^{-3} \text{ ev}$$

The energy represented by the sodium D lines ($\lambda \approx 6 \cdot 10^{-5}$ cm) is, however, about 2 ev, and so we may reasonably expect a doublet splitting of something like 1 part in 1000, which is almost exactly what it is observed to be.

The existence of this very strong field means that the orbital and spin motions of an electron in an atom are intimately coupled, so much so that they combine to define a resultant angular momentum, **J**, characterized by a quantum number j. The coupling takes place according to the rules of vector addition:

$$\mathbf{J} = \mathbf{L} + \mathbf{S}$$

The restrictions of quantum mechanics show, however, that j can only take on the values $l \pm \frac{1}{2}$. The quantum state of an electron is now described not by the four quantum numbers, n, l, m_l, m_s, but by an equivalent set n, l, j, m_j (where m with an appropriate suffix denotes the z component of the quantity designated). For given values of n, l, j the total angular momentum **J** can take up $2j + 1$ different orientations, which represent $2j + 1$ different quantum states of the system. For a given value of l, the total number of different quantum states is given by

$$[2(l + \tfrac{1}{2}) + 1] + [2(l - \tfrac{1}{2}) + 1] = 2(2l + 1)$$

We thus see that the introduction of j has not brought about any change in the total number of quantum states for the system, but only a reclassification of them.

8.3 SPECTROSCOPIC NOTATION

It is convenient at this point to present the notation by which the quantum states of individual electrons are described. In the absence of an external magnetic field, the energy level for a given quantum state is determined by the three quantities n, l, and j. It has become usual to identify the value of l by a letter code, as follows:

$l =$ 0   1   2   3   4   5   etc.
Designation s p d f g h etc.

The logic of the above scheme is not exactly obvious; we shall see later in this chapter (Section 7) how it came about. The state of an electron is then described basically by writing the value of the principal quantum number n, followed by the letter identifying l—e.g. $3p$ for $n = 3$, $l = 1$. The doublet fine structure makes this description incomplete, and so the value of j is written as a subscript to the l sym-

Some Applications of Quantum Mechanics 215

bol; and, as a reminder that the fine structure turns each level of a given l into a doublet, the figure 2 is written as a superscript just in front of the l symbol. Thus a level belonging to $n = 3, l = 1, j = \frac{3}{2}$ is written as $3\,^2p_{3/2}$. It might seem that the figure 2, denoting what is called the *multiplicity* of a level of given l, is really superfluous. This is true provided we are concerned with states of a single electron, since all levels are then doublets (unless $l = 0$). But, if we have a possible combined state of two or more electrons, the multiplicity of a level of given orbital momentum may assume other values.

In hydrogen and the alkali metals it is the single valency electron that is responsible, by its quantum jumps, for all the spectral lines in the visible region; and the values of orbital and total angular momentum for the whole electron structure of the atom are identical with those of the single electron. But to give expression to the fact that a quantum jump is in principle a change of state of the atom as a whole, it is written as a transition between states of the complete atom, and these are distinguished by writing the orbital momentum symbol with a capital letter, L. Thus, a sodium atom, for example, with its valency electron in a state with $n = 3, l = 1, j = \frac{1}{2}$, is designated by the notation $3\,^2P_{1/2}$.

When we come to atoms in which more than one electron contributes to the total state of the atom (e.g. Ca or Hg, each with two valency electrons) there arises a real distinction between the quantum numbers of the atomic state as a whole and the quantum numbers of the so-called "optical electrons" giving rise to it. We shall consider an example of this, but first we must introduce Pauli's exclusion principle, which dominates the permitted ways of coupling electrons together.

8.4 PAULI'S EXCLUSION PRINCIPLE

So far we have made no attempt to deal with systems containing more than one particle, but a very important part of wave mechanics is concerned with many-body problems. Let us consider first how we can describe a system of two electrons moving in the same potential. The same set of quantum states will be available to each; let us denote the wave functions describing these states by ψ_a, ψ_b, etc., and let us label the electrons as 1 and 2. Then, if one electron is in state a and the other in state b, we can try to set up a single wave function Ψ to describe the whole system, by putting

$$\Psi_\alpha = \psi_a(1)\,\psi_b(2) \quad \text{or} \quad \Psi_\beta = \psi_b(1)\,\psi_a(2)$$

The interpretation to be placed on a wave function of this form is that the quantity $|\Psi_\alpha|^2 \, d\tau_1 \, d\tau_2$, for example, is the probability that electron 1 is in state a and lies within a volume element $d\tau_1$ at $\mathbf{r}_1$, and at the same time electron 2 is in state b and lies within $d\tau_2$ at $\mathbf{r}_2$. Now Ψ_α (or Ψ_β) by itself is not an acceptable wave function, for it carries the implication that a given electron is labeled so that we can tell whether it is definitely in state a or definitely in state b, whereas, if the electrons are truly identical, this distinction is meaningless. We are therefore led to construct a wave function that expresses our ignorance in this matter:

$$\Psi = N[\psi_a(1)\,\psi_b(2) \pm \psi_b(1)\,\psi_a(2)]$$

If we take the positive sign, Ψ remains unchanged when we imagine an interchange between the two electrons; with the negative sign, Ψ is reversed in sign but retains the same magnitude. Thus in each case $\Psi^*\Psi$, which is the physically meaningful quantity, is unaffected—a condition that will not in general be satisfied for any other form of Ψ. The coefficient N that appears in Ψ is a normalizing factor. We require

$$\int_{\text{all space}} \Psi^*\Psi \, d\tau_1 \, d\tau_2 = 1$$

$$\therefore \quad 1 = N^2 \int [\psi_a^*(1)\,\psi_b^*(2) \pm \psi_b^*(1)\,\psi_a^*(2)]$$
$$\times [\psi_a(1)\,\psi_b(2) \pm \psi_b(1)\,\psi_a(2)] \, d\tau_1 \, d\tau_2$$
$$= N^2 \left(\int \psi_a^*\psi_a \, d\tau_1 \int \psi_b^*\psi_b \, d\tau_2 + \int \psi_b^*\psi_b \, d\tau_1 \int \psi_a^*\psi_a \, d\tau_2 \right.$$
$$\left. \pm \int \psi_a^*\psi_b \, d\tau_1 \int \psi_b^*\psi_a \, d\tau_2 \pm \int \psi_b^*\psi_a \, d\tau_1 \int \psi_a^*\psi_b \, d\tau_2 \right)$$

This simplifies drastically, because, if a and b represent eigenstates of one electron, we have

$$\int \psi_a^*\psi_a \, d\tau = \int \psi_b^*\psi_b \, d\tau = 1 \quad \text{(normalization condition)}$$

$$\int \psi_a^*\psi_b \, d\tau = \int \psi_b^*\psi_a \, d\tau = 0 \quad \text{(orthogonality condition)}$$

Hence $1 = 2N^2$, simply, and so

$$N = \frac{1}{\sqrt{2}}$$

We therefore arrive at two possible types of wave function for a system of two identical particles:

$$\Psi_S = \frac{1}{\sqrt{2}} [\psi_a(1)\,\psi_b(2) + \psi_b(1)\,\psi_a(2)] \quad \text{(symmetric)}$$

$$\Psi_A = \frac{1}{\sqrt{2}} [\psi_a(1)\,\psi_b(2) - \psi_b(1)\,\psi_a(2)] \quad \text{(antisymmetric)}$$

Some Applications of Quantum Mechanics 217

The titles "symmetric" and "antisymmetric" are a means of describing what happens to Ψ when the two particles are imagined to be interchanged.

Now there is a very obvious property of Ψ_A which proves to be of enormous importance: if we set $a = b$, i.e., if we suppose our two particles to be in the same quantum state, then Ψ_A vanishes. And this corresponds to a known property of systems of electrons: viz. that, when two electrons are moving in the same field of force (e.g. in the atom), certain states are missing, and they are precisely those states for which the two electrons would have the same set of quantum numbers and so be described by identical wave functions. This discovery was made by Pauli (1925) as a result of a study of spectral data, and was found to be of quite general application to systems of electrons. It has come to be known as the exclusion principle, and can be stated in two equivalent ways:

1. No two electrons moving in the same field of force can have the same complete set of quantum numbers (n, l, m_l, m_s) [or the alternative set (n, l, j, m_j)].

2. The wave function describing two electrons is antisymmetric; a symmetric function, though mathematically possible, is never found in nature.

It must be emphasized that the exclusion principle was first put forward as a purely empirical result. A number of years later Pauli (1940) analyzed the problem in a way that suggested an intimate connection between the symmetry of a possible wave function for two identical particles and the magnitude of the particle spin—whether it is an integer or an odd half-integer in units of $h/2\pi$.

8.5 TWO-ELECTRON SYSTEMS

To see how the Pauli exclusion principle applies to an actual physical system, let us consider an atom with two electrons in similar states—as for example in the alkaline earth metals, or zinc, or mercury, in which we know from chemistry that there are two valency electrons occupying a special outer position in the atom.

Suppose first of all that both electrons have $l = 0$ (and the same value of the principal quantum number n). Then each is restricted to the unique value $m_l = 0$ for the z component of l on any specified axis of quantization. Thus the three quantum numbers n, l, m_l are the same for both particles, and, according to the exclusion principle, they must differ in the remaining quantum number m_s. It follows that one electron must have $m_s = +\frac{1}{2}$ and the other must have $m_s = -\frac{1}{2}$. It really is not meaningful to speak of the separate electrons in this way, since they are indistinguishable; what we *can* do, however, is to give a definite value to the combined values of the orbital and spin momentum components. We designate these as M_L and M_S, and so for this case have

$$M_L = 0 + 0 = 0$$
$$M_S = \tfrac{1}{2} + (-\tfrac{1}{2}) = 0$$

No other combinations are possible. Now these M values are what would arise from a combined orbital momentum vector **L** equal to zero and a combined spin momentum **S** also equal to zero, and we infer

TABLE 8.1

$m_l(1)$	$m_s(1)$	$m_l(2)$	$m_s(2)$	M_L	M_S
1	$\frac{1}{2}$	1	$-\frac{1}{2}$	2	0
1	$\frac{1}{2}$	0	$\frac{1}{2}$	1	1
1	$\frac{1}{2}$	0	$-\frac{1}{2}$	1	0
1	$\frac{1}{2}$	-1	$\frac{1}{2}$	0	1
1	$\frac{1}{2}$	-1	$-\frac{1}{2}$	0	0
0	$\frac{1}{2}$	1	$-\frac{1}{2}$	1	0
0	$\frac{1}{2}$	0	$-\frac{1}{2}$	0	0
0	$\frac{1}{2}$	-1	$\frac{1}{2}$	-1	1
0	$\frac{1}{2}$	-1	$-\frac{1}{2}$	-1	0
-1	$\frac{1}{2}$	1	$-\frac{1}{2}$	0	0
-1	$\frac{1}{2}$	0	$-\frac{1}{2}$	-1	0
-1	$\frac{1}{2}$	-1	$-\frac{1}{2}$	-2	0
1	$-\frac{1}{2}$	0	$-\frac{1}{2}$	1	-1
1	$-\frac{1}{2}$	-1	$-\frac{1}{2}$	0	-1
-1	$-\frac{1}{2}$	0	$-\frac{1}{2}$	-1	-1

that the separate orbital and spin momentum vectors of the two electrons do indeed combine in this way and in no other. The resultant state is described as a "singlet" state ($2S + 1 = 1$) having $L = 0$, and in spectroscopic notation is written as $^1S_0$. As a suffix here we write the value of the total angular momentum J. (**J** = **L** + **S**.)

It should be noted that, if our two electrons did not have the same value of n, the values of m_s would not be restricted to being opposite, but could combine to make M_S equal to $+1$ or -1 as well as zero. These three possibilities would then correspond to the z components of a combined spin momentum equal to unity, so that, in addition to the state $^1S_0$, we could also have $^3S_1$ (multiplicity = $2S + 1 = 3$, $J = 1$). It was the complete absence of "triplet" states of this sort when n was

Some Applications of Quantum Mechanics

known to be the same for two s electrons that helped to lead Pauli to the discovery of the exclusion principle.

We have chosen a very simple first application of the Pauli principle; let us now see how it governs the possible states formed by the combination of two equivalent p electrons ($l = 1$) in the same atom. The problem at once becomes much more complicated, because the value of m_l for each electron may assume any of the three values ± 1 or 0. We can solve it, however, by systematically writing down the values of

TABLE 8.2

M_S	M_L	L	S
0	2, 1, 0, −1, −2	2	0
1 0 −1	1, 0, −1 1, 0, −1 1, 0, −1	1	1
0	0	0	0

m_l and m_s for the two electrons so as to exhaust all the different combinations that do not have $m_l(1) = m_l(2)$ and $m_s(1) = m_s(2)$ simultaneously. This is done in Table 8.1. Note that each particular combination $(m_l, m_s)_1$, $(m_l, m_s)_2$ appears once only; we do not obtain a new and distinct state by transposing the electrons, because they are indistinguishable.

We see that there are 15 different combined states. A little study shows that they can be classified as shown in Table 8.2.

We are thus able to infer that only three distinct types of state are possible:

Singlet D state: $^1D_2$
Triplet P state: $^3P_{0,1,2}$
Singlet S state: $^1S_0$

The subscripts indicate the values of J that can be formed by the combination of **L** with **S** in each case. Notice the conservation of the total number of states—in particular for the P states where we have

$$\text{Total number of states} = (2L + 1)(2S + 1) = \sum_{J=0,1,2} (2J + 1) = 9$$

A simple way of expressing these results is furnished by the concept of the vector addition of angular momenta. For two p electrons, the magnitude of the resultant orbital momentum vector **L** ($= \mathbf{l}_1 + \mathbf{l}_2$, with $l_1 = l_2 = 1$) is restricted to the values 0, 1, 2. The magnitude of the resultant spin **S** may be 0 or 1. Thus any possible combined state of the two electrons must belong to a singlet or triplet S, P, or D configuration. The detailed application of the Pauli principle then limits the allowed combinations of **L** with **S** as shown above if both electrons have the same n. This particular way of combining states, based on

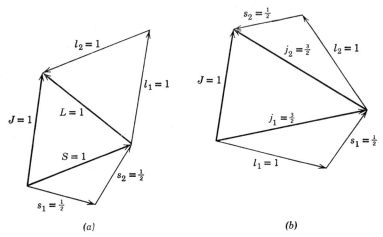

Figure 8.7. LS and jj coupling for a pair of p electrons ($l = 1$). (a) $L = 1$, $S = 1$, coupling to give $J = 1$. (b) $j_1 = \frac{3}{2}$, $j_2 = \frac{3}{2}$, coupling to give $J = 1$.

the (n, l, m_l, m_s) specification of the individual electron states, is called LS coupling (or Russell–Saunders coupling). We can go through an exactly similar procedure on the basis of the (n, l, j, m_j) classification, in which case we have what is called a jj coupling scheme (Fig. 8.7 illustrates the two types of coupling for a pair of p electrons). The choice of which to use is a genuine physical choice; it depends on whether the spin-orbit coupling of an individual electron (Section 2 of this chapter) is broken down or preserved when two electrons interact with each other in orbits of the same n. Usually the LS coupling is the more important for spectroscopy.

The principles that we have outlined can of course be extended to the case of many electrons, or to the vector combination of angular momenta for electrons having different values of l and j.

8.6 ATOMIC STRUCTURE AND THE PERIODIC TABLE

The major consequence of the exclusion principle (or rather of the property of electron systems that it expresses) is the building up of the

Some Applications of Quantum Mechanics 221

electron structure in a many-electron atom. To see how this comes about, let us consider first of all the sequence of levels for one electron moving in the field of a nucleus of charge Ze. The potential energy is given by

$$V(r) = -\frac{Ze^2}{r}$$

so that a section through the origin gives the appearance of a funnel-shaped potential well, as shown in Fig. 8.8. As we have seen, the

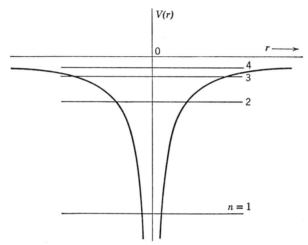

Figure 8.8. The radial variation of potential, and the first few energy levels, for a hydrogen-like atom.

energies of possible states are given to a first approximation by the equation

$$E_n = -C\frac{Z^2}{n^2}$$

where C is a constant of the dimensions of energy (actually $C = 2\pi^2 me^4/h^2 \equiv 13.5$ ev; cf. Chapter 5, Section 4). It is this result that gives special importance to n, the "principal quantum number," for it effectively determines the energy of a given state. But, when we consider the problem in more detail, we find that these levels split up into a number of others. It must be remembered that to each value of n there are n different values of l (from zero to $n - 1$), and to each value of l (except $l = 0$) there are two values of j. Thus for each n we can expect to find $2n - 1$ levels, because

1. Relativistic effects, as calculated by Sommerfeld (see Chapter 5, Section 9), cause a separation in energy of the states of different l.

2. The spin-orbit interaction leads to the doublet fine structure for each l (except for $l = 0$, when j can only be $\tfrac{1}{2}$).

The scheme of energy levels possible for one electron alone, moving in the Coulomb potential provided by a nucleus, thus becomes very complicated. We have mentioned how, in the Sommerfeld theory, the most eccentric orbits (i.e. those of the lowest l) lie lowest in energy. We could also argue that, for a given l, the level of smaller $j(= l - \tfrac{1}{2})$ must lie lower in energy than that for larger $j(= l + \tfrac{1}{2})$. For, if we return for a moment to the consideration of doublet structure, we see that the magnetic field **H** at the position of an electron is in the same direction as the orbital momentum vector **L**, since **L** is associated with a circulating positive charge. But the intrinsic magnetic moment μ of the electron is in the opposite direction to the spin momentum **S**, since in this case we have the circulation of a negative charge (Fig. 8.9). The magnetic potential energy $-\mu \cdot \mathbf{H}$ is negative when μ has a positive component along **H**, but this corresponds to the vector subtraction of **S** from **L**, and hence to the smaller value of **J**. Similarly, when **S** and **L** are oriented to give the larger value of **J**, the magnetic potential energy is positive.

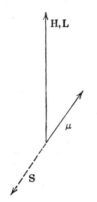

Figure 8.9. The sign of the spin-orbit interaction makes the magnetic potential energy negative, for a given value of l, when $j = l - \tfrac{1}{2}$. This case is shown in the figure.

When we come to consider a number of electrons attached to the same nucleus, the energy level scheme for any one of them is distorted. The principal reason for this is that the presence of other electrons leads to a partial shielding of the electric field due to the nucleus [as discussed in Chapter 5, Section 11-(3)]; and we can see, for example, that a nucleus Ze surrounded by $(Z - 1)$ electrons appears from a distance to be electrically identical with a proton. Thus one extra electron, added to make the structure electrically neutral, will behave like the electron in a hydrogen atom so long as it does not penetrate the charge cloud composed of the other electrons. Its potential energy for large r is thus not $-Ze^2/r$, but something approximating to $-e^2/r$ simply. This will be the case in the kind of circumstances described by a circular orbit of large n according to the old quantum theory (i.e., $l = n - 1$ in the wave-mechanical description). But, if we take a highly eccentric orbit (i.e. small l) for the same n, the electron can be thought of as penetrating through much of the electron cloud, and experiencing at some point in the orbit almost the full force of the nuclear attraction. This means a greatly enhanced negative contribution to the energy of the electron, so that, in a many-electron atom, the levels of different l are separated very distinctly—much more so, for a given value of n, than the corresponding levels for a single electron.

We can picture the development of the energy level structure in general terms by starting with the Bohr model, and then adding in turn the fine structure due to relativistic change of mass, the doublet structure due to spin, and the distortion due to electron shielding. Figure 8.10 is a strictly schematic representation for two successive values of the principal quantum number:

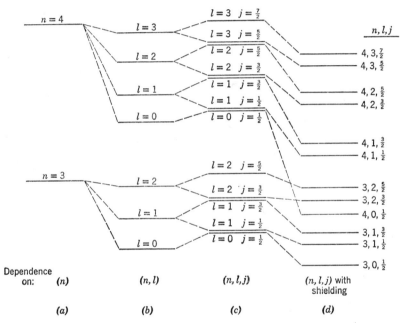

Figure 8.10. (a) Bohr levels for $n = 3$ and $n = 4$. (b) Splitting into levels of different l by Sommerfeld relativistic effect. (c) Doublet splitting of levels of given (n, l). (d) Displacement of levels as a result of shielding. The diagram is purely schematic; no attempt has been made to show anything except the order of the levels, and the energy spacings are quite unrealistic.

Two points of special interest are to be noted:

1. In the picture (c) the levels of the same (n, j) but different l are drawn close together. Until recently it was believed that they were strictly coincident, for, by what seemed a curious accident, the

positions of energy levels of a given j as calculated by wave mechanics (and taking into account the electron spin) coincided exactly with the levels of a given l, according to Sommerfeld's semiclassical calculation (without spin), provided one took $j = l + \frac{1}{2}$.

2. When electron shielding is taken into account, the levels do all appear as separate. Moreover, the great displacement in energy for penetrating orbits may break down the simple condition that all levels for a given n should lie lower in energy than all levels belonging to $n + 1$. In our diagram (d), for example, we show the level for $n = 4$, $l = 0$ falling below the two levels for $n = 3, l = 2$.

We are now in a position to discuss the building up of the periodic system of the elements. We imagine that we start with the potential well representing the Coulomb field of a nucleus Ze, and that we put into this one electron at a time until we have the full complement of Z electrons. The first electron will certainly fall into the lowest possible state, $n = 1, l = 0, j = \frac{1}{2}$. The second electron can also go into this state, since $2j + 1 = 2$—or in other words we have two distinct orientations of the vector $\mathbf{j}$ corresponding to the different values $m_j = \pm\frac{1}{2}$, and this means two recognizably different wave functions, identified by the set of quantum numbers (n, l, j, m_j). (In the absence of an external magnetic field the two states have the same energy.) But at this point we have exhausted the possibilities for $n = 1$, since by the Pauli principle we can associate only one electron with a given set of quantum numbers. We thus arrive at the concept of what is called a "closed shell." This term has a physical significance because the next electron to be added is not so firmly bound to the nucleus as the first two; it has to enter the next higher level $(n = 2, l = 0, j = \frac{1}{2})$, which corresponds classically to a larger radius of orbit $(r_n \propto n^2)$ and to a smaller ionization potential I_n $(I_n = -E_n \propto 1/n^2)$. Since $j = \frac{1}{2}$, we can again accommodate two electrons in this level; at this point we again fill a shell and must proceed to $n = 2, l = 1$. Here we have two close levels, belonging to $j = \frac{1}{2}$ and $j = \frac{3}{2}$. The first of these is filled by two electrons, and the second can accommodate four.

The complete development of this structure can now be argued with the help of one or two guiding principles. For given (n, l), the two levels of different j are close in energy, and the distinction between them is not particularly important. We can thus think of a closed shell as consisting of $2(2l + 1)$ electrons. The closed shell represents a chemically stable structure. The addition of one or two electrons beyond a closed shell gives a looser structure. It is rather easy to

Some Applications of Quantum Mechanics

detach the extra electrons, leaving a surplus of positive charge in the system as a whole; this is the basis of chemical reactivity and the formation of positive ions, and, in fact, we may expect electronic structures of this type to represent the metals. On the other hand, a structure that is short of a closed shell by one or two electrons will acquire further stability by completing the shell, thereby becoming a negative structure; here we have the possibility of forming negative ions from typical reactive nonmetals such as oxygen and the halogens. The closed shell structure by itself has its obvious expression in the inert gases.

By considering the progressive change of Z from 1 up to about 100, we can build up the periodic table in its entirety. We can illustrate the results by taking the first few shells, as shown in Table 8.3. The

TABLE 8.3

| Shell Being Filled | | Range | | |
n	l	of Z	Sequence of Chemical Elements	Total
1	0	1–2	H, He	2
2	0	3–4	Li, Be	
	1	5–10	B, C, N, O, F, Ne	8
3	0	11–12	Na, Mg	
	1	13–18	Al, Si, P, S, Cl, A	8
4	0	19–20	K, Ca	
3	2	21–30	Sc, Ti, V, Cr, Mn, Fe, Co, Ni, Cu, Zn	
4	1	31–36	Ga, Ge, As, Se, Br, Kr	18

horizontal lines mark the attainment of particularly stable configurations, viz. the inert gases He, Ne, A, Kr, etc. We may see the effect of penetrating orbits, with their depressed energies, in the fact that the shell for $n = 4$, $l = 0$ lies lower in energy than that for $n = 3$, $l = 2$, and so precedes it in the building of the electron system. This corresponds to what we indicated as a possibility in our schematic diagram (Fig. 8.10).

A full analysis of the periodic system in terms of electron quantum states brings to light a large number of finer details, but the present discussion contains most of the underlying features of the problem.

8.7 TERM DIAGRAMS FOR EXCITED ATOMIC STATES. SPECTRAL SERIES

Having considered the lowest possible state for an assemblage of electrons in an atom, let us now investigate the higher states that may be produced by excitation. To keep the discussion as simple as possible, let us limit ourselves to the one-electron excitations of an alkali metal atom. In this case we assume that the state of the whole atom

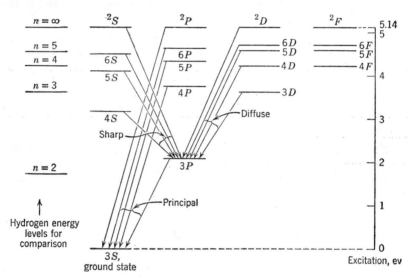

Figure 8.11. The main features of the term diagram for sodium (one-electron states). The corresponding levels for hydrogen are shown with their n-values. The D lines of sodium are the transitions $3P \to 3S$. The ionization potential of sodium (5.14 ev) is shown as the limit of excitation on the right-hand scale. Notice the drastic depression of the penetrating orbits compared to those of the same n for hydrogen.

is defined by the quantum state of the single valency electron; the remainder of the electrons pair off, as indicated in the previous section, to give a resultant angular momentum of zero.

The succession of levels of ascending n for an electron in the Coulomb potential has the characteristic feature that the levels crowd more and more closely together as the energy increases. The most deep-seated of the energy levels, representing the lowest energy state of the atom, has $L = l = 0$ for an alkali metal; i.e., it is an S state belonging to some particular value of the principal quantum number n. In sodium, for example ($Z = 11$), the first ten electrons take up all the states available for $n = 1$ and $n = 2$, and so the normal state of the atom is $3S$ (often, unfortunately, written as $1S$ because it is the lowest state involved in optical transitions). With $n = 3$ we may also have $l = 1$ ($3P$) and $l = 2$ ($3D$); these, because of their lesser penetration of the electron cloud, will be higher in energy than the $3S$ state (as explained in the previous section) and so will represent possible higher states of the atom. Still higher states become available if n itself is increased. Thus for $n = 4$ we can have $4S, 4P, 4D$, and $4F$; for $n = 5$ we can have $5S, 5P, 5D, 5F$ and $5G$. The highest states of all have $n = \infty$, and hence, according to the theory, their energy is zero for all l; at this stage the valency electron can just escape from the atom with zero kinetic energy. The largest energy release available as a quantum jump inside the atom is from such states to the $3S$ ground state.

The complete set of states can be presented on an energy level diagram. Figure 8.11 shows the more important features of the scheme for the sodium atom. It must be remembered that each level of given n and L is really a doublet (except for the S states), but this fine structure would not show up on the scale to which the diagram is drawn. Any quantum jump, giving rise to a spectral line, has an energy that is expressible as the difference between two energy levels in the diagram. Observation shows, however, that by no means all combinations of initial and final states are possible. We never, for example, find a transition between two levels of different n but the same L. Indeed, the choice is so restricted that transitions usually occur only under the following conditions:

(i) L must change by ± 1.
(ii) J may remain unchanged or else change by ± 1.

(The reasons for this type of restriction are discussed in the next two sections.) We at once see here the origin of various series in line spectra. Considering the sets of transitions that end on the levels $3S$ or $3P$, we may have

(a) $nP \to 3S$ $(n = 3, 4, 5, \cdots)$
(b) $nS \to 3P$ $(n = 4, 5, 6, \cdots)$
(c) $nD \to 3P$ $(n = 3, 4, 5, \cdots)$

Any one series of lines is expressible in the empirical form discovered by Rydberg (Chapter 5, Section 2), or more generally as follows:

$$k(n, L; n_0, L_0) = \frac{1}{hc}[E(n, L) - E(n_0, L_0)]$$

where k is the wave number of a line, $E(n, L)$ is the energy of the level at which the transition originates, and $E(n_0, L_0)$ is the fixed energy (for a given series) at which it ends. The fact that the wave number of any line is expressed as the difference of two terms has resulted in the use of the phrase "term diagram" to denote an energy level scheme such as Fig. 8.11. It is customary to express the energy in wave numbers (cm^{-1}). The great advantage of such a diagram (and the great achievement of empirical spectroscopy) is the recognition that a large variety of spectral line frequencies can be described and interrelated by the use of a quite limited number of terms.

Certain special features of the sodium scheme may be noted. The transitions $nP \to 3S$ provide the most energetic quanta (i.e. shortest wavelength) of any possible. The spectral lines of this type constitute the *principal series* (hence the use of the letter P for the levels $l = L = 1$ at which these transitions originate). The shortest wavelength possible (identifiable as the wavelength limit to which the series converges) corresponds to the full value of the ionization potential (5.13 volts) of the atom. The observed limit for the principal series in sodium has $k = 41,449.0$ cm^{-1}, $\lambda = 2412.6$ A, which is equivalent to 5.14 ev; the agreement is good.

The other two series of lines that we have specially mentioned both terminate on the same $3P$ level. This means that they have the same series limit, although they do not match anywhere else. The transitions $nS \to 3P$ form the *sharp series*, and the transitions $nD \to 3P$ form the *diffuse series;* these names serve to describe the general appearance of the lines, and have led to the use of the letters $S(L = 0)$ and $D(L = 2)$ to denote their parent levels. Many other details could be discussed, but they might only obscure the main features; we shall therefore leave the descriptive account of spectra at this point and turn to the underlying theory of radiative transitions.

8.8 RADIATIVE TRANSITIONS

A basic feature of Bohr's theory of the atom is the assertion that the emission of light happens only when an electron passes from one orbit to another, the frequency of the light being defined by

$$h\nu_{12} = E_1 - E_2$$

We must ask how the same result is reached through wave mechanics. Now classically the simplest motion of a point charge that can lead to radiation of energy is its acceleration along a straight line [Chapter 3,

Some Applications of Quantum Mechanics

Section 8(3)] and the simplest means of obtaining radiation of a given frequency is to make the charge undergo simple harmonic motion along a line:

$$x = x_0 \sin 2\pi\nu t$$

This oscillating charge is then equivalent to an electric dipole of varying magnitude $p = ex$, in terms of the displacement of the charge from some fixed origin O (Fig. 8.12). We shall therefore consider the question of evaluating the electric dipole moment of a system in wave mechanics.

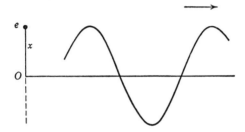

Figure 8.12. Radiation by an oscillating electron.

The description of an electron in an atom by a smooth wave function corresponds, as we have said, to a continuous charge distribution, and we can regard the value of $e\psi^*\psi$ at any point as being the local charge density ρ. In classical physics the electric dipole moment of a charge distribution is given by the equation

$$\mathbf{p} = \int \mathbf{r}\, \rho(\mathbf{r})\, d\tau$$

This is simply an extension of the equation $p = ex$ to three dimensions and to continuous (as distinct from discrete) charge distributions (Fig. 8.13). It is reasonable, then, to define a dipole moment in wave mechanics as

$$\mathbf{p} = \int \mathbf{r}\, e(\psi^*\psi)_{\mathbf{r},t}\, d\tau$$

$\psi(\mathbf{r}, t)$ or $\psi^*(\mathbf{r}, t)$ refers here to a stationary state of the system, which means that we can put

$$\psi(\mathbf{r}, t) = \psi(\mathbf{r}) \exp\left(-i\frac{2\pi E}{h} t\right), \qquad \psi^*(\mathbf{r}, t) = \psi^*(\mathbf{r}) \exp\left(+i\frac{2\pi E}{h} t\right)$$

(see Eigenfunctions, Chapter 7, Section 10). If ψ and ψ^* both refer to a definite stationary state n of an electron in an atom, we have $(\psi^*\psi)_{\mathbf{r},t} = \psi^*(\mathbf{r})\,\psi(\mathbf{r})$ simply, since the exponential factors containing E_n combine to give unity. Hence for a single eigenstate of an electron the dipole moment is independent of time, and this means that there is no radiation; it follows that the electron loses no energy so long as it remains in this state, and the objections that were leveled at Bohr's atomic model have no relevance here. (As a matter of fact, the dipole moment of an atom in a definite quantum state is not merely constant; it is zero. This follows from the geometrical symmetry of the wave functions.) But what we can now do is to ask if there exists a

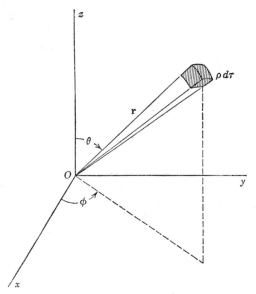

Figure 8.13. Volume element of continuous charge distribution in spherical polar co-ordinates.

kind of "mixed charge density" that is effective in the transition of an electron from one state to another. We simply pursue the strict logic of the argument for the static dipole moment, but this time we put

$$\rho_{12}(\mathbf{r}, t) = e\, \psi_2{}^*(\mathbf{r}, t)\, \psi_1(\mathbf{r}, t)$$

It will then follow that the effective dipole moment is given by

$$\mathbf{p}_{12} = \int \mathbf{r}\, e(\psi_2{}^*\psi_1)_{\mathbf{r},t}\, d\tau$$

i.e., $$\mathbf{p}_{12} = \exp\left[-i\,\frac{2\pi(E_1 - E_2)}{h}\,t\right] \int \mathbf{r}\, e\, \psi_2{}^*(\mathbf{r})\, \psi_1(\mathbf{r})\, d\tau$$

This automatically defines a dipole moment which is oscillating at the frequency characteristic of Bohr's theory and whose magnitude is

Some Applications of Quantum Mechanics 231

given by a certain integral that can be thought of as involving the degree of overlap of the initial and final wave functions, or of the corresponding charge distributions.

Under certain conditions the integral in the above formula is zero. In this case there is no possibility of a transition between states 1 and 2 as a result of electric dipole radiation. Such a transition is then said to be "forbidden," and there are mathematically well-defined conditions, known as *selection rules*, which decide whether a transition between two states is allowed or forbidden. The selection rules are intimately related to the quantum numbers of the initial and final states involved, and we shall illustrate them in their simplest form, taking as our basis the hydrogen atom without reference to the intrinsic spin of the electron. In this case the possible wave functions are those discussed in the first section of this chapter, and we can write them in the simplified form

$$\psi_{nlm}(r, \theta, \phi) = S_{nlm}(r, \theta)e^{im\phi}$$

We shall deal only with the selection rules depending on the quantum numbers m, and to do this explicitly we shall resolve the dipole moment $\mathbf{p}_{12}$ into its Cartesian components p_x, p_y, p_z. In terms of a spherical polar co-ordinate system (Fig. 8.13) we have

$$x = r \sin \theta \cos \phi, \qquad y = r \sin \theta \sin \phi, \qquad z = r \cos \theta$$

Then, setting aside the time dependence of $\mathbf{p}$, the effective component dipole moments for a transition from a state 1 to a state 2 are given by

$$p_x = \int x\, e\, \psi_2^*\psi_1\, d\tau = e \int r \sin \theta \cos \phi\; S_2^*(r, \theta)e^{-im_2\phi}S_1(r, \theta)e^{+im_1\phi}\, d\tau$$

$$p_y = \int y\, e\, \psi_2^*\psi_1\, d\tau = e \int r \sin \theta \sin \phi\; S_2^*(r, \theta)e^{-im_2\phi}S_1(r, \theta)e^{+im_1\phi}\, d\tau$$

$$p_z = \int z\, e\, \psi_2^*\psi_1\, d\tau = e \int r \cos \theta\; S_2^*(r, \theta)e^{-im_2\phi}S_1(r, \theta)e^{+im_1\phi}\, d\tau$$

For the volume element $d\tau$ we shall put

$$d\tau = r^2\, dr \sin \theta\, d\theta\, d\phi$$

and then we shall confine our attention to the integrals involving ϕ. This allows us to abbreviate the formulas for the components of $\mathbf{p}$ as follows:

$$p_x = A \int_0^{2\pi} \cos \phi\; e^{i(m_1-m_2)\phi}\, d\phi = \tfrac{1}{2} A \int_0^{2\pi} [e^{i(m_1-m_2+1)\phi} + e^{i(m_1-m_2-1)\phi}]\, d\phi$$

$$p_y = A \int_0^{2\pi} \sin\phi \, e^{i(m_1-m_2)\phi} \, d\phi = -\frac{i}{2} A \int_0^{2\pi} [e^{i(m_1-m_2+1)\phi} - e^{i(m_1-m_2-1)\phi}] \, d\phi$$

$$p_z = B \int_0^{2\pi} e^{i(m_1-m_2)\phi} \, d\phi$$

Now any integral of the form

$$I(\alpha) = \int_0^{2\pi} e^{i\alpha\phi} \, d\phi$$

where α is an integer, is zero unless α itself is zero. [For $\alpha = 0$, $I(\alpha) = 2\pi$.] We thus see that there are only *three* conditions under which **p** can be different from zero; hence for a given value of m_1 we are limited to just three effective values of m_2, as listed in Table 8.4.

TABLE 8.4

m_2	p_x	p_y	p_z
$m_1 + 1$	πA_+	$-i\pi A_+$	0
$m_1 - 1$	πA_-	$+i\pi A_-$	0
m_1	0	0	$2\pi B$

A more careful examination of these three forms for **p** shows that they correspond to the classical treatment of the Zeeman effect (Chapter 3, Section 3). The argument can be made as follows:

1. Let us suppose that a magnetic field is applied along the z direction. Then in consequence of the orbital magnetic moment the states of different m_2 lie at slightly different energies (Fig. 8.14).

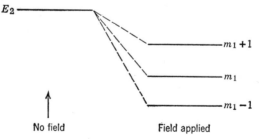

Figure 8.14. Energy levels in a magnetic field, leading to a Zeeman triplet.

2. We notice that, for $m_2 = m_1 \pm 1$, the values of p_x and p_y are equal in size, but that p_y has an additional factor $\pm i$. When we think in terms of simple-harmonic oscillations of the type $e^{i\omega t}$, we see that $\pm i \, [= \exp(\pm i\pi/2)]$ signifies that the oscillations of p_y either lead those

Some Applications of Quantum Mechanics 233

of p_x by 90° or lag behind them by 90°. But the combination of two vibrations at right angles when they differ in phase by 90° gives a circular motion in the xy plane.

3. Making use of the conclusions from (1) and (2), we can expect to describe three observed spectral lines as follows (for radiating atoms in an external magnetic field):

(a) Highest frequency: $m_2 = m_1 + 1$ Circularly polarized perpendicular to field
(b) Lowest frequency: $m_2 = m_1 - 1$ Circularly polarized in opposite sense to (a)
(c) Intermediate frequency: $m_2 = m_1$ Linearly polarized parallel to field

These correspond completely to the classical Zeeman triplet, which we can thus recognize as a special case of optical radiation according to wave mechanics. We shall come shortly to discuss the other types of Zeeman pattern that have no description in classical physics.

8.9 ELECTRON SPIN AND MAGNETISM

The fact of an electron spin, with an associated magnetic moment, became apparent from the study of doublets in line spectra. As we have seen, however, the value $s = \frac{1}{2}$ for the electron was superimposed

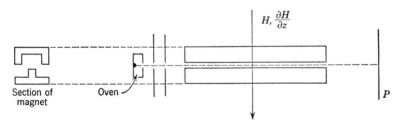

Figure 8.15. Schematic diagram of the Stern–Gerlach experiment.

arbitrarily on the existing atomic model, which had dealt hitherto solely in angular momenta measured in integral multiples of $h/2\pi$. The arbitrariness was made even greater because it appeared that the intrinsic magnetic moment of the electron was 1 Bohr magneton: i.e., as large as the magnetic moment associated with one whole unit of *orbital* angular momentum. The experimental evidence for this had really become available in 1921, when Stern and Gerlach passed a beam of silver atoms through a strongly inhomogeneous magnetic field transverse to their direction of motion (see Fig. 8.15). The idea was that, if the atoms possessed magnetic moments, they would first of all take up their quantized orientations with respect to the field direction, and would then experience the force characteristic of a magnetic dipole in a nonuniform magnetic field:

$$F = \mu_z \frac{\partial H}{\partial z} \qquad (\mu_z = \text{component of } \mu \text{ along } H)$$

Atomic magnets in passing through such a field would therefore be separated in position.

According to classical ideas, all orientations of μ would be possible, so that the "trace" received on a plate P would simply be broadened

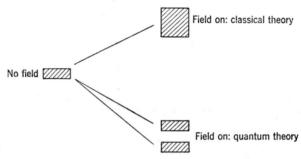

Figure 8.16. Idealized forms of trace caused by atomic beam on collector plate, according to classical and quantum theories.

when the field was switched on; but quantum conditions should allow only a limited number of components in the beam emerging from the field (Fig. 8.16). Stern and Gerlach found just two components in their silver beam, and from the amount of separation they were able to show that the effective magnetic moment was 1 Bohr magneton. Now the neutral silver atom has an odd number of electrons ($Z = 47$), and the structure consists of a set of closed shells with one last electron outside them in a state of $l = 0$. Thus it is the spin alone of this odd electron that is responsible for the angular momentum and magnetism of the atom as a whole; and we can interpret the results of the Stern-Gerlach experiment in the following way:

2 possible orientations in magnetic field: $\therefore \; s = \tfrac{1}{2}$

$\therefore$ Angular momentum along $H = \dfrac{1}{2} \cdot \dfrac{h}{2\pi} = \dfrac{1}{2}\hbar$

Some Applications of Quantum Mechanics 235

Effective magnetic moment = 1 Bohr magneton = $\dfrac{eh}{4\pi mc} = \dfrac{e\hbar}{2mc}$

$\therefore \dfrac{\text{Magnetic moment}}{\text{Angular momentum}} = 2\dfrac{e}{2mc}$

For ordinary orbital magnetism, the ratio of magnetic moment to angular momentum (which is called the *gyromagnetic ratio*) is simply $e/2mc$, only one half as large as for the spin. By taking account of this difference we are able to understand the so-called anomalous

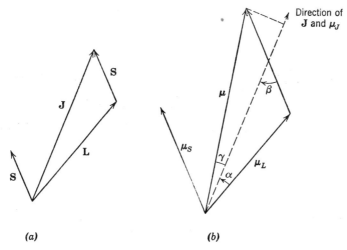

(a)   (b)

Figure 8.17. (a) The vector combination of **L** and **S** to give **J**. (b) The vector combination of μ_L and μ_S to give μ, which is then projected onto the direction of **J** to give the effective magnetic moment μ_J (i.e. $\mu_J = \mu \cos \gamma$).

Zeeman effect, which throws important light on electron configurations in atoms.

As we have seen in discussing doublet fine structure, the very strong internal magnetic fields in atoms give reality to an angular momentum vector **J** formed from **L** and **S**. This also means that an electron has a composite magnetic moment μ by the vector addition of μ_L and μ_S (see Fig. 8.17). The interacting magnetic dipoles exert a mutual torque, however, which has the consequence that the angular momentum vectors **L** and **S** precess about their resultant **J**. The rate at which they do this is of the order of 10^{11} revolutions per second, as we can estimate from the expression $\Delta W/h$, where ΔW is the doublet fine structure splitting expressed in ergs. Thus the direction of **J** becomes the only axis of physical significance in the system, so that effectively the magnetic moment becomes μ_J, where

$$\mu_J = \mu \cos \gamma = \mu_L \cos \alpha + \mu_S \cos \beta$$

In this formula we substitute the following:

$$\mu_L = \frac{e}{2mc} L = \frac{eh}{4\pi mc} [l(l+1)]^{1/2}$$

$$\mu_S = 2 \frac{e}{2mc} S = 2 \frac{eh}{4\pi mc} [s(s+1)]^{1/2} \qquad (s = \tfrac{1}{2})$$

$$\cos \alpha = \frac{J^2 + L^2 - S^2}{2JL}, \qquad \cos \beta = \frac{J^2 + S^2 - L^2}{2JS}$$

Then
$$\mu_J = \frac{eh}{4\pi mc} \cdot \frac{1}{2[j(j+1)]^{1/2}} \{[j(j+1) + l(l+1) - s(s+1)] + 2[j(j+1) + s(s+1) - l(l+1)]\}$$

$$= [j(j+1)]^{1/2} \frac{eh}{4\pi mc} \left[1 + \frac{j(j+1) + s(s+1) - l(l+1)}{2j(j+1)} \right]$$

i.e.,
$$\mu_J = [j(j+1)]^{1/2} g_j \frac{eh}{4\pi mc}$$

where g_j is a numerical factor (Landé's splitting factor) that is completely defined if l, s, and j are all given.

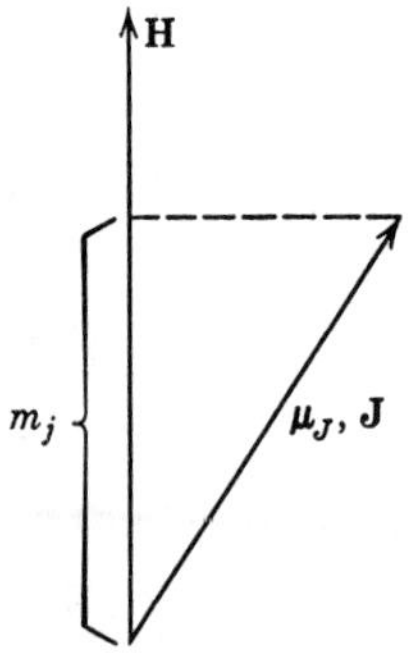

Now, if this whole system is put into an external magnetic field, the vector magnetic moment μ_J can take up $2j + 1$ different orientations characterized by the magnetic quantum number m_j (Fig. 8.18). The magnetic potential energy of a state of specified m_j is given by

$$\Delta W = -\mu_J \cdot \mathbf{H}$$

$$= -\mu_J H \frac{m_j}{[j(j+1)]^{1/2}}$$

Figure 8.18. Quantized orientation of resultant magnetic moment on an external field **H**.

or
$$\Delta W = -m_j g_j \frac{eh}{4\pi mc} H$$

If transitions occur between states $1(n_1, l_1, j_1, m_{j1})$ and states $2(n_2, l_2, j_2, m_{j2})$, we may then have a variety of lines of different frequencies. Figure 8.19, for example, shows the transitions between two groups of states belonging to $j = \tfrac{3}{2}$

Some Applications of Quantum Mechanics

and $j = \frac{1}{2}$. We might expect to find $(2j_1 + 1) \times (2j_2 + 1) = 8$ possible lines, but we discover that there are selection rules that limit the possibilities (cf. Sections 7 and 8 of this chapter). These selection rules, which have almost no exceptions in atomic spectroscopy, are as follows:

$$\left. \begin{array}{l} \Delta j = 0, \pm 1 \\ \Delta l = \pm 1 \\ \Delta m_j = 0, \pm 1 \end{array} \right\} \text{ for one-electron transitions}$$

Thus in the example we have chosen (which actually corresponds to

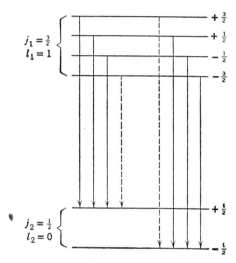

Figure 8.19. Transitions between magnetic sublevels in a magnetic field: the origin of the anomalous Zeeman effect. The diagram is drawn for the $3P_{\frac{3}{2}}$ and $3S_{\frac{1}{2}}$ levels of a one-electron atom (e.g. the sodium D_2 line). The broken lines represent forbidden transitions.

one of the D lines of sodium) there are six possible transitions, and these all show up in the Zeeman pattern when a magnetic field is applied. A principal difference between this analysis of optical transitions and the simpler one, disregarding spin, is that g_j is different for each (j, l) so that the magnetic splitting of levels in the initial state differs from that in the final state. All possible transitions then appear at *different* characteristic frequencies, whereas, if only l were operative, the spacing between successive magnetic levels would be independent of l, and several different transitions would occur with the same wavelength. (The picture would, in fact, be always that of a classical

Zeeman effect.) The ability of quantum mechanics to account for the varied and complex Zeeman patterns that are found in practice must count as one of its great successes.

We must remember, however, that our scheme depends entirely on the peculiar properties ascribed to the spinning electron. The apparently anomalous factor 2 appearing in the spin magnetism was explained by Thomas (1926) as a relativistic effect, which reduced the arbitrariness of the situation, although the half-integer spin quantum number of the electron remained a purely empirical fact. But then in 1928 Dirac succeeded in setting up a thorough-going relativistic wave equation, and from this the spin and magnetic moment of an electron emerged as an automatic and necessary consequence. It is curious to notice how, although the approach to wave mechanics was through special relativity, Schrödinger's equation is concerned with the nonrelativistic connection between energy and momentum:

$$T = \frac{p^2}{2m} \quad \text{where} \quad T = \text{kinetic energy}$$

If we adhere to a relativistic formulation, however, we have

$$p = \frac{m_0 v}{(1 - \beta^2)^{1/2}}, \quad E = \frac{m_0 c^2}{(1 - \beta^2)^{1/2}} \quad \text{where} \quad E = total \text{ energy}$$

It is easy to verify that the connection between energy and momentum in this case is

$$E^2 = p^2 c^2 + (m_0 c^2)^2 \qquad \text{(see Problem 3, Chapter 6)}$$

Dirac took this equation as basic, and invented a method by which it could be given expression in quantum mechanics. In more recent years some further refinements of Dirac's theory have been made, in consequence of the discovery that the anomalous factor in the gyromagnetic ratio for the electron is not *exactly* 2. [Precise and beautiful experiments by Lamb and Retherford (1947) led to the value 2.00232.] One consequence of this (and it was the basis of the Lamb–Retherford experiment) is that two levels of the same j but different l in a hydrogen-like atom do not occur at exactly the same energy, as they would according to all previous theories (cf. our discussion of the periodic table), and so it is possible to stimulate and observe the radiative transitions between them. The energy difference is exceedingly small, and corresponds to a long wavelength (some tens of centimeters) which lies in the 1000-Mc range of radio frequencies.

Some Applications of Quantum Mechanics 239

8.10 QUANTUM STATISTICS

In obtaining Boltzmann's formula (Chapter 4, Section 10) governing the distribution of a certain number of particles amongst a set of energy levels, we made two assumptions (one tacit, the other explicit) that must now be brought in question. These assumptions are:

1. That there is no restriction on the number of particles that can be accommodated in a given level.
2. That, although the permutation of particles within a level does not give rise to a new arrangement of the system, we do get a distinct new arrangement whenever particles belonging to different levels are interchanged.

If by a given level we understand a state labeled by its complete set of quantum numbers, it is clear that electrons, and any other types of particle that obey Pauli's exclusion principle, cannot be treated according to assumption 1. Moreover, in wave mechanics we cannot hold to assumption 2 because it implies our ability to distinguish in some way between particles that are identical. Quantum mechanics therefore calls for a new approach to the problems of statistical mechanics, and the starting point is the assertion that a wave function describing two identical particles is either symmetric or antisymmetric. As we have already seen, particles that are described by antisymmetric wave functions must conform to the exclusion principle, and electrons typify this class. On the other hand, we can well imagine particles that are described by symmetric wave functions; in this case the wave function for two particles in the same state does not vanish, which means that there is no limit on the number of particles per level. It appears that many species of neutral atoms and molecules, and also light quanta, belong to this category, and for most purposes the classical Boltzmann statistics can describe them adequately. It must be remembered, however, that there *is* a difference between the classical and the wave-mechanical enumeration of distinct arrangements for such particles, and under certain conditions it leads to an observable difference in physical behavior. We shall come to this question later (Section 13 of this chapter), but will first examine the much more drastic alterations brought about for particles that are governed by the exclusion principle.

Suppose that, in an energy interval $d\varepsilon_i$ at ε_i, there are g_i different quantum states available to an assembly of electrons, and suppose that n_i electrons lie within this range. Then each state contains either 0 or 1 electron, since any greater occupation would violate the exclusion principle. Thus $n_i \leqslant g_i$, and the situation is described very simply by saying that there are n_i occupied states and $(g_i - n_i)$ empty ones. The number of distinct arrangements that can be realized, for fixed values of g_i and n_i, is thus given by

$$p_i = \frac{g_i!}{n_i!(g_i - n_i)!}$$

Extending this to a whole set of possible energy states, we have the total number of arrangements given by

$$P = \prod_i \frac{g_i!}{n_i!(g_i - n_i)!}$$

The most probable configuration of the system is obtained by maximizing P, subject to the auxiliary conditions

$$\left. \begin{array}{l} \sum_i n_i = N \\ \sum_i \varepsilon_i n_i = E \end{array} \right\} \quad \begin{array}{l} \text{for a fixed number } N \text{ of particles} \\ \text{carrying a total energy } E \end{array}$$

We follow the same procedure as that for obtaining Boltzmann's distribution (Chapter 4, Section 10):

$$\delta(\log P) = -\sum_i \delta(\log n_i!) - \sum_i \delta[\log (g_i - n_i)!]$$

$$\approx -\sum_i \delta n_i \cdot \log n_i + \sum_i \delta n_i \cdot \log(g_i - n_i)$$

We now combine the conditions

$$\delta(\log P) = 0$$
$$\delta N = 0$$
$$\delta E = 0$$

Then

$$\sum_i \delta n_i \left[\log\left(\frac{g_i - n_i}{n_i}\right) - \alpha - \beta \varepsilon_i \right] = 0$$

whence

$$n(\varepsilon) = \frac{g(\varepsilon)}{e^{\alpha + \beta \varepsilon} + 1}$$

Some Applications of Quantum Mechanics

As in classical statistics, β can be identified with $1/kT$, so we may put

$$n(\varepsilon) = \frac{g(\varepsilon)\exp(-\varepsilon/kT)}{e^\alpha + \exp(-\varepsilon/kT)}$$

For T sufficiently large, this becomes equivalent to the Boltzmann distribution, since $\exp(-\varepsilon/kT)$ then becomes negligible compared to e^α. But, for $T \to 0$, it can be shown that the parameter α assumes the

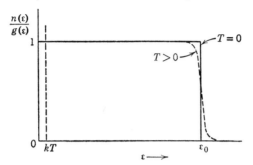

Figure 8.20. The Fermi–Dirac energy distribution function (probability of occupation of an individual state at a given energy). Observe the slight modification caused by change of temperature (assuming $\varepsilon_0 \gg kT$).

form $-\varepsilon_0/kT$, where ε_0 has the dimensions of energy, so for this case we have

$$n(\varepsilon) = \frac{g(\varepsilon)}{\exp[(\varepsilon - \varepsilon_0)/kT] + 1}$$

This function has remarkable properties, for, if we assume some very small value of $kT (\ll \varepsilon_0)$, the value of the exponential is practically zero for all $\varepsilon < \varepsilon_0$, but, as soon as ε exceeds ε_0, it becomes very large. Thus the value of $n(\varepsilon)/g(\varepsilon)$ remains very nearly equal to unity almost up to ε_0, and then falls rapidly to zero (Fig. 8.20). This then represents a totally different energy distribution of particles from what one finds using classical statistics, for under these conditions the bulk of the particles have energies far in excess of kT. This particular type of quantum statistics is known as Fermi–Dirac statistics, after its inventors.

8.11 ELECTRONS IN METALS

The behavior of electrons in metals provides a good example of the application and consequences of Fermi–Dirac statistics. To discuss this problem we regard the interior of a metal as representing a region of constant potential for an electron. The boundaries of the metal represent a high potential wall. Thus an electron inside a block of metal can be treated in wave mechanics as a particle in a box (Chapter 7, Section 11). This means that all its acceptable wave functions are in the form of plane waves with nodes at the walls. Each permitted wave function represents a distinct state of the spatial motion, and to find the number of states lying within a given energy interval we must count the corresponding number of stationary vibrations. But this is no new problem; it is precisely what we did for the standing waves of thermal radiation inside a cavity (Chapter 4, Section 9).

Let us suppose that we have N electrons confined in a cube of edge a. An electron of given energy has an associated de Broglie wave of wave number k, such that

$$k = \frac{mv}{h} = \frac{p}{h}$$

Thus the number $n(k)$ of possible eigenfunctions for wave numbers between zero and k is given by

$$n(k) = \frac{4\pi}{3} k^3 a^3$$

and the number within a range dk at k is given by

$$dn(k) = 4\pi V k^2 \, dk = \frac{4\pi V p^2 \, dp}{h^3}$$

by strict analogy with the radiation problem. Just as we had to bring in a further factor 2 to take account of the two states of polarization of photons, so here we bring in a factor 2 for the two possible spin orientations of an electron having a given space wave function. We therefore find

$$g(\varepsilon) = 2 \, dn(k) = \frac{8\pi V}{h^3} p^2 \, dp = \frac{8\pi V (2m^3)^{1/2}}{h^3} \varepsilon^{1/2} \, d\varepsilon$$

(We use the nonrelativistic relation $\varepsilon = p^2/2m$.)

Thus the complete formula for the energy distribution of electrons within a metal is

$$dn(\varepsilon) = \frac{8\pi V (2m^3)^{1/2}}{h^3} \cdot \frac{\varepsilon^{1/2} \, d\varepsilon}{\exp[(\varepsilon - \varepsilon_0)/kT] + 1} \qquad (kT \ll \varepsilon_0)$$

It is interesting to follow out some of the consequences of this result.

Some Applications of Quantum Mechanics — 243

For $kT \ll \varepsilon_0$, as we have seen, the denominator of the formula gives effectively a rectangle extending from $\varepsilon = 0$ to $\varepsilon = \varepsilon_0$. Supposing that this is exactly true, we can at once integrate the above formula to obtain the total number of particles:

$$N = \frac{8\pi V (2m^3)^{\frac{1}{2}}}{h^3} \int_0^{\varepsilon_0} \varepsilon^{\frac{1}{2}} d\varepsilon = \frac{16\pi V}{3h^3} (2m^3)^{\frac{1}{2}} \cdot \varepsilon_0^{\frac{3}{2}}$$

$$\therefore \varepsilon_0 = \frac{h^2}{8m} \left(\frac{3N}{\pi V}\right)^{\frac{2}{3}} = 5.84 \cdot 10^{-27} \left(\frac{N}{V}\right)^{\frac{2}{3}} \text{ ergs}$$

There is thus a unique connection between the maximum electron energy and the density of the free electrons in the metal. To find ε_0 in a typical case, let us consider silver, which as we have seen has just one odd electron per atom that can reasonably be assumed to act as a conduction electron in the metallic state. The density of silver is 10.5 g per cm^3, and its atomic weight is 108.

Thus $\dfrac{N}{V} = \dfrac{6.02 \cdot 10^{23} \times 10.5}{108} = 5.85 \cdot 10^{22}$ electrons/cm^3

From this we find $\varepsilon_0 = 8.8 \cdot 10^{-12}$ erg $\equiv 5.5$ ev. Now the work function ϕ of silver is about 4 volts. We infer from this that the most energetic electrons inside the metal lie at an energy $e\phi$ below the top of the potential barrier, so that our picture of a metal is a potential well of a total depth $\varepsilon_0 + e\phi$, and filled up to the level ε_0, as shown in Fig. 8.21. Because of the factor $\varepsilon^{\frac{1}{2}}$ in $dn/d\varepsilon$ the levels become more crowded toward the upper limit.

The application of Fermi–Dirac statistics resolves a seeming paradox in the electron theory of metals. We have seen (Chapter 3, Section 6) how the picture of a metal as a material containing a gas of free electrons gives a very satisfactory account of thermal and electric conductivity. On the other hand, it has been known for many years that this electron gas fails to contribute to the specific heats of metals. The specific heat of any crystalline solid should be $3R$ cal per mole per

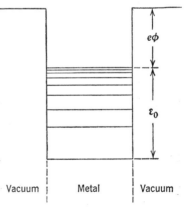

Figure 8.21. Electrons in the interior of a metal, showing the crowding of states toward the upper limit of the energy distribution.

°K, in consequence of the vibrations of atoms about their equilibrium positions. But a metal with, say, one free electron per atom should have in addition to this the specific heat of a monatomic gas, viz. $\tfrac{3}{2}R$ cal per mole per °K. Its failure to exhibit this behavior is understood when we realize that the electron gas does not remotely approach thermal equilibrium, in any ordinary sense, with the metal that contains it. At absolute zero the mean energy of the electrons is $\tfrac{3}{5}\varepsilon_0$,

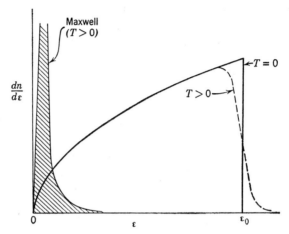

Figure 8.22. The energy distribution (number of particles per unit energy interval) according to Maxwell–Boltzmann and Fermi–Dirac, for $kT \ll \varepsilon_0$.

which we can take as equivalent to the mean gas-kinetic energy at some temperature T_0. That is, we put

$$\tfrac{3}{2}kT_0 = \tfrac{3}{5}\varepsilon_0$$

or
$$T_0 = \frac{2}{5}\frac{\varepsilon_0}{k}$$

Taking silver as an example once again, we find that this gives $T_0 \approx 25{,}000°\text{K}$. When the temperature of the metal is raised, there is no question of equipartition of energy between atoms and electrons. The energy spectrum of the electrons is only slightly affected (see Fig. 8.22); all that can happen is a slight displacement at the upper edge of the distribution, and the mean energy of the electrons as a whole is scarcely changed. But this is only another way of saying that the electron gas does not contribute appreciably to the specific heat of the structure. It is a very astonishing result, though there is no doubt of its correctness; what is equally surprising, in retrospect,

Some Applications of Quantum Mechanics 245

is that the use of Maxwell's distribution should have led so successfully to the Wiedemann–Franz law.

8.12 SPECIFIC HEATS OF SOLIDS

The study of the specific heats of solids provides one of the many interesting links between classical and quantum physics. The well-known law of Dulong and Petit (1819) recognized the fact that the specific heat multiplied by the atomic weight has about the same value (6 cal/°K) for all solid elements, although certain exceptions (e.g., boron, beryllium, and carbon in the form of diamond) fall far below this value. The measurements of specific heat on which the Dulong–Petit law was based were made at ordinary temperatures (say 0° to 100° C), and it was subsequently found that the anomalous substances exhibited a marked increase of specific heat with temperature, suggesting an ultimate approach to the more usual value. On the other hand, the specific heats of the supposedly "normal" substances were found to fall off when the measurements were made at lowered temperatures. We have already referred (Chapter 4, Section 9, and Section 11 of this chapter) to the expectation that a solid, regarded as an assembly of atomic oscillators, should have an atomic heat of $3Nk = 3R$ according to the principle of equipartition of energy. This would account satisfactorily for the figure of 6 cal per °K, but not for the temperature dependence. We must therefore examine the problem more carefully.

Figure 8.23 shows the general form of the variation of potential energy with distance for an atom or ion in the lattice structure of a solid. Classical mechanics would imply a state of equilibrium at $r = r_0$, and this would be a static situation for sufficiently low temperatures; the particle would have zero kinetic energy and hence a total (negative) energy equal to V_0. The points $r = 0$ and $r = 2r_0$ are the positions of neighboring atoms. (Our curve does *not* represent the mutual potential energy of two atoms or ions in isolation from all others; this is a related but distinct problem.) If thermal energy is now given to the particle, it can oscillate over a range of distance from r_1 to r_2 as shown. Provided the displacements are small, we can write, with good accuracy,

$$V(r) = V_0 + \tfrac{1}{2}\beta(r - r_0)^2$$

where β defines the nearly parabolic variation of potential energy with distance near $r = r_0$. Also, if the energy of the vibration is E and the kinetic energy is T, we have

$$T(r) = \tfrac{1}{2}M\dot{r}^2$$

and $$E + V_0 = T(r) + V(r) = \text{const}$$

Hence, by differentiation,

$$0 = M\dot{r}\ddot{r} + \beta(r - r_0)\dot{r}$$

or $$\ddot{x} + \frac{\beta}{M}x = 0$$

where $$x = r - r_0$$

This defines a simple-harmonic vibration of angular frequency $\omega = (\beta/M)^{1/2}$ of arbitrary amplitude, with two degrees of freedom. Since

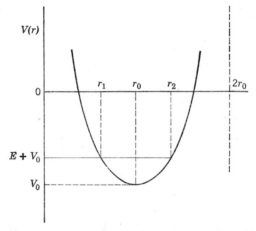

Figure 8.23. The potential seen by an atom in a solid lattice, with limits of oscillation defined (classically) by the total energy. The normal interatomic distance is r_0.

equivalent oscillations in two other independent perpendicular directions are also possible, the total energy at temperature T should be $3kT$ per atom, and hence $3RT$ per g atom of an element.

Quantum mechanics changes our thinking on this problem by denying that any arbitrary amplitude (with its associated energy) is possible. For one thing, there is bound to be a finite amount of vibration even at the absolute zero of temperature. This curious result is a manifestation of the uncertainty principle. For suppose we have an oscillation of amplitude x_0 and maximum momentum p_0. Then the particle in its vibration lies somewhere in the ranges zero to x_0 for displacement and zero to p_0 for momentum. Thus, by the uncertainty principle,

$$\Delta p \cdot \Delta x = p_0 x_0 \approx \frac{h}{2\pi} = \hbar$$

But the energy of the motion is given by

$$E_0 = \tfrac{1}{2}\beta x_0{}^2 = \frac{p_0{}^2}{2M} = \tfrac{1}{2} p_0 x_0 \left(\frac{\beta}{M}\right)^{\!\!1/2} = \tfrac{1}{2} p_0 x_0 \omega$$

Hence

$$E_0 \approx \tfrac{1}{2}\hbar\omega$$

The rigorous quantum-mechanical solution shows that this result is exact; there is a well-defined *zero-point energy* that can never be removed. The complete set of permitted eigenvalues of the energy in this "harmonic oscillator" potential is given by

$$E_n = (n + \tfrac{1}{2})\hbar\omega = (n + \tfrac{1}{2})h\nu$$

and so conforms to Planck's original quantum hypothesis for atomic oscillators generating cavity radiation. (The zero-point energy, being inaccessible, does not affect the situation and can be used to define the effective zero of the energy scale.)

Einstein, in 1907, applied the quantum theory to the problem of specific heats. For an assembly of N oscillators we shall have

$$E = \frac{3N\varepsilon_0}{\exp(\varepsilon_0/kT) - 1} \quad \text{with} \quad \varepsilon_0 = \hbar\left(\frac{\beta}{M}\right)^{\!\!1/2}$$

The factor 3 arises from the three-dimensional character of the vibration; apart from this the theory is exactly as already given for blackbody radiation (Chapter 4, Section 11). The specific heat C is then obtained as

$$C = \frac{dE}{dT}$$

Two extremes can be easily considered:

$kT \gg \varepsilon_0$: $\quad E \approx 3NkT, \qquad\qquad C \approx 3Nk = 3R$

$kT \ll \varepsilon_0$: $\quad E \approx 3N\varepsilon_0 \exp(-\varepsilon_0/kT), \quad C \approx 3R\left(\frac{\varepsilon_0}{kT}\right)^{\!2} \exp(-\varepsilon_0/kT)$

This shows that the Einstein formula predicts a rise of the specific heat from zero at $T = 0$ to a constant value of $3R$ at temperatures that are large compared with a characteristic temperature $\Theta\ (= \varepsilon_0/k)$

for a given substance. We see also that, if C is plotted against T/Θ, the specific heat curves for all solid substances should be the same, provided that Θ is correctly chosen for each.

Einstein's theory went a long way toward fitting the facts, but it was not perfect. The main objection to it is that it regards the atoms as vibrating independently of each other, whereas the field of force in which a given atom oscillates is provided entirely by its neighbors, and will change when they move. In other words there is a strong coupling between the motions of the different atoms; they behave more nearly like a line of point masses attached to a flexible string. In such a case the possible modes of vibration can be enumerated by considering the possible stationary waves that can be set up in the structure as a whole. This was done by Debye and by Born and von Kármán (1912).

Debye's treatment depends on the same analysis of standing waves that we have used in discussing the Rayleigh–Jeans law (Chapter 4, Section 9), but we are now dealing with elastic vibrations, and so one longitudinal disturbance, as well as two transverse vibrations, is allowed for any given value of the wave vector $\mathbf{k}$. The contribution to the energy in a frequency range $d\nu$ at ν is thus given by

$$dE = \left(\frac{8\pi V}{v_t^3}\nu^2\,d\nu + \frac{4\pi V}{v_l^3}\nu^2\,d\nu\right)\frac{h\nu}{\exp(h\nu/kT) - 1}$$

where V is the volume considered, and v_t, v_l are the velocities of transverse and longitudinal elastic waves in the substance. The total energy is obtained by integrating dE over the available range of frequencies. There cannot be a larger number of independent stationary waves than there are independent vibrations of the separate atoms; thus there is a maximum frequency ν_m such that the total number of waves in the range from zero to ν_m is equal to $3N$. This requires

$$3N = 4\pi V\left(\frac{2}{v_t^3} + \frac{1}{v_l^3}\right)\int_0^{\nu_m}\nu^2\,d\nu$$

whence
$$E = \frac{9N}{\nu_m^3}\int_0^{\nu_m}\frac{h\nu}{\exp(h\nu/kT) - 1}\nu^2\,d\nu$$

We can verify that this formula for E, like Einstein's, gives $E \approx 3NkT$ for T large. For very low temperatures, on the other hand, the upper limit ν_m, expressed as a multiple of kT/h, approaches infinity, and we have the same integral that appeared in the evaluation of Stefan's constant (Chapter 4, Section 11, and Appendix IV). Taking over that

Some Applications of Quantum Mechanics

result, we have

$$kT \ll h\nu_m: \quad E \approx 9NkT \left(\frac{kT}{h\nu_m}\right)^3 \int_0^\infty \frac{x^3\,dx}{e^x - 1}$$

i.e.,

$$E \approx \frac{3\pi^4}{5} RT \left(\frac{T}{\Theta}\right)^3$$

where $\Theta = h\nu_m/k$ is known as the *Debye temperature* and replaces the similar parameter ε_0/k of Einstein's treatment. By differentiation of

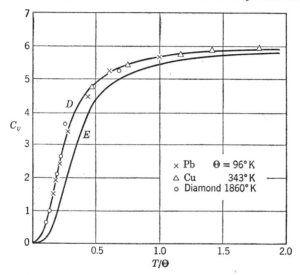

Figure 8.24. The specific (atomic) heats of several elements, plotted against T/Θ. The superiority of the Debye theoretical curve over Einstein's may be seen, especially for low temperatures.

the above formula we see that the specific heat at sufficiently low temperatures should vary as T^3; this has been verified by experiment and strongly favors Debye's theory over Einstein's. Figure 8.24 shows the convincing agreement between theory and experiment for several substances. It may be noted that the anomalous substances, such as diamond, differ from others merely in having a high characteristic temperature Θ, which is to be expected from the very high values of their elastic constants.

8.13 BOSE–EINSTEIN STATISTICS

We must draw attention to an objectionable feature of our derivation of the Debye formula. The mean energy of a standing wave of frequency ν is assumed to be given by

$$\bar{\varepsilon} = \frac{h\nu}{\exp\,(h\nu/kT)\,-\,1}$$

We mentioned in Chapter 5, Section 1, how this result was originally obtained by allowing a wave vibration of frequency ν to have energy in any integral multiple of $h\nu$, but that the photoelectric effect gave no grounds for this assumption. The value of $\bar{\varepsilon}$ is nevertheless correct, and the justification for it is to be found in another application of quantum statistics, this time for a system of particles (quanta of elastic or electromagnetic vibration) for which the wave function is symmetric with respect to interchange of any two particles (see Section 4 of this chapter). We have already pointed out that this permits any number of particles to occupy the same quantum state, which for light waves (photons) or for elastic-acoustic waves ("phonons") means the same frequency, direction of propagation, and state of polarization. By following out the consequences of this assumption, Bose (1924) was able to set up an acceptable form of statistics for light quanta, and Einstein (1925) extended the theory to include the behavior of gases.

The statistical mechanics of quanta is based upon three assumptions:

1. That any number of quanta can occupy the same quantum state.
2. That the total energy E is fixed.
3. That the total number of quanta is unrestricted. (This relaxed condition must not be used when the system considered is a gas of atoms or molecules instead of quanta.)

As in Section 10 of this chapter, we consider the number of distinct quantum states g_i in a range of energy $d\varepsilon_i$ at ε_i. The distribution of n_i quanta among these states can be obtained by a simple method described by Born.* We write down the number of a given state s followed by the identification of the quanta q occupying it, e.g.

$$s_1(q_1q_2q_3)\ s_2(q_4)\ s_3(0)\ s_4(q_5q_6)\ \cdot\ \cdot\ \cdot$$

Since there is no restriction on the number of quanta per state, and since the quanta are indistinguishable, we can run through all possible arrangements by naming the first state in the line (which can be done in g_i ways), and then permuting the number $(g_i + n_i - 1)$ that

---

* M. Born, *Atomic Physics*, 6th ed., London, Blackie & Son, 1957. The problem is also interestingly discussed in relation to the other forms of statistics (Boltzmann and Fermi–Dirac) by R. D. Cowan [*Am. J. Phys.* **25**, 463 (1957)].

Some Applications of Quantum Mechanics 251

remains. But the permutation of quanta within a state makes no physical difference; nor does the permutation of states among themselves. Thus the number of distinct arrangements at energy ε_i is given by

$$p_i = \frac{g_i(g_i + n_i - 1)!}{g_i! n_i!} = \frac{(g_i + n_i - 1)!}{(g_i - 1)! n_i!} \approx \frac{(g_i + n_i)!}{g_i! n_i!}$$

The total number of arrangements for all energies is thus

$$P = \prod_i \frac{(g_i + n_i)!}{g_i! n_i!}$$

Applying the standard maximizing procedure (as in Chapter 4, Section 10, and Section 10 of this chapter), we find

$$\delta(\log P) = 0 = \sum_i \delta[\log (g_i + n_i)!] - \delta[\log n_i!]$$

$$= \sum_i \delta n_i [\log (g_i + n_i) - \log n_i]$$

Also

$$\delta E = 0 = \sum_i \varepsilon_i \, \delta n_i$$

Hence

$$\sum_i \delta n_i \left[\log \left(\frac{g_i + n_i}{n_i} \right) - \beta \varepsilon_i \right] = 0$$

and so

$$n(\varepsilon) = \frac{g(\varepsilon)}{e^{\beta \varepsilon} - 1} \quad \text{with} \quad \beta = 1/kT$$

The number of quanta per state at energy $\varepsilon = h\nu$ is thus given by

$$\frac{n(h\nu)}{g(h\nu)} = \frac{1}{\exp(h\nu/kT) - 1}$$

and the amount of quantum energy per state at frequency ν can be written

$$\varepsilon_\nu = \frac{h\nu}{\exp(h\nu/kT) - 1}$$

We see, therefore, that the value of ε_ν is identical with the energy $\bar{\varepsilon}$ obtained from Planck's original formulation, but the artificial and incorrect features of the cavity radiation problem have at last been removed.

Problems

8.1. The lowest state of the hydrogen atom is discussed in Appendix VII. Consider in a similar way another possible state with $l = 0$, whose radial wave function is of the form

$$R(r) = (1 + \beta r)e^{-\alpha r}$$

Find the eigenvalue of the energy, and compare it to that of the lowest state. Draw the radial density function analogous to Fig. 8.3.

8.2. Carry out the complete enumeration of the states for two equivalent p electrons in jj coupling, and arrange them according to the component momenta j_1, j_2 and the resultant J. Verify that the total number of allowed states is the same as in LS coupling. What further states are possible if the electrons are not equivalent?

8.3. Using the term scheme shown in Fig. 8.11 for the sodium atom, make a table of the energies of the allowed transitions belonging to the sharp, principal, and diffuse series. Convert these to wavenumbers and to wavelengths, and from the latter construct the spectrum of one-electron transitions for sodium as it would be seen in a spectroscope. Identify the D lines (which appear as a single line in this simplified picture).

8.4. In a Stern–Gerlach experiment a beam of silver atoms ($A = 108$) is produced from an oven at a temperature of 1000° K and passes for a distance of 10 cm through an inhomogeneous magnetic field whose gradient is 10^4 gausses/cm. Find the splitting between the components of the trace if the beam continues for a further 10 cm before falling on a collector plate.

8.5. Figure 8.19 shows the energy levels and transitions that comprise the Zeeman effect for the D_2 line of sodium. With the help of Section 9 of this chapter, work out the relative magnitudes of the energy splittings ΔW. Make a scale drawing of the upper- and lower-level structures, and also a drawing of the D_2 line as it would appear in a magnetic field of 5000 gausses. On this latter diagram, show the line separations on a wavelength scale (taking 1 A as the unit of distance).

8.6. The quantization of angular momentum holds for the rotation of a diatomic molecule, just as for the orbital motion of an electron, and we can put

$$L = [J(J + 1)]^{1/2}\hbar$$

where J is the quantum number (an integer) for the rotation. If such a molecule has a moment of inertia I about its center of mass, obtain an expression for the energy of rotation as a function of J. Consider the molecule CO, in which the distance between the two nuclei is $1.1 \cdot 10^{-8}$ cm. Calculate the value of I, and hence the wavelengths associated with transitions between rotational states differing by one unit in J.

8.7. The density of gold is 19.3 g/cm^3, its atomic weight is 197, and its work function is about 4.5 volts. Assuming one free electron per atom, calculate the depth of the effective potential well in which the electrons can be assumed to move.

Some Applications of Quantum Mechanics 253

8.8. Using the data from the previous problem, calculate the approximate fraction of the electrons having energies greater than ε_0 by the amounts kT, $2kT$, $10kT$, $50kT$ for $T = 1250°$ K. Hence try to make a very rough estimate of the amount of thermionic emission current (in amperes/cm^2) that might be expected from gold at 1250° K.

8.9. The Debye temperature of copper is 343° K. Calculate the characteristic frequency associated with this, and the wavelength of the corresponding longitudinal waves. (Density of copper = 8.9 g/cm^3; Young's modulus = $1.2 \cdot 10^{12}$ dynes/cm^2.)

8.10. Using Θ/T as variable, make a graph of $E/3R\Theta$ against Θ/T for the total energy of elastic vibrations of a solid. [The first step is to construct a graph of the function $x^3/(e^x - 1)$ against x.] Hence find the theoretical value of the atomic heat C at $T = \frac{1}{2}\Theta$.

9 | *The Nucleus*

9.1 INTRODUCTORY COMMENTS

The subject of nuclear physics developed so much importance during the past few decades, and engaged the attention and time of so many investigators, that it came to dominate the whole field of atomic physics. In recent years the study of the solid state in its various aspects has begun to take on a comparable status for the same reasons —namely the possibility of rapid technical exploitation of the fundamental properties and behavior of matter in its various forms. The whole process has been magnified and accelerated by the ever-expanding scale of scientific and technical activity generally. One consequence is that it becomes difficult to present these more recent developments in proper perspective, and in proper relation to their antecedents. The discussion of nuclear physics in this book is an attempt, however inadequate, to deal with this problem, and to exhibit the study of the nucleus as part of a larger scheme. An inevitable feature of this approach is a somewhat ruthless limitation on the selection of topics, and the exclusion of a wealth of fascinating detail. The pages that follow do not and cannot purport to be an adequate presentation of nuclear physics, even at an elementary level; rather they are to be thought of as paving the way toward the study of nuclear physics as a subject in its own right. It is our hope that by this means the general sense of unity and balance in the study of modern physics can be preserved, albeit at the risk of a certain degree of superficiality.

9.2 THE NUCLEUS AS PART OF THE ATOM

We have discussed in Chapter 5, Section 3, the early work of Rutherford and his school, by which it was established that the entire positive charge and most of the mass of any atom are concentrated in a small volume at the center. All that was necessary for the setting up of Bohr's atomic theory was the knowledge that the nucleus could be considered a good approximation to a point charge; its exact dimensions were unimportant provided that they were small compared to the radii of the electron orbits—i.e. compared to about 10^{-8} cm for the case of hydrogen. It was known that the inverse-square law of Coulomb repulsion held good for the approach of an alpha particle to within about 10^{-12} cm of the center of a medium-sized nucleus (e.g. Cu), implying that the nuclear radius itself was smaller than this. It is now known that even the largest nuclei, such as uranium, have radii no greater than 10^{-12} cm. (We shall consider these things in more detail later.) One immediate deduction is that nuclear matter has no counterpart in our normal experience; it is almost unbelievably dense. Taking the uranium nucleus as an example, its mass is about $4 \cdot 10^{-22}$ g, and its volume about $4 \cdot 10^{-36}$ cm^3. Its density is therefore about 10^{14} g per cm^3, or 100,000 tons per mm^3. Evidently, then, the nucleus exists as something in its own right; we must be wary of trying to ascribe to it the kinds of properties that are characteristic of matter in bulk.

It is relatively seldom that we deal with nuclei isolated from all their surrounding electrons. The solitary electron attached to the proton in a hydrogen atom can indeed be removed fairly easily, and the outermost orbital electron in any other neutral atom can be detached if energy equivalent to about 10 ev is supplied. We say that the last electron has a certain *binding energy* with respect to the rest of the atom, and the energy needed to overcome it is measured in terms of the first ionization potential of the atom. (The values of such ionization potentials range from 3.87 volts for the readily ionized metal Cs to 24.46 volts for He.) The more deeply embedded electrons in the atom are, however, bound much more tightly, and, to take an extreme case, the innermost electrons of a uranium atom (those in the K orbit, with $n = 1$, $l = 0$) have a binding energy of the order of 100,000 ev or 100 kev. The inner electrons are therefore almost inseparable from the nucleus with respect to normal physical or chemical processes, and a moving atom, composed of the nucleus with its attendant electrons, may be likened to the sun with its planets moving through space. The comparison may also be used to drive home the extreme smallness of the nucleus in relation to the complete atom, for it so happens that the radius of the sun ($6.95 \cdot 10^5$ km) bears about the same ratio to the radius of the solar system ($5.9 \cdot 10^9$ km from the sun to the orbit of Pluto) as the radius of a nucleus bears to the orbital radius of its outermost electrons.

The close association of a nucleus with the electrons surrounding it has a number of consequences. One of them, stemming mainly from convenience, is that, when we wish to compare the masses of two nuclei, we normally do so in terms of the measured masses of the neutral atoms containing them. Only rarely is it necessary to evaluate the mass of a "bare nucleus." Other consequences, of considerable physical importance, arise from the mere presence of the electrons in the vicinity. Thinking of them as point charges describing orbits, as in the simple Bohr theory, we recognize that they will be the source of electric and magnetic fields at the position of the nucleus. And, when we adopt the wave-mechanical picture, we go one step further, for now we regard the electrons as a continuous distribution of probability density or charge, with a nonvanishing amplitude even inside the nucleus itself. We must not be surprised to discover, therefore, that the actual behavior of a nucleus may depend on the number and the disposition of the electrons surrounding it.

9.3 NUCLEAR CHARGE AND MASS

The science of chemistry had revealed an orderly progression of the elements, with groupings according to chemical properties, that is expressed in the periodic table. The application of the principles of atomic mechanics, and in particular the Pauli principle, had made this sequence intelligible, and confirmed that the ordinal number of an element in the periodic table (i.e. the atomic number) is to be equated to the total positive charge on the nucleus, measured as a multiple of the electronic charge. The historic work of Moseley in 1913 on the frequencies of the characteristic X rays of the elements [see Chapter 5, Section 11(3)] had in fact established this latter conclusion in the early days of the Bohr theory. It must be remembered, however, that the attempt to set up a periodic table of the elements was first made in terms of the chemical atomic weights, and that an arrangement of the atoms in ascending atomic weight differs in only one or two particulars from the sequence in terms of the atomic number Z (see Fig. 9.1). It was also recognized that the atomic weights themselves, referred to oxygen as a standard (16.000) tended to fall close to whole numbers, and gave grounds for the belief that nuclei were built up from smaller structural units.

There were, of course, exceptions such as chlorine with an atomic weight close to 35.5 that made generalizations difficult. Nevertheless Rayleigh, as early as 1901, had given it as his conviction that the near-integral values were not merely the result of chance.

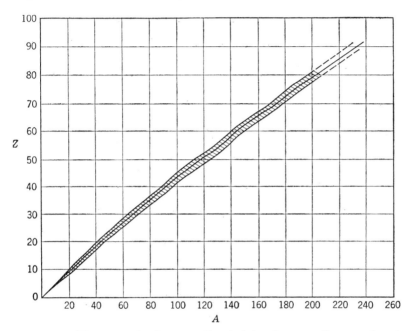

Figure 9.1. The connection between Z and A for the naturally occurring elements. The shaded part indicates the approximate region of stability.

The evidence for a discontinuous structure of nuclei was greatly reinforced by the discovery of *isotopy**; i.e., the existence of two or more discrete values of the atomic weight for a given chemical element. This phenomenon had first been recognized in natural radioactivity, when it was found that certain radioactive atoms might, by the successive emission of an alpha particle and two electrons, undergo a loss of mass equivalent to a neutral helium atom and yet preserve their chemical identity. Evidence to this effect became conclusive in 1913, which was a kind of *annus mirabilis* for atomic physics. [In it came

* This word, meaning "same place" (in the periodic table) was coined by F. Soddy, who played a prominent part in the discovery of the effect.

also the publication of Bohr's theory of the nuclear atom, and the experimental confirmation of the theory by the alpha-particle scattering experiments of Geiger and Marsden (Chapter 5, Section 3) and by Moseley's X-ray studies just mentioned.] At about the same time, too, J. J. Thomson obtained the first direct evidence for isotopy amongst the normal stable elements. His method had a good deal in common with his earlier determination of e/m for electrons, but merits a separate discussion.

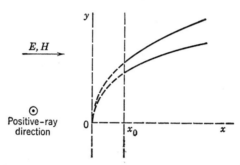

Figure 9.2. Diagram for the consideration of J. J. Thomson's positive-ray parabolas. There is no trace for $x < x_0$; this corresponds to a maximum energy $\frac{1}{2}Mv^2$ for the accelerated ions (assuming them to be singly charged) that is the same for all M.

An electric discharge at low pressure was set up between two electrodes in a bulb containing a substance of interest in gaseous or vapor form. Positive ions were accelerated to the cathode, and a fraction of them escaped through a fine canal, emerging as "positive rays." They were subjected to electric and magnetic fields acting in the same direction (perpendicular to the line of flight) over a short distance l, and then traveled in straight lines through a field-free region until they struck a photographic plate that recorded their arrival. Let us suppose that ions of a certain type have mass M, charge ne, and velocity v. Then, if we choose co-ordinates x, y (Fig. 9.2) to describe the displacements parallel and perpendicular to the common direction of the applied fields, we have

$$\frac{d^2x}{dt^2} = \frac{neE}{M}$$

$$\frac{d^2y}{dt^2} = \frac{nevH}{Mc}$$

The time of flight through the fields is l/v, so that the transverse velocities acquired are given by

$$u = \frac{neE}{M} \cdot \frac{l}{v}$$

$$w = \frac{nevH}{Mc} \cdot \frac{l}{v}$$

The final displacements on the photographic plate, distance $L(\gg l)$ away, are thus given by

$$x = \frac{neE}{M} \cdot \frac{lL}{v^2}$$

$$y = \frac{neH}{Mc} \cdot \frac{lL}{v}$$

For ions of a given type, regardless of velocity, we therefore have

$$y^2 = \frac{lL}{c^2} \cdot \frac{H^2}{E} \cdot \frac{ne}{M} x$$

which represents a parabolic trace on the photographic plate. By this means Thomson was able to analyze the positive rays into the various mass and charge components, and in a famous investigation, reported in 1913, he showed that with neon in his discharge tube he always obtained a faint parabola corresponding to $M = 22$, together with a strong one for $M = 20$. This was in clear though qualitative accord with the fact that the chemical atomic weight of neon is just over 20 (20.18, to be exact).

It came to appear very likely, therefore, that the masses of individual atoms are, to a close approximation, integral multiples A of some basic unit. This mass number A, together with the charge number Z, then characterizes any given atomic species X in terms of the mass and charge of its nucleus. It is customary to write down this identification in the form $_ZX^A$. The proton, with $Z = 1$, $A = 1$, is an obvious choice as a unit in the structure of heavier nuclei. We know that this is not the whole story, because Z is less than half of A for most nuclei. When the proton and the electron were the only elementary particles known, it was tempting to suppose, by economy of hypotheses, that a given nucleus (Z, A) contained A protons (to provide the bulk of its mass) and $A - Z$ electrons (to balance out the surplus charge). This picture has been rejected. It is our confident present belief that a nucleus of charge Z contains precisely Z protons; the balance of the mass, with no addition to the charge, is then provided by

$A - Z$ neutrons. The reasons for this change of view will be brought out in the next sections.

9.4 NUCLEAR EFFECTS IN LINE SPECTRA

In Chapter 8, Section 2, we discussed the fine structure of atomic energy levels arising from the "spin-orbit interaction" between the electron magnetic moment and the magnetic field resulting from its orbital motion. This fine structure could in certain cases be readily observed in spectra (e.g. the D lines and similar doublets in the sodium spectrum) with a simple prism spectroscope. But the development of high-resolution spectroscopy showed that the individual "lines" of a fine structure pattern often possess a complex structure themselves. From the measured wavelengths we can infer that the individual optical transitions begin or end on members of a very close group of energy levels, spaced by perhaps about 1/100,000 of the energy of the main quantum jump. This is hyperfine structure (often referred to as HFS), and it can arise for two quite distinct reasons. The first reason, which itself has two aspects, is the existence of isotopes.

The effective value of the Rydberg constant for a hydrogen-like atom with a nucleus of mass M is given (see Chapter 5, Section 6) by the equation

$$R_M = \frac{M}{M+m} R_\infty$$

where m is the electron mass. If we consider two isotopes, of masses M and $M + \Delta M$, we can readily verify that the fractional difference between their Rydberg constants is given approximately by the equation

$$\frac{\Delta R}{R_\infty} = \frac{m}{M} \cdot \frac{\Delta M}{M}$$

Thus, for example, with $M = 10$, $\Delta M = 1$ (which would correspond to the two boron isotopes B^{10} and B^{11}), we should have $\Delta R/R_\infty$ equal to about 1/185,000; this would lead to a doublet hyperfine structure for all the lines in the boron spectrum. Since the effect varies inversely as the square of the mass (approximately) it becomes a splitting of about 0.1% between corresponding lines for light hydrogen ($M = 1$) and heavy hydrogen ($M = 2$). Indeed, it was by this that H^2 (deuterium) was discovered—a research that resulted in the award of a

The Nucleus 261

Nobel prize in 1934 to H. C. Urey. It is, however, clear that the $1/M^2$ variation makes the finite mass effect unobservably small for heavier atoms; the above formula shows that $\Delta R/R_\infty$ is down to about 10^{-7} for $M = 70$, $\Delta M = 1$. Nevertheless these heavier atoms do exhibit a well-defined isotopic effect with line splittings of the order of 1 in 10^5. It appears that this is a consequence of the finite nuclear size, which (when we are concerned with very small effects) makes it no longer correct to treat the nucleus as a point charge (see Problem 2 at

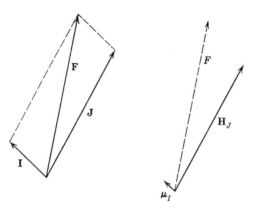

Figure 9.3. The production of magnetic hyperfine structure by interaction of the nuclear magnetic moment μ_I with the electronic field H_J.

the end of this chapter). Different isotopes have nuclear charge distributions that vary in a more or less systematic way with the isotopic mass; the trend is for a heavier isotope to produce a spectral line of lowered frequency, which is just the opposite of the finite mass effect in the Rydberg constant.

But, when all allowance has been made for isotopic effects, there may still remain a complexity in spectral line structure. This is what we may call the true hyperfine structure, in that it arises from the interaction of a single nuclear species with its orbital electrons. Pauli (1926) suggested that it was due to the existence of nuclear magnetic moments with associated angular momenta. If we suppose a nuclear angular momentum **I** (numerically equal to $[I(I + 1)]^{1/2}$ in units of $h/2\pi$) and an associated magnetic moment μ_I, these will couple, through the magnetic interaction, with the corresponding quantities **J** and μ_J of the atomic electrons. The theory is exactly analogous to the treatment of electron spin and magnetism (Chapter 8, Section 9) and leads to the introduction of a total angular momentum $\mathbf{F}(= \mathbf{I} + \mathbf{J})$ for the nucleus and electrons taken together. F will then have $2I + 1$

or $2J + 1$ different values (whichever is less), and each value of F corresponds to a distinct energy of the atom as a whole. If the magnetic field caused by the orbital electrons at the position of the nucleus is H_J, the energy of a given state of the atom is modified by the amount

$$W = -\mu_I \cdot \mathbf{H}_J$$

The angle between μ_I and $\mathbf{H}_J$ is set by the vector addition law for $\mathbf{I}$ and $\mathbf{J}$ (see Fig. 9.3):

$$\cos \widehat{(\mathbf{IJ})} = \frac{I(I+1) + J(J+1) - F(F+1)}{2[I(I+1) J(J+1)]^{1/2}}$$

It follows, therefore, that we can put

$$W = \frac{a}{2}[F(F+1) - I(I+1) - J(J+1)]$$

where a is a constant for the system. Since F is quantized, the change of W between two successive values of F is given by

$$\Delta W(F, F-1) = aF$$

The constant a is thus a suitable unit in terms of which to express the hyperfine separation of atomic energy levels resulting from nuclear magnetism.

9.5 NUCLEAR MAGNETISM AND SPIN

Observation of the multiplicity and the *relative* spacings in the hyperfine structure can be used to determine I; this is basically just a matter of counting and shows that I is always some multiple of $\frac{1}{2}(\times h/2\pi)$ or else is zero. If some estimate of H_J is available, then the *absolute* spacing of the HFS gives a value for μ_I. Calculation shows that H_J tends to be in the range from 10^5 to 10^7 gausses, depending on the electron configurations. Despite these immensely strong fields the hyperfine splittings are exceedingly small. To illustrate the order of magnitude, let us evaluate μ_I on the assumption that $H_J = 10^6$ gausses and that the HFS separations are about $1/100{,}000$ of the wavelength for a spectral line in the visible at about 6000 A. Since this wavelength corresponds to a quantum energy of about 2 ev or about $3 \cdot 10^{-12}$ erg, we have

$$\Delta W \approx 3 \cdot 10^{-17} \text{ erg}$$

Hence, with the assumed value of H_J, we find (very roughly indeed)

$$\mu_I \approx 10^{-23} \text{ emu} \approx 10^{-3} \text{ Bohr magneton}$$

All observed nuclear magnetic moments are of this order. Now if, in the theoretical expression for the magneton, we replace the electron mass by the proton mass, we obtain a unit that is entirely appropriate to the situation; this is the *nuclear magneton*, and it is given by

$$\mu_M = \frac{eh}{4\pi Mc} = \frac{\text{Bohr magneton}}{1836} = 0.505 \cdot 10^{-23} \text{ emu}$$

These facts suggest that the magnetism of nuclei can be plausibly accounted for by the orbital motion of protons inside them, together possibly with intrinsic magnetic moments of comparable size, analogous to the spin magnetic moment of the electron.

From the above discussion it becomes a very strong presumption that nuclei do not contain electrons; otherwise we should certainly expect to observe magnetic moments of the order of magnitude of the Bohr magneton, at least in those nuclei for which $A - Z$ is odd (and which would therefore, on a proton–electron model, contain an odd number of electrons). This conclusion is reinforced by the systematic properties of I. Whenever A is odd, I is an odd multiple of $\frac{1}{2}$, and, whenever A is even, I is either zero or an integer.

Now if we consider a nucleus having A even and Z odd, this would, on the proton–electron picture, have an *even* number A of protons and an *odd* number $(A - Z)$ of electrons. Thus, regardless of whether the proton spin is an integer or half of an odd integer, such a nucleus would have a resultant I that is "half-integral" (by which is meant half of an *odd* integer). This is completely at variance with observation; N^{14}, for example, has $I = 1$. We are thus led naturally to the proton–neutron model; and the fact that I is always half-integral for odd A is in itself clear evidence that the individual proton and neutron spins are half-integral. For, if Z (now identified as the number of protons) is odd, the neutron number $N(= A - Z)$ is even in an odd-mass nucleus; and, if Z is even, then N is odd. The nuclei $_3\text{Li}^7$ and $_4\text{Be}^9$ (both having $I = \frac{3}{2}$) typify these two possibilities. It is known from spectroscopic and other evidence that the proton spin is just $\frac{1}{2}$, as for the electron. It has also been established by methods and arguments that we shall not go into that the neutron spin has this same value. Upon these primary data we can proceed to build a detailed theory of the magnetic moments and total angular momenta of individual nuclei.

We have discussed these matters at some length because of their important bearing on the problem of nuclear constitution, which is crucial to all that follows. Even so, we have concentrated on a single source of evidence, namely the emission spectra (line spectra) of single atoms. It is only right to point out, therefore, that a great deal of what we know about spins and magnetic moments of nuclei has in fact been discovered in other ways, particularly from the study of molecular spectra (band spectra) and from a diversity of beautiful experiments employing atomic beams subjected to magnetic fields (outgrowths of the Stern–Gerlach type of experiment). These matters are fully discussed in many texts.

9.6 NEUTRONS AND PROTONS

The neutron has become a very familiar particle during the past two decades; for this reason we have felt free to mention it in the earlier sections of this chapter without giving it a formal introduction. But we must now describe its properties more fully (and compare them to the properties of the proton) if we are to have some clear picture of the structure and behavior of nuclei.

The idea that there might exist a neutral particle of almost the same mass as the proton was given public utterance by Rutherford and others in 1920. It was then regarded as a possible close combination of a proton and an electron—an idea that has been definitely abandoned. Such a particle, it was recognized, would be able to pass almost unimpeded through matter, being immune from the Coulomb forces that scatter and retard all other types of particle. The existence of the neutron was finally established by Chadwick (1932) from a study of the balance of energy and momentum when beryllium was bombarded with alpha particles. It became clear that the bombardment resulted in the emission of a neutral particle with a mass closely equal to that of the proton. (This is a typical example of a nuclear reaction, a topic that we shall come to later—Section 18 of this chapter.) More refined measurements showed that the neutron is slightly heavier —by about 0.1%—than the proton.

We have already seen that the proton and neutron both have spin $\frac{1}{2}$. It is natural that the proton, regarded as a spinning charge, should have an associated magnetic moment. What is surprising is that the neutron also has a magnetic dipole moment, of a comparable size but of opposite sign.† These magnetic moments have been determined

---

† The magnetic moment is reckoned as positive if the magnetic dipole vector points in the same direction as the angular momentum vector. The proton has a positive moment on this convention, whereas the electron and neutron moments are negative.

The Nucleus

with great accuracy by directly measuring the energy of a quantum that will reverse the direction of the dipole with respect to an applied magnetic field of a known strength. If the magnetic moment is μ, the energy change for reversal in a field H is $2\mu H$. (With $s = \frac{1}{2}$, only these two extreme orientations are possible.) A transition from one orientation to the other can therefore be stimulated by radiation of frequency ν such that

$$h\nu = 2\mu H$$

It has been possible to obtain a direct comparison of the neutron and proton magnetic moments by this means, keeping H constant and adjusting ν to its resonant values for the two types of particle.

Neither the neutron nor the proton is found free in nature, except in such transient forms as cosmic radiation. But, whereas the proton acquires stability once it has captured an electron into an orbit around it, the neutron exists only in the interior of nuclei. Moreover, the proton will preserve its identity so long as it is kept apart from other matter. The neutron, on the other hand, is spontaneously radioactive; if held in a high vacuum, it can be observed to emit an electron (called a beta particle in this context) and to be converted into a proton in the process. This is the prototype of beta decay (of which more later—Sections 13 and 14 of this chapter).

In Table 9.1 we summarize those properties of neutron and proton that we have mentioned.

TABLE 9.1

Property	Neutron	Proton	Unit of Measurement
Mass	1.00898	1.00759	amu* $= 1.66 \cdot 10^{-24}$ g
Charge	0	+1	$e = 4.80 \cdot 10^{-10}$ esu
Magnetic moment	-1.9135	$+2.7927$	$\mu_M = 0.505 \cdot 10^{-23}$ emu

* amu = atomic mass unit.

9.7 NUCLEAR MASSES AND BINDING ENERGIES

The first result of measuring the masses of individual isotopes was to show that nuclear masses are very close to integral multiples of a basic unit. But, as soon as more precise measurements became possible (starting with Aston in 1925), it was found that there were deviations from a whole number rule. The deviations are very slight, being always less than 1%, and in most cases not more than a tenth of this. The detailed study of isotopic masses is the subject of mass spectroscopy; it has been brought to an astonishing degree of perfection (by

Dempster, Bainbridge, Mattauch, Nier, and others), and in certain cases an accuracy of a few parts per million has been achieved, making it possible to give the *difference* between isotopic mass M and mass number A with an accuracy of between 0.1% and 1%. This has been done by exploiting the focusing properties of suitably shaped electric and magnetic fields, and is a specialized subject in its own right which we shall not discuss here.

If we regard a nucleus as composed of neutrons and protons, we can define a binding energy B, which is the energy that would be required to dismember the nucleus into its separate neutrons and protons, removed completely from each others' influence. Making use of Einstein's mass-energy equivalence, we have

$$B = c^2 \Delta M$$

where
$$\Delta M = ZM_p + NM_n - M(Z, N)$$
$$= ZM_p + (A - Z)M_n - M(Z, N)$$

It will also prove to be useful to evaluate the ratio B/A, which is the mean binding energy per particle in the nucleus. This is called the mean binding energy *per nucleon*, the word "nucleon" having been coined to signify a neutron or a proton without discrimination. In the above equations $M(Z, N)$ is the mass of the *nucleus* (Z, N). Let us add and subtract the mass of Z electrons to the above equation for ΔM. This gives

$$\Delta M = Z(M_p + m) + NM_n - [M(Z, N) + Zm]$$

Now the electrons in an atom of charge Z have a certain total binding energy $b(Z)$, so that we can put

$$M_p + m = M_H + b(1)/c^2$$
$$M(Z, N) + Zm = M(Z, A) + b(Z)/c^2$$

where M_H is the mass of a neutral hydrogen atom and $M(Z, A)$ is the mass of a neutral *atom* of charge Z and mass number A. Hence we have

$$\Delta M = ZM_H + NM_n - M(Z, A) - \delta M$$

where
$$\delta M = \frac{b(Z) - Zb(1)}{c^2}$$

In practice δM is an almost negligible contribution to ΔM for most purposes, and it becomes an excellent approximation to write

$$\Delta M = ZM_H + NM_n - M(Z, A)$$

so that the nuclear binding energy is given in terms of the atomic mass only. The electronic binding correction is, however, available if needed, and is given semiempirically by the equation

$$b(Z) = 15.73 Z^{7/3} \text{ ev}$$

To illustrate the order of magnitude of B and B/A, let us take the case of $_8O^{16}$, the standard of isotopic masses. We have $Z = N = 8$, and also

$$M_H = 1.008145$$
$$M_n = 1.008986$$
$$M(O^{16}) = 16.000000$$

Hence
$$\Delta M = 8 \times 1.008145 + 8 \times 1.008986 - 16.000000$$
$$= 0.137048 \text{ amu}$$

We wish to convert this into a binding energy ($c^2 \Delta M$) measured in convenient units. Now we have

$$1 \text{ amu} = 1.6598 \cdot 10^{-24} \text{ g} \equiv 1.4916 \cdot 10^{-3} \text{ erg}$$
$$\equiv 9.312 \cdot 10^8 \text{ ev}$$

It is convenient, therefore, to introduce the Mev (million electron volts) or Bev (billion‡ electron volts) as a unit, so that we put

$$1 \text{ amu} = 931.2 \text{ Mev} = 0.931 \text{ Bev}$$

So far as nuclear binding energies are concerned, the Mev is the more suitable choice, and the binding energy of $_8O^{16}$ is given in these terms as

$$B(O^{16}) = 127.6 \text{ Mev}$$

We can see that this is vastly greater than the electronic binding energy, for we have

$$b(8) \approx 2 \text{ kev}$$

a correction of only about 0.001%. The binding energy per nucleon for $_8O^{16}$ is equal to $B/16$ or 7.98 Mev.

The variation of B/A with A for the stable isotopes of the elements is of great importance for the understanding of nuclear structure. It is shown in Fig. 9.4. We must remember that B is not uniquely defined by A, since it varies with Z for a fixed value of A; but, since Z and A go hand in hand, or very nearly so (Fig. 9.1), we can overlook this complication for the moment. We see from the figure that the lightest

‡ 1 billion = 10^9.

nuclei tend to be rather loosely bound, but that, for A greater than about 10, the binding energy per nucleon remains almost constant in the neighborhood of 8 Mev. This is in remarkable contrast to the mean binding energy per orbital electron, which rises as $Z^{4/3}$ and so increases by a factor of about 500 between the beginning and the end of the periodic table.

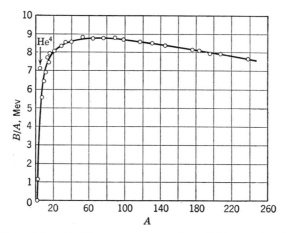

Figure 9.4. Mean binding energy per nucleon, B/A, as a function of A.

Since the energy equivalent of the mass of a single nucleon (average of neutron and proton) is about 939 Mev, we see that the binding energy of a nucleus is typically about 0.9% of the total mass.

9.8 NUCLEAR RADII

The experiments on the scattering of alpha particles by nuclei showed, as we have seen, that the Coulomb law of repulsion was operative between the whole charge of the alpha particle and the whole charge of the nucleus so long as the distance between their centers was more than about 10^{-12} cm. For collisions involving a still closer approach, however, deviations from the Rutherford scattering law appeared; the exact distance at which this so-called "anomalous scattering" set in depended on the particular nucleus involved. Two possible causes of the anomaly can be recognized:

(a) If the electric charge distributions of two particles begin to interpenetrate, the Coulomb force between them is no longer given by the inverse-square law evaluated as if the total charges were concen-

*The Nucleus*_____269

trated at their centers. It falls below this value in a manner prescribed by Gauss's theorem: viz., that, at a given distance r from the center of a cloud of electric charge, only that part of the charge contained within radius r will contribute to the electric field.

(b) There may be a specifically nuclear force that comes into play only when neutrons and protons, or nuclei composed of them, come within a distance that is a small multiple of 10^{-13} cm. We believe that the range of any such force must be short, because the Rutherford law describes the nuclear scattering so perfectly for all but the closest collisions. (Although this conclusion is correct, the argument as stated is not really convincing. To illustrate its shortcomings, one may consider the interactions between neutral molecules of a gas. The forces between such molecules are of extremely short range, falling off as a high inverse power of the distance; yet the ultimate origin of the forces is the long-range Coulomb interaction between individual charges.)

In the scattering of alpha particles the more important of the above two possibilities is in fact the second—the existence of a characteristic field of force around a nucleus that reaches out for a very limited distance. Granted that the neutrons and protons composing a nucleus are contained within a small volume, which we shall assume to be a sphere of a certain radius, it is plausible that the nuclear forces extend to a somewhat larger radius, and hence that the anomaly (b) makes itself felt at larger distances than (a). It is appropriate at this point to ask what we really mean by "the radius" of a nucleus. The answer will depend on the precise way in which we put this question to test, and many different methods have been devised, of which we shall mention only two.

(1) The Scattering of Fast Neutrons

A neutron is in many ways an ideal tool for investigating nuclei. Being uncharged, it has virtually no interaction with the atomic electrons and is not deviated at large distance by the nucleus. We thus have the possibility of a simple "hit or miss" situation, in which the neutron can be considered to strike a nucleus if their centers come within a distance equal to the sum of their radii (Fig. 9.5). If we denote the nuclear radius by R, and the neutron radius by r, we thus have an effective collision area given by

$$\sigma_c = \pi(R + r)^2$$

By measuring the attenuation of neutrons in passing through a layer of material containing a known number of target nuclei per cubic centimeter, it is possible to infer the value of the cross section σ_c, and hence the value of R if r is known.

We have described the collision as though it takes place between rigid spheres, but this leaves out of account the wave-mechanical features of the encounter. The incident neutron is described by a de Broglie wave of a certain wavelength λ. The collision process is then to be thought of as one in which the incident neutron plane wave is partly absorbed and partly diffracted by the nucleus (just like light falling on a small obstacle). If the nuclear circumference $2\pi R$ is appreciably greater than λ, it proves to be legitimate to replace r by $\lambda/2\pi$ (written as $\lambdabar$). Furthermore, the fact of diffraction, in addition to absorption, can be shown to lead to a doubling of the effective collision area for removal of neutrons from the incident beam. Thus we arrive at a total collision cross section given by

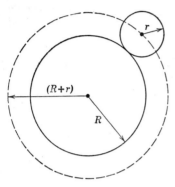

Figure 9.5. Diagram to illustrate the cross section available for collision between two spheres.

$$\sigma = 2\pi(R + \lambdabar)^2$$

and it is this total cross section that is inferred from experiment.

In order to satisfy the condition $\lambdabar < R$, it is necessary to use quite energetic neutrons. If the neutron energy E is expressed in Mev, we have

$$\lambdabar = \frac{4.55 \cdot 10^{-13}}{E^{1/2}} \text{ cm}$$

Hence, for $E = (4.55)^2 = 20.7$ Mev, the value of $\lambdabar$ is exactly 10^{-13} cm (the corresponding neutron velocity is about one-fifth the speed of light); this value of $\lambdabar$ is suitably small. Figure 9.6 shows the results of an experiment of this type, and it immediately brings out a conclusion that is obtained in all studies of nuclear size—the nuclear radius is proportional to the cube root of the mass number A. (We should perhaps make it clear that this result was established, e.g. by studies of anomalous scattering of alpha particles, long before the neutron cross-section measurements were made, but the neutron experiments commend themselves by being so directly interpretable.) The fit to the data of Fig. 9.6 can be made by putting

$$R(A) = r_0 A^{\frac{1}{3}}$$

with $\qquad r_0 \approx 1.4 \cdot 10^{-13}$ cm

We see that for $A \approx 250$, corresponding to the heaviest nuclei known, R is about $8.5 \cdot 10^{-13}$ cm.

A very simple statement of the relation between R and A is that the nuclear volume is proportional to A; i.e. to the total number of nucleons: or, in other words, that the density of all nuclei is the same.

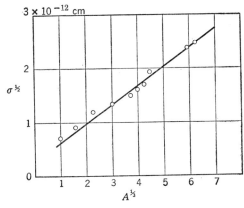

Figure 9.6. A graph of the square root of the collision cross section versus $A^{\frac{1}{3}}$ for 20-Mev neutrons. [After Day and Henkel, *Phys. Rev.*, **92**, 358 (1953)]

The value of R obtained in this way is a "nuclear force radius"; it represents the distance from the center of the nucleus at which an external uncharged nucleon first feels its influence. This is what we might call an operational definition of R, and if we had considered a proton instead of a neutron as the probing particle, we might have concluded by a comparable argument that the effective value of R was in this case infinite, since the Coulomb field causes a disturbance of the proton trajectory out to any distance (or at least to the point where screening of the nucleus by the orbital electrons becomes effective). The evaluation of a useful radius from neutron scattering is possible simply because the nuclear force becomes vanishingly small within a finite distance of the order of 10^{-13} cm.

(2) The Scattering of Fast Electrons

The interaction between an electron and a nucleus is limited almost entirely to the electric forces. Thus an electron, if it may be regarded as a point charge, can be used to map out the nuclear charge distribution through a study of the Coulomb scattering. This will then be tantamount to a knowledge of the radial distribution of the protons within the nucleus. We can conceive of the electron as passing right through the nuclear charge cloud in undergoing a large-angle deflection (Fig. 9.7). If the charge density is $\rho(r)$ at radius r, the total charge

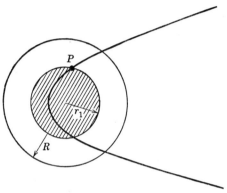

Figure 9.7. An electron passing through a nuclear charge cloud.

effective in acting on the electron when it is at point P of its trajectory is given by

$$q(r_1) = \int_0^{r_1} 4\pi r^2 \, \rho(r) \, dr$$

Since $q(r_1)$ is less than the total charge Ze for penetrating orbits of this type, the electron is deflected less than it would be by a point nucleus, and the discrepancy will be most marked for the largest deflections, since these correspond to the closest collisions. The relative probability of scattering $w(\theta)$ can be calculated as a function of θ for a point nucleus and for charge clouds of various types and sizes. Figure 9.8 shows an example of the results obtained, with an excellent fit between theory and experiment for a uniform charge distribution of a certain radius. The correct calculations are of course wave-mechanical; the particle-trajectory picture provided by classical physics does not stand up to detailed analysis.

Experiments of this type have been carried out by Hofstadter and others. They are done with electrons of very high energy, since the structure of the nuclear charge will not be revealed unless the value of $\lambda/2\pi$ for the electrons is small compared to the nuclear radius. To see how to estimate the electron wavelength, consider the relativistic equation (Problem 3, Chapter 6) connecting energy and momentum

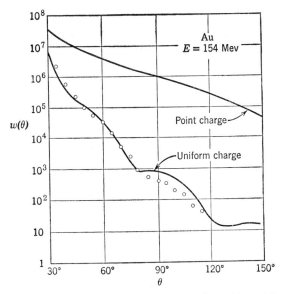

Figure 9.8. The scattering of 154 Mev electrons by gold nuclei. [After Hofstadter et al., *Phys. Rev.* **95**, 512 (1954), and Yennie et al., *Phys. Rev.*, **95**, 500 (1954)]

for a particle:
$$E^2 = p^2c^2 + (m_0c^2)^2$$

If $E \gg m_0c^2$, we have
$$p \approx \frac{E}{c}$$

simply, and hence
$$\lambda = \frac{h}{p} \approx \frac{hc}{E}$$

It is convenient to express E as a multiple ε of the rest energy m_0c^2; in this case we can put
$$\lambda = \frac{1}{\varepsilon} \cdot \frac{h}{m_0c}$$

But we recognize h/m_0c as the Compton wavelength, $2.4 \cdot 10^{-10}$ cm, and so we have
$$\lambdabar = \frac{\lambda}{2\pi} \approx \frac{4 \cdot 10^{-11}}{\varepsilon} \text{ cm}$$

Hence, for $\varepsilon = 400$ (which means an electron energy of about 200 Mev, since $m_0c^2 \equiv 0.51$ Mev), $\lambdabar$ is about 10^{-13} cm. Under such conditions the electron therefore becomes a suitably fine instrument for exploring nuclear structure.

The result of the electron-scattering experiments is to show once again a proportionality of R to $A^{1/3}$. But the proportionality constant r_0 is found to be about $1.1 \cdot 10^{-13}$ cm: i.e. about 20% less than that deduced from neutron scattering. Thus the "charge radius" and the "nuclear force radius" of a given nucleus are not the same thing by any means, but, as we have pointed out, this need cause no surprise.

9.9 THE SEMIEMPIRICAL MASS FORMULA

We now have the picture of a stable nucleus as possessing (a) a well-defined and uniform density, (b) a radius R proportional to $A^{1/3}$, (c) a binding energy per nucleon that is nearly constant at about 8 Mev, and (d) a ratio of protons to neutrons that is defined within narrow limits for a given value of A and is close to unity for small A (falling progressively below unity as A increases).

The first three facts suggest a certain resemblance between a nucleus and a liquid droplet, and can be taken to imply that a given nucleon, like a molecule in the interior of a liquid, interacts only with its nearest neighbors. The forces of attraction then provide a negative contribution to the potential energy of the nucleon considered, and this energy is not increased by the addition of further layers of nucleons on the outside. So long as a nucleon has its quota of nearest neighbors, the potential energy per nucleon from this cause is thus constant. The fact (d) implies some kind of symmetry in the nuclear structure, a symmetry that is broken down by the Coulomb repulsion between the protons—which will clearly act in such a way as to favor an increase of N at the expense of Z.

It proves possible to set up, on the basis of the above considerations, a semiempirical formula for the mass or binding energy of a nucleus. Only one other feature of major importance has to be added, and this follows naturally from what we have already deduced. It is the existence of a surface energy, just as on a liquid or solid surface, arising from the unsatisfied attractions of nucleons lying on the periphery of a nucleus. This will lead to an increase of the energy or mass, proportional to the surface area. We thus have the following contributions to the energy of a given nucleus:

The Nucleus

(i) A negative "volume" energy proportional to A.
(ii) A positive surface energy proportional to $4\pi R^2$.
(iii) A positive Coulomb energy equal to $\tfrac{3}{5}(Z^2 e^2 / R)$. (See Problem 1 at the end of this chapter.)
(iv) A positive asymmetry energy that would make $Z = N = A/2$ if the Coulomb forces were absent. To give the simplest possible expression to the fact that a departure of Z, *in either direction*, from the value $A/2$ will cause an increase of potential energy, we suppose the asymmetry energy to be proportional to $(Z - \tfrac{1}{2}A)^2$. More sophisticated arguments suggest that it is also proportional to $1/A$.

Putting these together, we therefore have

$$B(Z, A) = c_1 A - c_2 \cdot 4\pi R^2 - \frac{3}{5}\frac{Z^2 e^2}{R} - c_3 \frac{(Z - \tfrac{1}{2}A)^2}{A}$$

with $\qquad R = r_0 A^{1/3}$

Hence the total mass of the atom is given by

$$M(Z, A) = Z M_\mathrm{H} + N M_n - c_1 A + c_2 \cdot 4\pi R^2 + \frac{3}{5}\frac{Z^2 e^2}{R}$$
$$+ c_3 \frac{(Z - \tfrac{1}{2}A)^2}{A}$$

For a given value of A, the most probable value of Z is that which makes M a minimum. Mathematically, this means solving the equation $(\partial M / \partial Z)_A = 0$; physically, it implies that an atom of mass number A and of arbitrarily assigned Z may adjust itself by means of the reversible (and charge-conserving) reaction

$$_0 n^1 \rightleftharpoons {}_1 p^1 + e^-$$

until it arrives at the most stable apportionment of neutrons and protons. Now

$$M(Z, A) = [A M_n - c_1 A + c_2 \cdot 4\pi R^2] - Z(M_n - M_\mathrm{H})$$
$$+ \frac{3}{5}\frac{Z^2 e^2}{R} + c_3 \frac{(Z - \tfrac{1}{2}A)^2}{A}$$

where the square bracket is independent of Z. Hence

$$\left(\frac{\partial M}{\partial Z}\right)_A = -(M_n - M_\mathrm{H}) + \frac{6}{5}\frac{Z e^2}{R} + \frac{2 c_3 Z}{A} - c_3$$

so that Z defines itself by the equation

$$Z(A) = \frac{A}{2} \cdot \frac{1 + (M_n - M_H)/c_3}{1 + (3e^2/5R)/(c_3/A)}$$

In practice, this equation will be used to determine the value of c_3 from the known variation of Z and A for the stable isotopes (Fig. 9.1). We can then go back to the more complicated equation for M or B so as to solve for c_1 and c_2 from the knowledge of isotopic masses. The most immediately interesting quantity is c_1, the volume energy per nucleon. The best values of the constants appear to be

$$c_1 = 14.1 \text{ Mev}$$

$$4\pi r_0{}^2 c_2 = 13 \text{ Mev} \qquad (r_0 = 1.45 \cdot 10^{-13} \text{ cm})$$

$$c_3 = 76 \text{ Mev}$$

It should be remarked that the mass formula, in which the constants can be established with reference to only a few values of A, provides an excellent account of the binding-energy curve (Fig. 9.4) over its whole range.

9.10 NUCLEAR FORCES

From the evidence already presented it begins to appear that we can make some specific statements about the properties of nuclear forces:

1. They are primarily attractive. This is clear from the fact that nuclei hold together, despite the disruptive effect of the Coulomb repulsions between the protons.

2. They seem to be essentially the same for neutrons and protons, since there is a clear tendency for a nucleus to have $Z = N$. If the binding energy per proton, for example, were greater than that per neutron, we should expect to find that light nuclei, at any rate, would have Z greater than N.

3. The range of nuclear forces is not greater than about $2 \cdot 10^{-13}$ cm. One indication of this is that the nuclear force radius of a proton, as inferred from fast neutron scattering, is about $1.5 \cdot 10^{-13}$ cm. Many other lines of evidence point to the same conclusion.

4. Within their short range the nuclear forces are very strong; they give rise, as we shall see, to a mutual potential energy of about 30 or 40 Mev. (For nucleons inside a nucleus, just as for electrons in

atomic orbits, this potential energy is partly compensated by kinetic energy so that the binding energy is numerically much smaller than the potential energy.) The electrostatic potential energy of two protons, regarded as point charges, separated by $1.5 \cdot 10^{-13}$ cm is about 1 Mev, and so is quite small compared with the nuclear potential energy at the same distance. For many purposes it can be regarded as a mere correction or "perturbation" upon the purely nuclear force system.

5. The nuclear forces have the property of *saturation:* that is to say, a given nucleon is able to interact with only a limited number of others. This is apparent, as we have remarked earlier, from the fact that the mean binding energy per nucleon, after rising rapidly among the very lightest nuclei, remains almost constant for all further increase of A.

Point 5 merits a more extended discussion. The reason for saturation may be sought in part, but only in part, from the Pauli exclusion principle. Protons and neutrons, like electrons, are subject to the exclusion principle. This has been established by indirect evidence and conforms to Pauli's suggestion (Chapter 8, Section 4) that groups of similar particles of half-integer spin must have antisymmetric wave functions. The practical consequence of this is that only two protons and two neutrons can be accommodated in the same state of spatial motion in a nucleus (the factor 2 coming from the two possible spin orientations). The nucleus $_2\text{He}^4$ represents just such an assemblage, and an extra proton or neutron can be added only in the next orbit, as it were. But this might result in a very feeble attraction of the $_2\text{He}^4$ "core" for the extra nucleon—and in fact both $_3\text{Li}^5$ and $_2\text{He}^5$ are completely unstable. It has been conclusively proved, however, that the Pauli principle by itself cannot lead to the saturation behavior and to the near constancy of B/A if the nuclear forces are purely attractive, "ordinary" forces (also called Wigner forces). To meet this difficulty the concept of "exchange forces" has been developed.

We do not as yet understand the nature of the interaction between two nucleons, although we know a good deal about its properties. But we do know that a strong interaction occurs, and one way of accounting for it is to suppose that some attribute or entity is constantly being handed back and forth between two nucleons. We have a kind of pattern for this behavior in the H_2^+ ion, which holds together as a result of the continual exchange of the single electron between the two protons. We can imagine that, in the interaction of a neutron with a proton, for example, such properties as charge and spin are interchanged through some agency or other. And this, together with the

Pauli principle, can lead to a type of saturation property that would not arise through the operation of the Pauli principle alone. To see how it might come about, let us suppose that there is some kind of exchange interaction that leads to the exchange of charge between nucleons, and let us consider how this might affect the addition of a proton to a deuteron to form $_2\text{He}^3$.

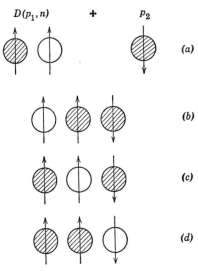

Figure 9.9. The violation of the Pauli exclusion principle as a consequence of exchange interactions. In (d) the charge transfer from proton 2 to the neutron calls for having two protons in the same space and spin state.

The deuteron is known to consist of a neutron n and a proton p_1 aligned with their spins parallel and in a state of zero relative orbital momentum. The second proton p_2, if it is to be added in the same space state as proton 1, must be given the opposite spin direction. Figure 9.9a shows the original configuration as prescribed by the exclusion principle. Figures 9.9b, c, and d now show the consequences of the three possible charge exchange interactions: viz. $p_1 - n$, $p_1 - p_2$, and $n - p_2$. The first two of these bring about no significant change in the situation, but the third (d) results in giving us two protons with parallel spins (and in the same space state). Hence the Pauli principle is violated, and physically this manifests itself as a repulsion, because it means that this particular configuration of nucleons in one place is not allowed. The type of exchange force that we have naively illustrated here was introduced by Heisenberg (1932); other possible types are obtained by considering exchange of spin only (Bartlett, 1936) or of spin *and* charge (Majorana, 1933).§ Each of these will convert the attractive poten-

§ We must caution the reader that our definition of exchange operations will appear to differ from the usual mathematical statements of them. The reason is that we describe the exchange in terms of the transfer of the *attributes* (spin and/or charge) from one position to another, whereas the mathematical exchange operators, as normally written, express the transfer of the *particles* from one position or spin state to another. Thus, for example, the charge exchange $p_2 - n$ in Fig. 9.9d can be described by saying that the proton p_2 (identifiable as such) arrives at the original position of the neutron n and also acquires its spin state. The Heisenberg exchange operation can therefore be written as one that exchanges both space and spin co-ordinates of the supposedly labeled particles.

The Nucleus

tial to a repulsion under certain conditions, and so can contribute to the saturation properties of the forces.

One is bound to ask whether it is meaningful to introduce these peculiar exchange interactions. A very definite affirmative answer to this question has been given by experiments on nucleon–nucleon scattering. When a collimated beam of extremely energetic neutrons, with kinetic energy of the order of 100 Mev, falls upon a target containing hydrogen, there is scattering of the neutrons by the effectively free protons (the chemical binding forces are negligible). Let us suppose that the mutual potential energy of a neutron and a proton is a function only of the radial distance between their centers, and that it takes the form shown in Fig. 9.10. Then the force exerted between the two particles is given by $-dV/dr$, and we have zero force for $r < R_0$ or for $r > R_0 + a$. But between $r = R_0$ and $r = R_0 + a$ we have a constant force F, directed toward the origin, and of magnitude equal to V_0/a. Now let us imagine that a fast neutron, of mass M and speed v, passes across this potential as shown in Fig. 9.11. During

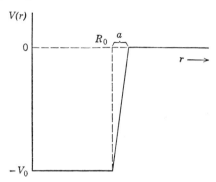

Figure 9.10. Schematic representation of a nuclear potential well with a boundary layer of thickness a.

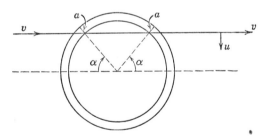

Figure 9.11. Passage of a particle through the potential of Fig. 9.10, resulting in acquisition of a transverse component of momentum.

its two traversals of the thin layer in which $V(r)$ is changing, the neutron will receive a transverse impulse, which will give it a transverse velocity u. With the angle α as shown, we shall have

$$Mu = \text{(transverse force)} \times \text{(time)} = \frac{V_0}{a} \sin \alpha \frac{2a \sec \alpha}{v} = \frac{2V_0 \tan \alpha}{v}$$

(Notice that the thickness a of the transition layer drops out; only the total well depth V_0 matters.) We can put $\alpha = 45°$ to characterize an average type of collision, and for such a collision the angle θ through which the neutron would be deflected is given roughly by

$$\theta = \frac{u}{v} = \frac{2V_0}{Mv^2} = \frac{V_0}{T}$$

where T is the initial kinetic energy of the neutron. Hence for $T \approx 100$ Mev, $V_0 \approx 30$ Mev, we should have $\alpha \approx 20°$. We might

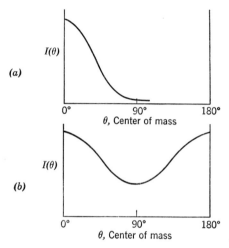

Figure 9.12. Types of angular distribution of scattered intensity expected in scattering of neutrons by protons: (*a*) with ordinary forces, (*b*) with a mixture of ordinary and charge-exchange forces.

therefore expect to see an angular distribution of scattering like that in Fig. 9.12*a*. (This shows the scattering as seen in a co-ordinate system attached to the center of mass of the neutron–proton system. See also Problem 8 at the end of this chapter.) But what is observed in practice is a curve with a backward peak also, again as seen in the center of mass system (Fig. 9.12*b*). The backward scattering is a clear demonstration that the neutron, in being deflected through some angle θ, can acquire the charge of the proton as it passes by. The original proton then appears as a neutron in the direction $\pi - \theta$. The two types of collision, with or without charge exchange, are depicted in Fig. 9.13.

Granted the reality of exchange interactions between nucleons, it is a matter of extreme interest to know what precisely is being exchanged. There seems no doubt that certain types of mesons fill this role. In

The Nucleus 281

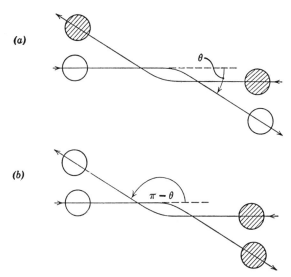

Figure 9.13. Pictorial representation of neutron–proton scattering (as seen in the center of mass system): (a) with no charge transfer, (b) with charge transfer. The same type of encounter appears to deflect the neutron (incident from the left) through θ in (a) and through $\pi - \theta$ in (b).

particular the π meson (or pion), which has a mass of about 270 electron masses and can exist in positive, negative or neutral forms, is believed to play the most important part. But we shall not attempt to go into this problem here.

9.11 NUCLEAR POTENTIAL WELLS. ENERGY LEVELS

We had occasion in the previous section to use a very schematic version of the type of potential that might represent the interaction between a proton and a neutron. We regarded it as the source of a force acting on a point particle, which was not altogether unreasonable in considering the scattering of 100-Mev neutrons ($\lambda \approx 5 \cdot 10^{-14}$ cm). More correctly and generally, however, we should study the neutron–proton system as a problem in solving the Schrödinger wave equation for this particular shape of potential well. This has of course been done in great detail with a view to understanding the so-called "nuclear two-body problem," and so making quantitative inferences about nuclear forces and the exact shape of the potential function. Evidently two possible classes of solutions exist:

(a) States of negative total energy, which thus represent bound states of the neutron–proton system (or of the comparable neutron–neutron and proton–proton systems) with quantized "eigenvalues" of the energy (see Appendix VIII).

(b) States of positive total energy, which are unbound and correspond physically to the scattering of neutrons by protons.

We know that there is one state of type (a), namely the deuteron ground state, which is a stable combination of neutron and proton with a binding energy of 2.19 Mev. There are no other bound states of the

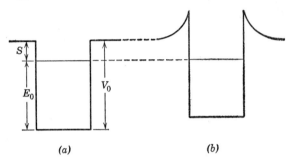

Figure 9.14. The nuclear potential well as seen by (a) a neutron and (b) a proton. E_0 represents the highest energy level normally occupied, and S the extra "separation energy" required to extract a nucleon from the nucleus.

two-nucleon system. From the detailed analysis of the deuteron problem and from measurements on nucleon–nucleon scattering at a variety of energies, it is possible to arrive at more or less consistent conclusions regarding the range, depth, and shape of the nuclear potential—and to choose an appropriate admixture of ordinary and exchange forces. It is from studies of this sort that we can infer a nuclear potential well depth of about 30 Mev (Section 10, point 4, in this chapter). Some elegant and sophisticated theoretical techniques have been developed in the attempt to systematize this information.

A less fundamental but extremely interesting system is that represented by a nucleus containing many nucleons, in which the finer details of the forces become secondary. We are then able to regard a nucleus as a more or less featureless potential well in which the individual nucleons move. In the simplest possible model, the potential is assumed to remain at a constant negative value from the center of the nucleus up to a radial distance R. A neutron inside the nucleus can thus be considered to find itself in a three-dimensional analog of the rectangular potential well (which has been considered in Chapter 7, Section 11). A section of such a potential is shown in Fig. 9.14a. A proton will be subject to the additional potential provided by the Coulomb forces, so that the walls of its potential well can extend to high positive energies. It is customary in simple treatments to suppose that this Coulomb potential is cut off at the nuclear radius R, so that the floor of the proton potential well, also, is regarded as flat (Fig. 9.14b).

We now have the problem of filling these two potential wells with neutrons or protons, subject to the dictates of the Pauli exclusion principle, until we have N neutrons and Z protons present. This situation has a great deal in common with the problem of free electrons in a metal, which we discussed in Chapter 8, Section 11. If we assume that the nucleons occupy the lowest possible states, leaving no vacancies, we have a maximum kinetic energy E_0 given by

$$E_0 = \frac{h^2}{8M}\left(\frac{3N}{\pi V}\right)^{2/3}$$

which represents the electron-energy formula with the nucleon mass replacing the electron mass. We now substitute

$$N \approx \frac{A}{2}$$

$$V = \tfrac{4}{3}\pi R^3 = \tfrac{4}{3}\pi r_0^3 A$$

$$\therefore \frac{N}{V} = \frac{3}{8\pi r_0^3}$$

and

$$\left(\frac{3N}{\pi V}\right)^{2/3} = \frac{1}{4r_0^2}\left(\frac{9}{\pi^2}\right)^{2/3} \approx \frac{1}{4r_0^2}$$

Hence

$$E_0 \approx \frac{h^2}{32 M r_0^2}$$

We notice that the result is independent of A, and, when the values of h, M, and r_0 are substituted, we find

$$E_0 \approx 30 \text{ Mev}$$

(the precise figure depending on the value assumed for r_0). When account is taken of the fact that $Z < A/2 < N$ in the heavier nuclei, we see that E_0 for neutrons is expected to be somewhat greater, and E_0 for protons somewhat less, than the value given above.

We can pursue the analogy with the electron gas problem a little further. Just as we have a characteristic work function ϕ for removal of an electron from a metal, so we have a separation energy S (see

Fig. 9.14a) that represents the least energy needed to remove a nucleon from the nucleus. We must be careful not to press this comparison too far, however, because the removal of one nucleon from a nucleus causes a non-negligible change in the nuclear radius and in the energy levels of the remaining particles. The strict definition of the separation energy is given in terms of the binding energies of the initial and final nuclei; thus for removal of a neutron we have

$$S_n = B(Z, A) - B(Z, A - 1)$$

The value of S can be derived directly from the values of B in a given case, or indirectly from the semiempirical mass formula. Its value is between about 6 Mev and 10 Mev for either a neutron or a proton in most nuclei. We thus deduce

$$V_0 = E_0 + S \approx 38 \text{ Mev}$$

This is the well depth measured from the zero potential energy level; for protons we must add to it the Coulomb barrier height Ze^2/R to obtain the over-all height of the potential discontinuity at the nuclear surface.

Given a "rectangular" nuclear potential well of known depth and radius, we can proceed to find the permitted values of energy, angular momentum, etc., corresponding to the stationary states of the system. This can be done in the first instance by regarding the nucleons as independent. We then have a fairly close parallel to the orbital electron problem (Chapter 8, Section 6), leading naturally to the expectation that there should be a regular shell structure, with particularly stable arrangements analogous to the inert gas configurations of the orbital electron system. Such a scheme was propounded by Bartlett as long ago as 1932, but it did not gain much credence because the predicted values of N and Z to give nuclei of greater stability (as judged, for example, by especially large values of the binding energy) did not agree with the facts. The matter was more or less dropped until 1949, when Maria Mayer (following a suggestion of Fermi) and Haxel, Jensen, and Suess examined the consequences of a very strong spin-orbit coupling in nuclei. Such a coupling would make j, rather than l, the important quantum number.

The possibility of a significant spin-orbit splitting (equivalent to the doublet fine structure of electronic levels) had been very reasonably set aside because the nucleon magnetic moments that might be expected to give rise to it were known to be so small. If we suppose that a nucleon in an orbit in the nucleus finds itself in a magnetic field H, given in order of magnitude by a nuclear magneton at a distance r_0, we shall have (cf. Chapter 8, Section 2)

$$H \approx \frac{2\mu_M}{r_0^3} = \frac{10^{-23}}{(1.2 \cdot 10^{-13})^3} \approx 6 \cdot 10^{15} \text{ gausses}$$

The doublet splitting for a proton magnetic moment in such a field would then be given by

$$\Delta W = 2\mu_p H$$

$$= 2 \times 1.4 \cdot 10^{-23} \times 6 \cdot 10^{15} \text{ ergs}$$

$$\approx 0.1 \text{ Mev}$$

This is small compared to the calculated spacings between levels of different l. But, if we ignore this difficulty and assume a spin-orbit splitting of the order of 50 times the above calculated value, then the nuclear shell model becomes quite spectacularly successful.

The levels are classified according to (a) an ordinal number n (which does not have quite the same significance as the principal quantum number of electronic levels), (b) the orbital angular momentum l, and (c) the total angular momentum j. The spin-orbit coupling is such as to make the state of higher $j(= l + \frac{1}{2})$ lie lower in energy than the state of lower $j(= l - \frac{1}{2})$ for a given value of l; this is just the opposite of the electronic situation. The size of the spin-orbit splitting is taken to increase as l increases, by analogy with atomic fine structure. This leads to a clustering of the levels in such a way as to produce significant energy gaps at certain stages of the level structure. The filling of a level with its complement of $2j + 1$ neutrons or protons just below such a gap will give a closed shell of an especially prominent kind, and the next nucleon to be added will tend to be weakly bound. The essential behavior is indicated in Fig. 9.15. It may be noted that the heaviest nuclei known

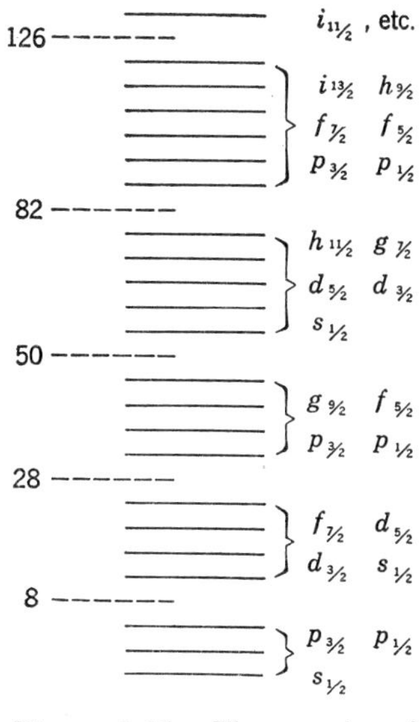

Figure 9.15. The grouping of nuclear levels leading to significant closed-shell configurations at the magic numbers shown on the left. Within each group there is no attempt to show the true ordering or spacing of levels. (But see Problem 10 at the end of this chapter.)

do not go much beyond $Z \approx 90$, $N \approx 140$, so that it is sufficient for most purposes to carry the energy level scheme up to levels with $l = 6$: i.e. $i_{13/2}$ and $i_{11/2}$, as shown.

Among the major successes of the shell model in its revised form are the following:

1. It correctly predicts the so-called "magic numbers": i.e., those values of N and Z at which there are clearly defined discontinuities of binding energy and stability corresponding to shell closures. (The most important of these are $Z = 28, 50, 82$; $N = 28, 50, 82, 126$.)

2. For nuclei of odd A, which therefore contain either one odd proton or one odd neutron, the model predicts with almost complete success the value of the nuclear spin I, regarding this as being equal to the j value of the odd nucleon, the other nucleons being paired off to give zero resultant angular momentum.

3. Not only does the model predict the value of the angular momentum for odd A; it also requires that **j** should be a particular vector sum of the spin **s** with a specified orbital momentum **l**. This means that the sign and size of the nuclear magnetic moment are determined without room for ambiguity (see Chapter 8, Section 9). The predictions are largely borne out by the measured values. This is really a very stringent test of the general correctness of the shell picture.

We do not wish to suggest, however, that the shell model provides a complete explanation of nuclear structure. We have glossed over a large number of details, and the fact remains that there is much that is empirical in our interpretations. There is still not a good understanding of nuclear forces as such, and the magnitude of the spin-orbit force is a considerable mystery at the present time (1958).

9.12 ISOTOPIC ABUNDANCES. NUCLEAR STABILITY

The resolution of a given chemical element into its various isotopes reveals not only the various isotopic species that are present, but also their relative abundances in the normal chemical substance. The mass spectroscope is the most direct and accurate means of obtaining this information.

The first and simplest observation we can make is on the existence or nonexistence of an isotope of a certain A for a given value of Z. This at once reveals some very striking statistics. Amongst the naturally occurring elements there are 284 different nuclear species (or "nuclides"). Of these 166 (i.e. well over half) have both Z and N

even. Only 8, of which two are radioactive and two are exceedingly rare, have both Z and N odd. (The four prominent nuclei in this class are $_1H^2$, $_3Li^6$, $_5B^{10}$ and $_7N^{14}$—all very light nuclei, it may be noted.) The remaining 110 nuclides (i.e. all those with odd A) divide themselves almost equally into odd Z with even N, and even Z with odd N. These facts are in accord with the idea that nucleons of the same type

TABLE 9.2

Mass number, A	112	113	114	115	116	117	118	119	120	121	122	123	124
% relative abundance	0.95	0	0.65	0.34	14.24	7.57	24.01	8.58	32.97	0	4.71	0	5.98

will tend to pair off with each other and that in so doing they will arrive at a state of minimum energy and maximum stability.

When we examine the isotopic abundances themselves, this impression is reinforced. An excellent example is provided by the isotopic composition of the element tin ($Z = 50$). The percentages of the different isotopes are as shown in Table 9.2.

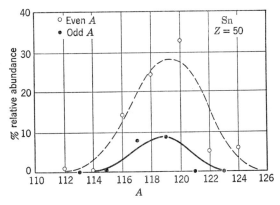

Figure 9.16. The isotopes of tin and their relative abundances.

These same data are shown graphically in Fig. 9.16. The smooth curves are sketched in merely as an aid to visualizing the trend of the values, and are not to be thought of as "best fits" to the experimental points (which are in fact determined with great accuracy). We can see how the isotopes of even A (which thus have both Z and N even) are spread over a wider range of A than the isotopes of odd A and are in aggregate about five times as abundant. All this suggests very strongly that in the process of element building, whatever it might have been, there existed the possibility of adjustments and regroupings of the nucleons so as to favor particular combinations of Z and N: namely those that minimized the total energy of the system.

We have another line of approach to the question of nuclear stability. Just as we can recognize a most probable value of A for a given value of Z, so we can recognize a most probable value of Z for a given value of A. Indeed, this is one of the concerns of the binding-energy formula (Section 9), which would define a unique Z for each A. Nature allows a little more variety than this—but not very much. In a few instances there are as many as three values of Z for a given A, but broadly speaking we can say that

(a) If A is odd, only one value of Z is found among the naturally occurring nuclides.

(b) If A is even, there may be either one or two values of Z. (Note: Nuclei of the same mass number A but different Z are called *isobars*.)

In order to understand these facts, we bring into our consideration of binding energy an effect that we omitted from the mass formula. This

TABLE 9.3

Z	N	Correction to B
Even	Even	$+\delta$
Even	Odd	0
Odd	Even	0
Odd	Odd	$-\delta$

is the stabilizing effect of nucleon pairing that we have already referred to. There appears to be a characteristic pairing energy of about 1 Mev when two nucleons of the same kind are brought together. Thus, if we take two possible isobars having the same *even* value of A, such that one has both Z and N even, and the other has both Z and N odd, then the first has an extra binding energy of about 2 Mev with respect to the second after the differences of Coulomb energy and asymmetry energy have been allowed for. (The even–even nucleus has both a proton pair and a neutron pair in excess of the odd–odd nucleus.) If, on the other hand, we take two isobars having the same *odd* value of A, the odd Z, even N and the even Z, odd N combinations have the same number of paired nucleons; so there is no jump in the binding energy in going from one to the other. The pairing effect can be incorporated

The Nucleus

in the semiempirical mass formula by adding a term to the binding energy $B(Z, A)$ as shown in Table 9.3.

We are now in a position to understand in outline the systematics of the naturally occurring nuclei. First of all, we see how the pairing effect will favor the production of even Z, even N nuclides above all others and will discourage the odd–odd types. We see that the even–odd and odd–even nuclei (in terms of Z and N) are similar to each other and have an intermediate status. This fits in well with the facts about the numbers of nuclides of the four different types. And now, to see

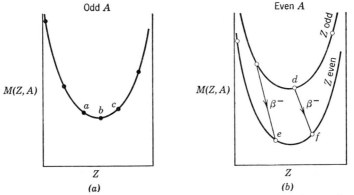

Figure 9.17. The curves of $M(Z, A)$ as a function of Z for a given value of A (a) if A is odd and (b) if A is even.

how the possible values of Z for a given A are defined, let us consider the curves of Fig. 9.17. Here we show the variation of atomic mass with Z for a given value of A according to the mass formula, regarding Z as a continuous variable. For odd A this gives a single curve, whereas for even A the pairing effect gives rise to two curves separated by an energy difference 2δ. Mathematically the value of Z is given by the minimum of the curve in each case, but in general this will not be an integer and so we choose the nearest integral value that is available.

Now for odd A there will always be a single value of Z that defines an atom of lower mass than either of its neighbors (unless by some freak we have the same value of M for two successive values of Z). But for even A we may have the situation shown in Fig. 9.17b, where an odd–odd combination d lies near the minimum of its mass curve but yet is heavier than the even–even combinations e and f to either side of it. In this case we need not be surprised to find two stable even–even nuclides of the same A, differing in Z by two units. Of course, if the mass formula gives a minimum for Z close to an even integer on the *lower* curve of Fig. 9.17b, we should expect to find just this one nuclide for the given value of A. This makes intelligible, in general terms, the facts (a) and (b), above, about the association of Z with A.

9.13 BETA DECAY AND THE POSITRON

We have mentioned (Section 6 of this chapter) the fact that the neutron may decay spontaneously into a proton and an electron (beta particle). We have referred to this again (Section 9) as a possibly reversible reaction through which an atom with a given value of A could adjust itself so as to arrive at its least massive and hence most stable state. Let us now consider this explicitly with the help of Fig. 9.17. The point b on the curve for odd A represents an atom (Z, A). The point a represents a heavier atom $(Z - 1, A)$. Thus a neutron in a might be able to turn into a proton and an electron with the release of energy; in this process the nuclear charge would be raised by one unit, and the extra orbital electron required for b would (in principle at least) be automatically supplied from the nucleus. This, basically, is the process of nuclear beta decay, except that the electron ejected from the nucleus normally escapes from the atom completely, and some other electron is subsequently captured from outside to restore neutrality to the atom. It is a fact of observation that beta decay will take place in this way whenever it is energetically possible. (The identification of beta rays as electrons was made by Becquerel in 1900.) Hence, if a nucleus is ever formed with too few protons for its particular value of A, it will undergo a series of beta-particle (i.e. electron) emissions until it reaches the stable state. (This is important in the products of nuclear fission.)

But what if we have an atom with an excessive ratio of protons to neutrons? Point c in Fig. 9.17a represents such an atom $(Z + 1, A)$. Will it be possible for the nucleus of this atom to capture one of its own orbital electrons, thereby turning into b (Z, A) with a release of energy? The answer is "yes," and the most common process of this type is K capture, in which the captured electron is provided from the innermost orbit. (We have here a good example of the nucleus behaving as a part of the atom, and not in isolation; cf. Section 2 of this chapter.)

But orbital electron capture may not be the only possible process. In 1932, C. D. Anderson discovered in the cosmic rays a particle having the same mass as the electron and with a charge of $+e$. This is the *positron*, and, except for the obvious differences that arise from the reversed charge, it behaves exactly like an ordinary electron. (It is now beginning to appear that every particle known to us has a corresponding "antiparticle"; the electron–positron relationship was the first of this type to be discovered, and had in fact been called for by Dirac's relativistic wave equation, referred to in Chapter 8, Section 9.) And a positron may be ejected from a nucleus of charge $Z + 1$, converting it to charge Z, if enough energy is available.

For positron emission from a nucleus to occur, the mass difference between the atoms involved has to be greater than about 1 Mev. Writing $M(Z, N)$ to denote the mass of a *nucleus* (as in Section 7 of this chapter), we see that positron emission is possible if

$$M(Z + 1, N) > M(Z, N + 1) + m$$

since the rest-mass energy of the positron has to be created. Now, if we ignore the electron binding energy, we have

$$M(Z + 1, N) = M(Z + 1, A) - (Z + 1)m$$

and $$M(Z, N + 1) = M(Z, A) - Zm$$

Hence the condition for positron emission to be energetically possible can be written in terms of the masses of the neutral atoms as follows:

$$M(Z + 1, A) > M(Z, A) + 2m$$

The rest-mass energy of an electron or a positron is 0.511 Mev; thus positron emission can occur only if the parent atom $(Z + 1)$ has a surplus energy of at least 1.02 Mev with respect to the daughter (Z). If this mass excess is available, positron emission and orbital electron capture will provide alternative modes of decay for the heavier atom; if it is not, the orbital capture will continue to provide a means by which the nuclear transformation can occur, no matter how small the mass difference between the atoms may be.

Our discussion here has been based on Fig. 9.17a for odd A, but it applies equally to even A. We may draw attention to one interesting possibility, however, that is peculiar to even A. If the masses of the atoms of different Z are as shown in Fig. 9.17b, with an isobar of odd Z near the minimum of its mass curve, this atom (d in the figure) may be able to decay both by ordinary electron (β^-) emission to f, and by positron (β^+) emission to e. This dual process has actually been observed in certain cases.

9.14 BETA SPECTRA. THE NEUTRINO

Our discussion of beta decay so far has regarded it merely as a means of converting one nuclide into another in such a way as to minimize the potential energy of the system. Let us now look at some details of the process itself.

The phenomenon of beta emission was discovered in the heavy radioactive elements (in uranium and thorium minerals) by Becquerel and others about 1899. In more recent times beta-active nuclei have been produced in great variety by bombarding stable nuclei with

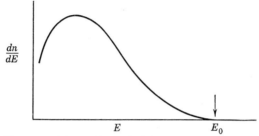

Figure 9.18. The general form of a typical beta-ray spectrum. E is the electron kinetic energy.

neutrons. Capture of a neutron by a nucleus will tend to raise the neutron–proton ratio above the value required for stability, and the nucleus may try to restore the situation by emitting an electron. Quite generally, beta-particle emission converts an atom of well-defined mass and energy into another atom of smaller mass and energy. We should therefore expect a definite amount of energy to be carried away by the electron. What happens is totally different. If the beta-particles emerging from a radioactive source are analyzed, by magnetic deviation or some other means, so as to obtain their energy spectrum, a curve of the kind shown in Fig. 9.18 will typically be obtained. The kinetic energy E_0 is the available energy as deduced from the mass difference of the initial and final atoms. This total energy is almost never found on the electron; an indefinite fraction of it seems to disappear in the decay process. And energy is not the only thing that apparently fails to be conserved. We have noted that the proton, the neutron, and the electron are all particles of spin $\frac{1}{2}(\times h/2\pi)$; this being so, it is impossible for spin angular momentum to be conserved if a neutron decays into a proton and an electron and nothing else, since the vector sum of proton and electron spins can only be 0 or 1. These facts created a cruel dilemma for physicists around 1930. To abandon the treasured laws of conservation of energy and angular momentum seemed unthinkable; yet the only way to avoid it was to postulate that, in the process of each beta decay, an unobserved (and apparently unobservable) particle carried away the balance of energy and momentum. The lesser of the two evils was chosen, and so the "neutrino" was born (Pauli, 1931).

Apart from being a vehicle for angular momentum and energy, the neutrino can have almost no attributes. It is certainly uncharged, or it would interact with matter and so be detectable. Its mass is probably zero, since the upper limit E_0 of a beta-particle spectrum always coincides exactly with the mass difference of the transition and leaves nothing over for the rest mass of the neutrino. Its spin has to be $\frac{1}{2}$. With its zero rest mass it must always travel at the speed of light.

In 1934 Fermi built a quantitative theory of beta decay on the electron–neutrino hypothesis, and was able to account for the shape of the electron energy spectrum. The theory was highly successful in other respects, too, although it seemed to be based on a virtual fiction. But finally, in 1956, Cowan, Reines, et al. succeeded in pinning down the neutrino as something that really existed. They did this by exposing a large tank of material to the flood of neutrinos emerging from a nuclear reactor (and produced there by the beta-decays that always accompany fission processes). Some of the neutrinos were effective in causing a kind of beta decay in reverse, by which the capture of a neutrino led to the ejection of a detectable beta-particle (actually a positron). The difficulty of this splendid experiment can perhaps be judged from the fact that the probability is only about 1 in 10^{12} that a neutrino will interact with anything in passing right through the earth!

All beta-decay processes can be ultimately referred back to the prototype beta decay, which we now write in the form

$$_0n^1 \rightarrow {_1p^1} + e^- + \nu$$

where ν is the neutrino.

The mass difference between neutron and proton is equivalent to 1.293 Mev, and the rest mass of the electron is 0.511 Mev. Thus the energy release available in the beta-decay of the free neutron is 0.782 Mev. The neutron-decay process has been studied in detail in a beautiful experiment by Robson (1951); besides measuring the energy spectrum, he found the half-life of the neutron to be about 13 min—i.e., out of any sample of neutrons, one half will undergo decay in this time. We are tempted to say that, since the electron is observed to leave the neutron (or, more generally, a parent nucleus), it must have been there beforehand. As we have seen, however (Section 5 of this chapter), the evidence is strongly against this. We must conceive of the electron being created in the act of the beta-decay process, just as we think of a photon as being created in a radiative transition and having no prior existence.

9.15 EMISSION OF HEAVY PARTICLES FROM NUCLEI

(1) The Criteria for Nuclear Stability

Any nucleus that occurs in nature is stable against complete disruption into its separate neutrons and protons. It does not follow, however, that it is necessarily stable in any absolute sense. We have seen, for example, that two atoms with the same odd value of A, and with Z values differing by unity, cannot both be stable; the heavier will be able to transform into the lighter by β^- decay or by orbital electron capture. But, if the mass difference is exceedingly small, the detailed theory of beta decay shows that the rate of transformation will be very slow, and may even occur on a time scale comparable with the age of the earth (about $5 \cdot 10^9$ years). If this happens, it is quite possible that both types of atom will occur naturally, and it may be a difficult matter to discover which of the two is in fact unstable. Several instances of this are known.

The question can now be enlarged to include stability of a nucleus against the loss of one, or perhaps several, of its constituent nucleons. And it is here that we find the basic reason for the occurrence in nature of the heavy radioactive elements. Let us consider what is probably the most famous of all such atoms, $_{92}U^{238}$, the heaviest nuclide found in nature. Its binding energy is very large—over 1800 Mev—and corresponds to a mean binding energy per nucleon of 7.6 Mev. With the help of the empirical mass formula we shall estimate the energy needed to remove a neutron, a proton, or an alpha particle ($_2He^4$) from this nucleus. We can discuss this in terms of binding energies. If an atom $M(Z, A)$ is to be separated into two parts, $M_1(Z_1, A_1)$ and $M_2(Z_2, A_2)$, we have

$$Z_1 + Z_2 = Z$$
$$A_1 + A_2 = A$$
$$M(Z, A) = ZM_H + (A - Z)M_n - B(Z, A)$$

and hence

$$M_1 + M_2 - M = B - (B_1 + B_2)$$

Thus, if the total binding energy of the initial atom is greater than the sum of the binding energies of the two parts, energy must be supplied to cause the separation; if it is less, the separation is energetically possible in a spontaneous fashion. Taking values of the constants as given in Section 9 of this chapter, we have

$$B(Z, A) = 14.1A - 13A^{2/3} - 0.595\frac{Z^2}{A^{1/3}} - 76\frac{(Z - \tfrac{1}{2}A)^2}{A} + f(\delta) \quad \text{(in Mev)}$$

where $f(\delta)$ is the pairing energy contribution.

In Table 9.4 we show the computation of the approximate total binding energies of $_{92}U^{238}$ and of the three atoms that would result from it by removal of $_0n^1$, $_1H^1$, or $_2He^4$. It is interesting to note the relative importance of the different contributions to B (all in Mev).

TABLE 9.4

Z, A	Volume, 14.1A	Surface, $-13A^{2/3}$	Coulomb, $-0.595Z^2/A^{1/3}$	Asymmetry, $-76(Z - \tfrac{1}{2}A)^2/A$	Pairing, $f(\delta)$	Total Binding, $B(Z, A)$
92, 238	3356	−499	−812	−233	+1	1813
92, 237	3342	−498	−814	−225	0	1805
91, 237	3342	−498	−796	−243	0	1805
90, 234	3300	−494	−782	−237	+1	1788

Now the binding energy of an individual neutron or hydrogen atom is zero, by definition. Thus we see that about 8 Mev (1813 minus 1805) would have to be supplied to remove a neutron or a proton from the U^{238} nucleus; it is stable against the emission of these particles. But when we consider the possible removal of an alpha particle, we find that more than enough energy is already available. The total binding energy of $_2He^4$ is about 28 Mev. Hence the emission of an alpha particle by U^{238} should occur with the release of about 3 Mev of energy (1788 + 28 − 1813). (The precise figure, obtained by observation on the alpha particles themselves, is 4.2 Mev.) We see here how the stability or instability of a nucleus can be defined only with respect to some specified mode of breakup.

We can express the above results in simple pictorial fashion by imagining that a particle trying to escape from a nucleus will find itself with a certain energy in a potential that is determined by itself and the rest of the nucleus. Figure 9.19 shows the same uranium nucleus as seen in this way by a neutron, a proton, and an alpha particle. We

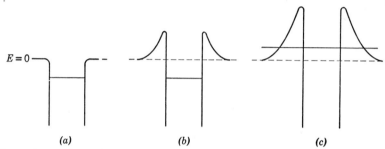

Figure 9.19. A uranium nucleus as it might appear to (a) a neutron, (b) a proton, (c) an alpha particle. The neutron and the proton are bound, but the alpha particle has a positive total energy and can escape.

have included here the effect of the Coulomb barrier, which for the alpha particle is about twice as high as for the proton.

(2) Alpha Decay

When the stability of heavy nuclei such as uranium with respect to emission of light nuclei is considered in more detail, it is found that the alpha particle is the only one for which a surplus of energy is available, permitting appearance of the alpha particle outside the nucleus with a positive kinetic energy. The reason for this is to be found in the abnormally large binding energy of $_2$He^4 compared to that of neighboring nuclides. And this, essentially, is why alpha decay is the starting point for the radioactive transformations of the heavy naturally occurring nuclides.

To understand the process in detail we must go to wave mechanics. We have indicated the nature of the problem in Chapter 7, Section 12, where we gave a qualitative discussion of the penetration of potential barriers by the tunnel effect. Alpha-particle emission is an excellent example of the effect, and the explanation of it in these terms by Gamow (1928) and by Condon and Gurney (1929) was a triumph for quantum mechanics.

We shall not attempt to give the details of the theory, but the essentials can be understood from Fig. 9.20. An alpha particle leaving a nucleus of total charge Ze will experience a Coulomb potential given by

$$V(r) = \frac{2(Z-2)e^2}{r}$$

This comes simply from considering the division of the total charge into two parts; r is the distance between the center of the alpha particle

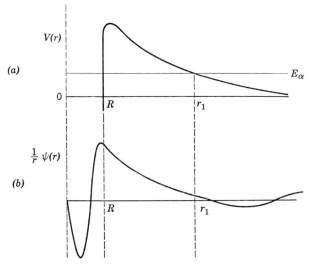

Figure 9.20. (a) A nuclear potential barrier in alpha emission. (b) The corresponding type of wave function.

and the center of mass of the remaining nucleons. In the simple picture that we have used throughout, the Coulomb potential is considered to be sharply cut off at $r = R$ where the nuclear forces take over. The maximum height of the Coulomb barrier is obtained at this radius, and so we have

$$V_{max} = V(R) = \frac{2(Z-2)e^2}{r_0 A^{1/3}} \quad \text{approximately}$$

Substituting numerical values, we find for uranium

$$V_{max} \approx 29 \text{ Mev}$$

The alpha particle cannot appear outside the nucleus with positive kinetic energy until it reaches a radial distance r_1 such that

$$E_\alpha = V(r_1)$$

A typical energy for natural alpha particles is about 5 Mev; this would define

$$r_1 = \frac{2(Z-2)e^2}{E_\alpha} \approx 5 \cdot 10^{-12} \text{ cm}$$

Since $R \approx 10^{-12}$ cm for any of the heavy radioactive nuclei, we see that the barrier is very thick on the nuclear scale, besides being very high. This drastically attenuates the wave-function amplitude (Fig. 9.20b) in passing from inside to outside the nucleus, making the penetrability of the barrier exceedingly small for the energies of most natural alpha particles, and so making the process of alpha decay very improbable.

The probability of decay per unit time is described quantitatively by a *decay constant* λ, a quantity, measured as the reciprocal of a time, which can be defined for any spontaneous decay process. If N_0 unstable nuclei of the same type are present in a system at time zero, the number remaining at time t is given by

$$N(t) = N_0 e^{-\lambda t}$$

This is derived in the same way, and on the same basic assumptions of probability theory, as the free path formula of Chapter 1, Section 7. Its applicability to nuclear decay implies that the breakup or decay of nuclei is a random and statistical process; it is impossible to predict the instant at which any particular nucleus will disintegrate, and a group of similar unstable nuclei all formed at the same instant will break up or decay at different times in such a way that, statistically, the exponential law just given is obeyed.

The *half-life* $\tau_{1/2}$ (mentioned in connection with neutron beta-decay, Section 14 of this chapter) is defined through the equation

$$N(\tau_{1/2}) = N_0 \exp(-\lambda \tau_{1/2}) = \tfrac{1}{2} N_0$$

Hence
$$\tau_{1/2} = \frac{\log_e 2}{\lambda} = \frac{0.693}{\lambda}$$

The half-life of U^{238} is $4.5 \cdot 10^9$ years, which probably implies that about half the uranium present when the earth was formed is still in existence today. We must dispel the idea, however, that alpha-particle emission is bound to be a slow process, even in heavy nuclei with high potential barriers. The quantum-mechanical theory predicts an enormously rapid variation of decay constant with alpha-particle energy. A *very rough* statement of the theoretical result for emission from nuclei with $Z \approx 90$ is

$$\log_{10} \lambda \approx 60 - \frac{150}{E_\alpha^{1/2}} \qquad (E_\alpha \text{ in Mev})$$

This would give values of decay constant and half-life as in Table 9.5 (all figures rounded off to powers of 10). This energy dependence

TABLE 9.5

$E_\alpha^{1/2}$	E_α	λ	$\tau_{1/2}$
2.0 (Mev)$^{1/2}$	4.0 Mev	10^{-15} sec^{-1}	10^8 years
2.5	6.25	1	1 sec
3.0	9.0	10^{10}	10^{-10} sec

seems almost unbelievably extreme; yet, when a graph is made of $\log_{10} \lambda$ against $E_\alpha^{-1/2}$ for the natural alpha emitters, we find it confirmed.

(3) Radioactive Series Decay

All nuclides having $Z > 83$ are unstable with respect either to alpha or to beta emission, and most have extremely short half-lives; yet many of these occur naturally in radioactive minerals. How is this possible? The answer lies in the existence of just three nuclides, $_{92}U^{238}$, $_{92}U^{235}$, and $_{90}Th^{232}$, which happen to have half-lives for alpha decay that are comparable with the age of the earth (or perhaps, more properly, of the universe). These three nuclides act as the parents and controlling agents of three separate chains of radioactive decay processes; each chain terminates with a stable isotope of lead ($Z = 82$: a "magic number"). All the intermediate members of a given chain are in what is called "secular equilibrium"; each member of the series is decaying at almost exactly the same rate at which it is formed, and the over-all effect of the process is simply to convert uranium or thorium into lead. Now the rate at which the atoms of a given member of the chain are decaying is given by

$$\frac{dN}{dt} = \frac{d}{dt}(N_0 e^{-\lambda t}) = -\lambda N$$

and this represents the rate of formation of the next member of the series, since the act of decay converts the first type of atom into the second. Thus the secular equilibrium is defined by the equation

$$\lambda_1 N_1 = \lambda_2 N_2 = \cdots \approx \text{const}$$

where N_1, N_2, etc. are the numbers of the different types of atoms present at a given instant. The numbers of atoms of the different types present in equilibrium are therefore inversely proportional to their decay constants, or directly proportional to their half-lives. And on the long geological time scale the magnitudes of N_1, N_2, etc. decline according to the gradual disappearance of the parent material. To give one example, the element radium ($_{88}Ra^{226}$) in the U^{238} series has a half-life of 1620 years. In secular equilibrium, therefore, it is present in the ratio $1.6 \cdot 10^3/4.5 \cdot 10^9$ relative to the uranium. This is about 1 part in 3 million, or about 1 g of radium in 3 tons of uranium.

It is a fascinating thing that this array of unstable elements, running through 10 units of Z, should be sustained by a few nuclides that are themselves unstable. It is equally fascinating to consider the violence of the impulse that causes U^{238}, for example, to jettison 10 units of charge and 32 units of mass (by 8 alpha emissions and 6 beta decays) in order to achieve final stability.

(4) Fission

The story of the discovery of nuclear fission in 1939 is a familiar and dramatic one—of how the attempt to synthesize elements beyond uranium led first to perplexity, and then finally to the realization that what had been done was to divide the nucleus instead of adding to it. We shall not try to trace the development of the subject here, but will simply consider some of the principles underlying it.

First of all, what *is* fission? It is the division of a heavy nucleus into two parts of comparable but not necessarily equal size. It can happen spontaneously in a few of the very heaviest nuclides, but is vastly stimulated by the provision of extra energy to the nucleus (notably by neutrons). The energy release when fission takes place is very large—about 200 Mev—and most of it goes into the kinetic energy of the separating fragments. The process is accompanied by debris in the form of two or three neutrons (of vital importance to chain-reacting systems); it is followed by several beta-decay processes in the fission fragments before a final state of equilibrium is reached.

Some of these facts can be understood directly by referring to the curves of binding energy and Z as functions of A (Figs. 9.4 and 9.1). The mean binding energy per nucleon in U^{238} is close to 7.6 Mev; the mean binding energy per nucleon for a stable nucleus of half this mass is about 8.5 Mev. Hence there would be a gain of binding energy of about 0.9 Mev per nucleon if U^{238} could divide into two equal parts of this sort, and the net release of energy would be about $0.9 \times 238 = 214$ Mev. The actual fission process will not be able to do this in one step, because the stable value of Z corresponding to $A = 119$ is 50 rather than 46 ($= 92/2$); the initial division would give a value of Z to each fragment that was too small by 4 units, and the fragments would each be expected to undergo four successive β^- decays before reaching a stable state.

The amount of energy released in the division process itself can be estimated for symmetric fission from the empirical mass formula. From the discussion of "Criteria for Nuclear Stability" earlier in this section it is easy to deduce that the energy release Q is given (if pairing energy is ignored) by

$$Q = 2B(\tfrac{1}{2}Z, \tfrac{1}{2}A) - B(Z, A)$$
$$= -13(2^{1/3} - 1)A^{2/3} + 0.595(1 - 2^{-2/3})Z^2/A^{1/3}$$

or $\quad Q = 0.220\,(Z^2/A^{1/3}) - 3.38 A^{2/3}$ Mev

For $Z = 92$, $A = 238$, this gives $Q \approx 170$ Mev, which is all but about 40 Mev of the total available energy. The balance goes into kinetic energy of the ejected neutrons, energy of beta-decay transformations, etc.

The discussion of fission cannot be given in quite the same terms as that of alpha decay, because the fission process involves a serious distortion of the whole nucleus. It is not realistic to think of one fragment providing a static potential well in which the other fragment moves. Instead, we look to the liquid drop model and consider the possible distortion of an initially spherical drop which, it must be remembered, has a more or less uniform charge density. The effect of an initial distortion from sphere to ellipsoid (from i to ii in Fig. 9.21a) is two-fold:

(a) The surface area, and hence the surface energy, increases.
(b) The electrostatic potential energy falls.

The first effect outweighs the second, and so the total potential energy begins to rise as the amount of distortion increases. For larger deformations (such as iii in the figure), the calculation of the precise energy becomes difficult, but we know that it must ultimately fall away as the fragments separate to large distance; thus the potential energy of the system must have the general form shown in Fig. 9.21b. Detailed calculation shows that ΔV, the height of the potential bump above the normal energy of the system, is only a few Mev for nuclei such as uranium.

Now, even though we cannot really visualize the fission fragments as individual particles traversing the potential barrier, the fact remains that a state of positive kinetic energy exists for them for any value of r greater than that corresponding to configuration (iv) in Fig. 9.21, and we can still, in a sense, regard this as an example of the tunnel effect. The probability that the tunneling will take place is exceedingly small, and we have here the basic theoretical interpretation of spontaneous fission. (If U^{238} decayed only by spontaneous fission, its half-life

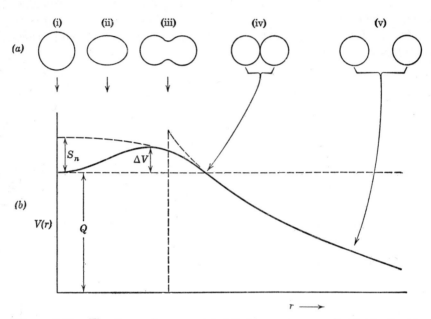

Figure 9.21. The stages of a symmetrical fission process according to the liquid drop model. (a) The configuration of the system. (b) The associated potential energy diagram. Q represents the relative kinetic energy of the fragments at infinite separation.

would be 10^{16} years.) The situation may be completely changed, however, if a few extra Mev of "activation energy" are added to the undistorted nucleus. We may be able to make the total energy of the system equal to or greater than the maximum potential energy $Q + \Delta V$, and then there is nothing to prevent the fragments from flying apart almost instantaneously. This is regarded as the explanation of induced fission. To take the best-known example, when U^{235} captures a slow neutron, it becomes U^{236}. Now, if U^{236} existed as a normal nucleus, it would have a neutron separation energy of about 6 Mev, which means that a neutron "falling into the nucleus," as it were, gives up this 6 Mev in reaching its normal level in the structure. It so happens that this amount of energy exceeds, by about 1 Mev, the height of the potential energy bump ΔV. The fission is therefore possible if the extra energy is retained within the U^{236} nucleus. This form of the theory of fission was first given by Bohr and Wheeler in 1939, almost as soon as the facts of fission were known.

9.16 EXCITED STATES OF NUCLEI. GAMMA RADIATION

Until the last paragraph of the previous section, we have made no mention of the possibility of raising nuclei above their lowest states (although we have discussed nuclei that are unstable even in such "ground states"). But, in describing the fission of U^{235} by slow neutrons, we pointed out that the added neutron makes available its binding energy for excitation of the system. The study of the excited states of nuclei is a large part of the entire subject of nuclear physics, both experimental and theoretical, and from it has come a great deal of what we know about nuclear structure and nuclear forces. Much of this knowledge has been embodied in our account of the nucleus, and we do not propose to discuss it separately in detail. Some general comments should, however, be made.

Whether the nucleus is regarded as a liquid drop, or as a Fermi gas, or as a collection of nucleons moving in definite orbits in a potential well, it can always be described as a system subject to the laws of quantum mechanics. Being a system within more or less well-defined boundaries, it will have a variety of distinct quantum states, characterized by eigenvalues of energy, angular momentum, etc. And, since it contains electric charges, there will in general be the possibility of radiative transitions between the different energy levels, according to the same quantum condition that holds for electronic transitions:

$$h\nu_{12} = E_1 - E_2$$

We may in certain cases be able to identify a certain quantum jump as due to a single nucleon in a nucleus, just like a one-electron transition in ordinary spectroscopy. On the other hand, we may find that the whole nucleus is behaving like an oscillating or rotating charge cloud. But in any case there will be the possibility of observing line spectra of electromagnetic radiation from any nucleus that has been raised above its ground state. This is *gamma radiation*, first observed as the penetrating component, undeviated by magnetic fields, in the radiations from radioactive substances. The energies of gamma-ray quanta range from about 10 kev to 10 Mev; the corresponding wavelengths, given by hc/E, are from about 10^{-8} cm to 10^{-11} cm, and so overlap the wavelength range for X rays. The proof that gamma rays are electromagnetic radiation was first obtained by Rutherford and Andrade (1914), who showed that the rays could be diffracted by crystals. They can also act on the individual orbital electrons of an atom to cause a photoelectric effect (Chapter 5, Section 1) or Compton scattering (Chapter 5, Section 13).

The most direct information that gamma rays can give about nuclei concerns the positions and spacings of the nuclear excited levels.

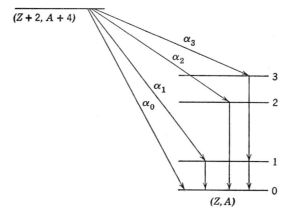

Figure 9.22. A complex alpha-particle spectrum, corresponding to transitions from the ground state of a parent nucleus to various states of a residual nucleus.

This was known from the study of natural radioactivity, where it was found that the formation of a particular nucleus by alpha or beta decay of some preceding nucleus in a radioactive series did not necessarily produce it in its lowest possible state. In alpha decay, for example, there might exist the kind of scheme shown in Fig. 9.22, with several discrete energies of alpha particle. The highest energy alpha particle leads to the lowest state of the nucleus (Z, A). The lower-energy alpha emissions lead to higher states (1, 2, 3, etc.) which then de-excite, either directly or in "cascade," by gamma radiation to the ground state. The alpha-particle energy, plus the total energy of any gamma-ray quanta following it, is a constant. When the excited states are formed as a result of beta decay, it is the maximum energy of the electron spectrum (Fig. 9.18), representing the total available energy of a particular transition, that takes the place of the alpha-particle energy in the energy balance of the process.

9.17 LEVEL WIDTHS. COMPETING PROCESSES

Gamma radiation is nearly always possible as a means for a nucleus to get rid of surplus energy. There are quantum-mechanical "selection rules" for gamma emission, just as for optical spectra (Chapter 8, Sections 7–9), but they are far less restrictive. Virtually all optical transitions correspond to oscillations of classical electric dipoles, but gamma radiation can arise from a variety of more complicated motions of the nuclear charge system, and can take place for almost any difference of angular momentum between initial and final states (whereas electric dipole emission is impossible for an angular momentum change of more than one unit of $h/2\pi$). Important information about the quantum numbers of excited states of nuclei can be obtained by studying the rate at which the states emit their gamma radiation. Two factors enter here:

1. For a given value of the gamma-ray energy, the rate of radiation is least when the angular momentum change between initial and final states is largest.

2. For a given value of the angular momentum change, the rate of radiation is greatest when the energy of the emitted radiation is greatest.

(These are general trends, and must not be considered binding in individual cases. They are, however, trends that apply also to the emission of alpha or beta rays, although the details of the energy and angular momentum dependence are different for all three.) The question naturally arises—how do we measure the rate of radiation by a nucleus? The answer is that, if it is not measured directly in terms of the mean lifetime of the radiating state, it may be inferred with the help of Heisenberg's uncertainty principle, which for gamma radiation at least can be discussed in almost classical terms.

Let us consider the hypothetical case of a nuclear level A that can decay in two ways—either directly to the ground state C, or indirectly through an intermediate level B. So far as level A is concerned, the only relevant processes are the emissions of the quanta γ_1 or γ_2 (Fig. 9.23). These are mutually exclusive, and there will be an independent probability of decay per unit time for each. The *total* probability of decay per unit time, p, is thus given by

$$p = p_1 + p_2$$

The reciprocal of p is a characteristic time for decay, and, by analogy with the mean free path problem (Chapter 1, Section 7), it will be realized that this time is the mean life (τ_m) of the exponentially decaying excited state. But the existence of a non-infinite τ_m implies a finite spread of frequency $\Delta\nu$ in the radiations (as discussed in Chapter 7, Section 9) such that

$$\Delta\nu \cdot \tau_m = \frac{1}{2\pi}$$

Hence
$$\Delta\nu = \frac{1}{2\pi\tau_m}$$

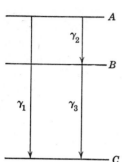

Figure 9.23. Alternative modes of de-excitation of a nuclear level A by gamma rays of different energies.

Let us now multiply this equation by h on both sides:

$$h\,\Delta\nu = \frac{h}{2\pi\tau_m} = \frac{\hbar}{\tau_m}$$

The quantity $h\,\Delta\nu$ has the dimension of energy, and represents a finite uncertainty in the energy of the radiating state as a result of its finite life. This is called the *width* Γ, and we have

$$\tau_m = \frac{\hbar}{\Gamma}, \qquad p = \frac{\Gamma}{\hbar}$$

The energy of any gamma quantum originating in the level A will have at least this uncertainty—and, if it terminates on a level that is itself subject to decay, there will be a further spread. The gamma-ray spectrum, plotted on an energy scale, will thus have lines with characteristic widths Γ. (The same applies in principle to optical spectra, of course, but here the widths are negligibly small as a rule.) In order to identify the probability per second of emission of a particular gamma ray, we introduce what are called "partial widths," defined as follows:

$$\Gamma_1 = \hbar p_1, \qquad \Gamma_2 = \hbar p_2, \qquad \Gamma_1 + \Gamma_2 = \Gamma$$

$$\frac{\Gamma_1}{\Gamma} = \frac{p_1}{p}, \qquad \frac{\Gamma_2}{\Gamma} = \frac{p_2}{p}$$

The term "partial width" is somewhat unfortunate, since each individual gamma-ray line has the full energy spread Γ, and quantities such as Γ_1 and Γ_2 merely express the *probability* that the decay will take place by the various possible paths.

It will now be profitable to consider what happens as a given nucleus is raised to higher and higher levels of excitation, starting from its ground state. There are two main effects:

1. The average number of levels per unit energy interval increases. This corresponds to the rapid variation with total energy (much stronger than linear) of the total number of different modes of motion for a collection of nucleons, and is to be expected on general grounds for any many-body system.

2. The average level width Γ increases. This follows directly from what we have said about gamma radiation, because (a) the individual partial widths tend to increase as the energy above the ground state increases, and (b) more and more different paths for de-excitation become available.

For a typical stable nucleus these two effects go hand in hand until, at a certain stage, the nucleus acquires enough energy to become unstable with respect to emission of heavy particles (neutron, proton, alpha particle, etc.). This happens, of course, each time the separation energy for a particular mode of breakup is reached, and so will occur at about 6 to 10 Mev excitation for emission of a single neutron or proton. The opening of this new avenue for de-excitation leads to an enhanced increase of Γ with energy. It is customary to describe the situation by saying that heavy-particle emission now *competes* with quantum emission. This is true in that the ratio Γ_p/Γ_γ of the partial widths for particle and gamma-ray emission rises rapidly from zero toward a large number as the excitation above the heavy-particle "threshold" is increased, but it tends to suggest that the onset of the heavy-particle emission actually reduces the absolute probability per second of quantum emission, which is not so. Nevertheless, the gamma emission tends to be effectively suppressed in the final result. The explanation of this gives some useful insight into the nature of nuclear processes.

Let us suppose that we have a nucleus excited above the neutron separation threshold. Then some neutron in the nucleus can be thought of as finding itself in the potential shown in Fig. 9.24. This is similar to the potential step discussed in Chapter 7, Section 12. Each time the neutron strikes the potential discontinuity from the inside, there will be a certain probability for it to escape. This probability is given by the ratio of transmitted current to incident current in the wave-mechanical problem of Chapter 7, Section 12, and in the notation of that problem we have

$$\text{Transmission probability} = \frac{v_2|C|^2}{v_1|A|^2} = \frac{|A|^2 - |B|^2}{|A|^2} = \frac{4K_1K_2}{(K_1 + K_2)^2}$$

where
$$K_1 = \frac{(2ME_1)^{1/2}}{\hbar}, \qquad K_2 = \frac{(2ME_2)^{1/2}}{\hbar}$$

If the kinetic energy E_2 with which the neutron emerges is much less

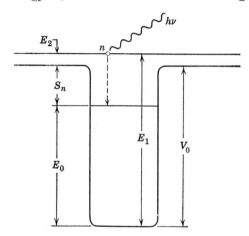

Figure 9.24. The competition between gamma radiation and neutron emission for a nucleus excited above the neutron separation energy.

than E_1 (which will be true in any case of interest), we can put $K_2 \ll K_1$, and so

$$\text{Transmission probability} \approx \frac{4K_2}{K_1} = 4\left(\frac{E_2}{E_1}\right)^{1/2} \approx 4\left(\frac{E_2}{V_0}\right)^{1/2}$$

According to this, therefore, the probability per unit time that the neutron will be emitted is proportional to its velocity, and we can put

$$\Gamma_n \propto E_2^{1/2}$$

The partial width for gamma emission, on the other hand, is a function of $(S_n + E_2)$, the total de-excitation energy. So long as we have $E_2 \ll S_n$, therefore, the value of Γ_n rises rapidly with E_2, whereas Γ_γ is not much affected. And what happens in practice is that Γ_n, which is bound to be less than Γ_γ for very small values of E_2, overtakes and surpasses it for higher excitations of the nucleus.

The same behavior, in essence, occurs for emission of other particles, although we must now consider the effects of the Coulomb barrier. Thus at sufficiently high excitation the total level width Γ becomes nearly equal to the partial width Γ_p for particle emission (including all possibilities). The mean life of the excited nuclear state is thus nearly equal to $\hbar/\Gamma_p$; the probability per second of gamma emission is equal to $\Gamma_\gamma/\hbar$ throughout this time. Hence the probability that the excited state will decay by gamma emission at all is given by $(\Gamma_\gamma/\hbar) \times (\hbar/\Gamma_p)$ = Γ_γ/Γ_p, which will, for the reasons given, become much less than unity for sufficiently high excitation.

A good example of this competitive behavior is found in the fission of U^{235} by slow neutrons. The excited state of U^{236} formed by the neutron capture can do three things:

(a) Since a neutron is able to enter in the first place, there must be some probability that a neutron (not necessarily the same one) will be able to escape, leaving the U^{235} nucleus in its original state.

(b) There may be emission of gamma rays, bringing the U^{236} nucleus to its lowest state; this is called "radiative capture" (of the neutron).

(c) There may be fission.

Experiment shows that the probability of (a) is negligibly small, and that (b) takes place in about 15% of all cases. Thus we can say that the ratio of the partial widths for radiative capture and for fission is $\Gamma_\gamma/\Gamma_f = 15/85 \approx 0.18$.

9.18 NUCLEAR REACTIONS AND SCATTERING

(1) The Compound Nucleus. Resonance Processes

In 1919 Rutherford found that, when alpha particles from a radioactive source were allowed to pass through air, they somehow ejected protons. It was a very inefficient process: only about one alpha particle in 300,000 succeeded in causing it. Nevertheless, this was a historic discovery, for it was the first recognized case of a nuclear reaction. Further investigation proved that the reaction was:

$$_2He^4 + {_7N^{14}} \rightarrow {_8O^{17}} + {_1H^1}$$

We can readily understand such a process in terms of our present picture of the nucleus. We recognize that it is basically what may be called a "rearrangement collision"; the final result has been to transfer one proton and two neutrons from the helium nucleus to the nitrogen. The reaction takes place with the conservation of momentum, energy (in all forms), charge, and the numbers of neutrons, protons, and electrons. This bare account leaves a great deal unsaid, however; let us now examine the situation a little more critically.

Suppose we look at the reaction from the common center of mass of the colliding and separating particles. Then we shall see an alpha particle and a nitrogen nucleus approaching from opposite directions. They strike each other and form an agglomerate of nine protons and nine neutrons which is momentarily at rest from our chosen point of view. An instant later it explodes, and we see a proton and an O^{17} nucleus receding in opposite directions—but not necessarily along the line of motion of the initial pair of particles. In the intermediate stage all the kinetic energy of relative motion (as measured in the center of mass) of the colliding particles is stored up; what can this imply about the details of the process?

A system of nine protons and nine neutrons is, at least in principle, a form of $_9F^{18}$, which can therefore be considered to constitute an intermediate stage in the nuclear reaction. And Bohr, in 1936, suggested that it is a vitally important stage, at least for nuclear reactions in which the colliding particles have energies not more than a few Mev. His main reason for being able to assert this was the knowledge that nuclear reactions exhibit well-defined *resonances* in this region, by which it is meant that the yield of the end products of the reaction is a sensitive function of the kinetic energy of the reacting particles, and passes through sharp maxima (see Appendix III). This resonance behavior is more or less implicit in our discussion of excited states of nuclei (Sections 16 and 17 of this chapter). For, if a nucleus has a series of well-defined energy levels, we shall expect to be able to stimulate transitions between them whenever the energy that is put into the system becomes equal to a characteristic energy difference between two levels. The probability of absorbing the excitation energy will remain high over a range of energy about equal to the width Γ of the state being formed. It is in precisely this way that we describe the resonance absorption of light by atoms, as in the Fraunhofer lines. From this point of view, the collision of two nuclei (such as $_2He^4$ and $_7N^{14}$) is only one of many possible ways of forming an excited state of a *compound nucleus*. Once this has been achieved, we consider the various modes of decay of the excited state as a separate problem.

The physical importance to be attached to the compound nucleus depends primarily on its mean lifetime. If this is very long compared to the time that it would take for the colliding particles to pass by each other within the range of nuclear forces, then the initial and final states of the system are effectively isolated from each other by the compound state, and an intelligible account of the nuclear reaction is impossible without it. On the other hand, if the lifetime is exceedingly short, it may only complicate and confuse matters to bring in the compound nucleus. Now the velocity of an alpha particle or a proton with an energy of a few Mev is of the order of 10^9 cm per sec and the sum of the diameters of two colliding nuclei is about 10^{-12} cm, so that the duration τ of a passing encounter would be of the order of 10^{-21} sec. This, it will be realized, is a very short time indeed, and corresponding to it there is an energy $\hbar/\tau \approx 1$ Mev. Unless the bombarding particle energy is made quite high, the observed resonance widths do not remotely approach this figure; so the compound nucleus picture is valid. Figure 9.25 shows two resonances (with the suggestion of a

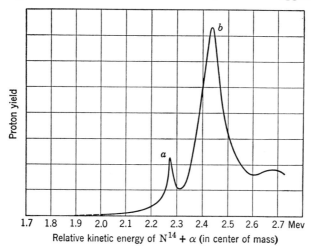

Figure 9.25. Resonances in the reaction $N^{14}(\alpha, p)O^{17}$ [after Heydenburg and Temmer, *Phys. Rev.*, **92**, 89 (1953)].

third) for emission of protons when N^{14} is bombarded with alpha particles of an accurately controlled and variable energy, and we can conclude that the F^{18} nucleus has a significant existence as a compound state in this region of excitation.

Figure 9.26 shows about twenty of the known excited states in the level structure of F^{18}. The excitation energies at which the system first becomes able to break up in various ways are also indicated. The levels marked a and b are the two resonance levels of Fig. 9.25. It may be noticed that the reaction $N^{14}(\alpha, p)O^{17}$, around which we are building this whole discussion, is *endo-ergic;* i.e., a certain minimum collision energy has to be put into the $N^{14} + \alpha$ system before it can re-form itself as $O^{17} + p$. The exact amount needed can be readily calculated from the known atomic masses or binding energies. The inverse reaction, $O^{17}(p, \alpha)N^{14}$, can evidently take place with the liberation of this same energy, and so will be energetically possible whenever a proton strikes an O^{17} nucleus.

If we consider the bombardment of N^{14} with alpha particles of such high energy that the F^{18} compound state has about 10 Mev of excitation, we see from Fig. 9.26 that it may de-excite by emission of a neutron, a proton, a deuteron, or an alpha particle. (It may also, in principle, emit gamma radiation, but from what we have said in the last section we should expect the chance to be slight.) It is quite

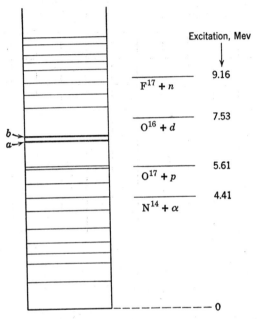

Figure 9.26. Energy levels in F^{18}, showing energies for separation into various pairs of nuclei. The heavily marked levels a and b are the pronounced resonance levels of Fig. 9.25.

possible that the breakup process may lead to excited states of the residual nuclei (F^{17}, O^{17}, O^{16}, N^{14}), in which case the kinetic energy spectrum of the ejected light nuclei will show a number of discrete values, just as in Fig. 9.22. Thus, from the resonance behavior we can learn about the levels of the compound nucleus, and from the energy distribution of the emitted particles we can learn about the levels of the possible residual nuclei.

The study of nuclear reactions has given powerful support to the idea that the specifically nuclear forces exerted by protons and neutrons are the same. By producing excited states of two nuclei with the same total number of nucleons, but with different Z (e.g. $_6C^{13}$ and $_7N^{13}$), and by mapping out their separate energy level structures, it

The Nucleus 313

has been shown that such nuclei have a far-reaching correspondence in their quantum states. Investigations of this type are made possible through the great variety of rearrangement collisions that can be brought about by nuclear bombardments. In this way, too, we have come to acquire a quite detailed knowledge of large numbers of nuclei that have no stable existence in nature.

(2) The Optical Model. Scattering

If we look again at Fig. 9.25, we can see that, although the resonances are unmistakable, they do overlap appreciably. For lower energies of excitation of the F^{18}, because of the tendency toward lower values of Γ, the resonances would become much more clearly separated. Conversely, at higher energies, the increase of Γ and the reduction in mean spacing of levels will combine to blur the picture, so that we are always dealing with an ill-defined mixture of levels. Under these latter conditions it becomes physically reasonable, and theoretically convenient, to regard any nuclear collision as involving an average over so many compound levels that we can revert to the picture of a nucleus as a more or less structureless potential well (cf. Section 10 of this chapter). This simplified picture has proved particularly useful for the description of scattering processes at high energy; it has been named "the optical model."

In first introducing Schrödinger's equation (Chapter 7, Section 7), we pointed out how it was equivalent to the equation of a wave in a medium whose refractive index varies from place to place. A particle coming to a region of negative potential will have its kinetic energy increased and so will find its de Broglie wavelength shortened. The passage of a particle across such a region will therefore be analogous to the passage of light waves through a layer of material with refractive index greater than 1. When this picture is extended to three dimensions, we see that the behavior of a neutron (to take the simplest case), when it encounters the potential well of a nucleus, can be compared to the behavior of a plane wave of light falling on a small sphere of a refracting substance. This is the basis of the optical treatment of nuclear scattering processes.

To obtain some idea of the conditions, let us consider the scattering of neutrons of about 80 Mev. If we take the depth of the nuclear potential well to be about 40 Mev, the kinetic energy inside the nucleus of a neutron being scattered is about 120 Mev. Thus the neutron wavelength inside the nucleus is less, by a factor $\sqrt{3/2}$, than the wavelength outside; the nuclear material has an effective refractive index equal to this ratio: i.e. about 1.22. The neutron wavelength λ outside the nucleus is close to $3 \cdot 10^{-13}$ cm, which is appreciably less than (but still comparable with) the diameter of most nuclei. It is to be expected, therefore, that diffraction effects will be important (cf. Chapter 7, Section 4).

Figure 9.27 shows the Fraunhofer type of diffraction pattern for scattering by copper of neutrons of about the energy we have considered. Now, for copper, $A = 63$, $A^{1/3} \approx 4$, and so the nuclear diameter d is close to $1.2 \cdot 10^{-12}$ cm. The detailed theory of the scattering shows that it is nearly correct to put

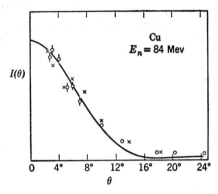

Figure 9.27. The optical type of diffraction scattering for 84-Mev neutrons scattered by copper [from Bratenahl et al., *Phys. Rev.*, **77**, 597 (1950)].

$$I(\theta) \propto \left[\frac{\sin (\pi d \sin \theta/\lambda)}{(\pi d \sin \theta/\lambda)} \right]^2 \quad (\theta \text{ small}, \lambda \ll \pi d)$$

where $I(\theta)$ is the intensity of neutrons scattered through an angle θ. This is precisely the same angular dependence as for diffraction by a slit of width d, and the first zero of the pattern would occur when $d \sin \theta = \lambda$. With our values for λ and d, this gives

$$\theta \approx \sin \theta = \frac{\lambda}{d} = \tfrac{1}{4} \text{ radian} \approx 15°$$

It may be seen that the experimental results bear this out very well.

The optical model (which might equally well be called "the acoustical model," since it owes an acknowledged debt to Rayleigh's treatment of the scattering of sound) has also been found useful for nuclear collision problems at much lower energies, provided that we deliberately average the observed effects over a large number of neighboring resonances. This carries the implication that the classification of nuclear reactions into two distinct types, which we have made in this section, is not so clear-cut as one might have supposed. As with so many physical theories, each picture is rather accurate and convincing over a certain range of conditions, but tends to become confused or modified outside that range.

Problems

9.1. Regarding the nucleus as containing a total charge Ze uniformly distributed within a sphere of radius R, show that: (a) The total electrostatic potential energy is $\frac{3}{5}Z^2e^2/R$. (Consider building up the charge distribution by bringing up successive concentric shells.) (b) The electrostatic potential at a point inside the charge cloud is given by

$$V(r) = \frac{Ze}{2R^3}(3R^2 - r^2)$$

9.2. Consider the electrostatic potential energy of a K electron attached to a nucleus of charge Z and radius R. The problem is to evaluate a quantity W defined by

$$W = \int_{r=0}^{\infty} e\, V(r)\, p(r)\, dr$$

where $p(r)\, dr$ is the probability of finding the electron between r and $r + dr$ (and is given in Appendix VII, provided that a_0 is replaced by a_0/Z). $V(r)$ is given by Problem 1b above for $r \leq R$, and is simply Ze/r for $r \geq R$. Carry out the integrations over these two regions separately, using as variable the dimensionless quantity $x = 2r/a$, where a is the radius of the first Bohr orbit of the atom in question ($= a_0/Z$). Approximate your answer by using the fact that $R/a \ll 1$, and hence show that

$$W(R) \approx \frac{Ze^2}{a}\left(1 - \frac{2}{5}\frac{R^2}{a^2}\right)$$

9.3. Make use of the result of Problem 9.2 to evaluate the difference of W for two isotopes having $Z = 62$ (Sm) and $A_1 = 145$, $A_2 = 155$. Convert this into a difference of wave numbers (cm^{-1}). This illustrates the type of hyperfine structure due to the isotopic effect (Section 4 of this chapter).

9.4. Calculate the resonant frequencies needed to cause "spin flip" of neutrons and protons in a magnetic field of 5000 gausses.

9.5. Making use of the binding energy curve (Fig. 9.4), calculate the total amount of energy that would be made available if (a) eight atoms of deuterium ($_1H^2$) could be combined into $_8O^{16}$, (b) two atoms of $_{10}Ne^{20}$ could be combined into $_{20}Ca^{40}$. Express the answers in kilocalories per g atom of the product, and compare the magnitude of the answers with the energy releases typical of chemical reactions.

9.6. Alpha particles with a speed of $2 \cdot 10^9$ cm/sec are used to bombard gold nuclei ($Z = 79$) which can be regarded as infinitely heavy. (a) Find the closest distance of approach of an alpha particle to a gold nucleus in a head-on collision. (b) Find the angle through which an alpha particle is deflected in a single encounter with a gold nucleus if the continuation of its initial path would, in the absence of any deflection, miss the center of the nucleus by $5 \cdot 10^{-13}$ cm. (Refer to Appendix V.)

9.7. Making use of the parameters quoted in Section 9 of this chapter, make a graph of the mathematically defined atomic number Z_A as a function of A. (Remember that R, in the formula for Z_A as given in the text, is itself a function of A.) Compare the result with Fig. 9.1. Mark in on your graph a selection of points representing known isotopes. ($M_n - M_H = 0.782$ Mev.)

9.8. In Section 10 of this chapter, we discuss the deflection through an angle of the order of V_0/T for an energetic particle traversing a nucleus. If θ is the deflection, we have, approximately,

$$\theta = \frac{V_0}{T} \tan \alpha \qquad \text{(Fig. 9.11)}$$

It is instructive to carry the problem further to arrive at a differential cross section for scattering into unit solid angle at θ (cf. Appendix V). We have, approximately,

$$\frac{d\sigma}{d\Omega}(\theta) = \frac{2\pi p \, dp}{2\pi \theta \, d\theta}$$

where p is the impact parameter and is equal to $R \sin \alpha$. Proceed from this to the result

$$\frac{d\sigma}{d\Omega}(\theta) \approx R^2 \frac{(V_0/T)^2}{[(V_0/T)^2 + \theta^2]^2}$$

and make a graph of this as a function of θ for $T = 2V_0$.

9.9. The separation energies S_p and S_n for neutrons and protons from many nuclei are nearly equal at about 8 Mev. Make use of this fact to draw the nuclear potential as seen by the most energetic proton and neutron in a nucleus with $Z = 50$, $A = 120$. Take account of the difference between neutron and proton densities N/V and Z/V in the theory of Section 11.

9.10. For nucleons in a rectangular potential well one can say that, in general, (a) for a given value of the principal quantum number, the levels of larger l have more negative energies than those of smaller l, (b) the spin-orbit splitting may be assumed proportional to $(2l + 1)$, (c) the level of larger j lies lower for a given value of l. With the help of these principles, arrange the groups of levels in Fig. 9.15 in a plausible manner. [Then check the result with a discussion of the shell model—e.g. M.H.L. Pryce, *Repts. Progr. in Phys.*, **17**, 1 (1954).]

9.11. Construct a table similar to Table 9.4 of the text, but taking $_{92}U^{235}$ instead of $_{92}U^{238}$ as the assumed parent atom. Check your findings against a standard compilation of nuclear data.

9.12. A radioactive source is known to contain a mixture of two unrelated radioactive substances. The total counting rate as a function of time is given in the accompanying table. By making a suitable semilogarithmic plot of

t, min	Count	t, min	Count
0	20,000	8	2200
1	14,260	10	1420
2	10,380	12	932
3	7,720	14	617
4	5,840	16	408
5	4,500	18	273
6	3,510	20	183

these data, find the decay constant and half-life for each substance, and the counting rate due to each at $t = 2.5$ min.

9.13. Calculate the energy release at the initial and subsequent stages of a fission of U^{238} that breaks up unsymmetrically into $_{39}Y^{96}$ and $_{53}I^{140}$, with the simultaneous emission of two neutrons. (Use the empirical mass formula.)

9.14. The following are some atomic mass values in amu:

$_0n^1$	1.008986	$_9F^{18}$	18.006646
$_1H^1$	1.008145	$_9F^{19}$	19.004448
$_1H^2$	2.014740	$_{10}Ne^{19}$	19.007945
$_2He^4$	4.003874	$_{10}Ne^{20}$	19.998769
	$_8O^{16}$ (standard) = 16.000000		

Discuss what nuclear reactions may occur when protons of variable energy are used to bombard a target of F^{19}.

Appendix

THE EVALUATION OF MEAN VELOCITY AND MEAN-SQUARE VELOCITY IN KINETIC THEORY

The Maxwell velocity distribution can be written

$$f(v) = Av^2 \exp(-mv^2/2kT)$$

where $f(v)\,dv$ is the number of molecules between v and $v + dv$, irrespective of direction. The average of any power of the velocity is therefore defined as follows:

$$\overline{v^n} = \frac{\int_0^\infty v^n \cdot v^2 e^{-\alpha v^2}\,dv}{\int_0^\infty v^2 e^{-\alpha v^2}\,dv} \qquad (\alpha = m/2kT)$$

Let us look first at the denominator of this expression. Integrating by parts, we have

$$\int_0^\infty v^2 e^{-\alpha v^2}\,dv = \left[-\frac{1}{2\alpha} v e^{-\alpha v^2}\right]_0^\infty + \frac{1}{2\alpha}\int_0^\infty e^{-\alpha v^2}\,dv$$

$$= \frac{1}{2\alpha}\int_0^\infty e^{-\alpha v^2}\,dv \quad \text{simply}$$

The integral $I = \int_0^\infty e^{-\alpha v^2}\,dv$ is the "error integral" of probability

Appendix I. 319

theory and is evaluated by a special device. We put

$$I = \int_0^\infty e^{-\alpha x^2}\, dx = \int_0^\infty e^{-\alpha y^2}\, dy$$

where x and y are independent of each other and can be used separately in the evaluation of the definite integral. It is then possible to write

$$I^2 = \int_{x=0}^\infty \int_{y=0}^\infty e^{-\alpha(x^2+y^2)}\, dx\, dy$$

We treat x and y as rectangular co-ordinates spanning the positive

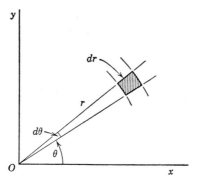

Figure A1.1. Two-dimensional space spanned by rectangular or polar co-ordinates.

quadrant (Fig. A1.1). The same space can be spanned by introducing polar co-ordinates (r, θ) as shown, with the substitutions

$$x^2 + y^2 = r^2$$

$$dx\, dy \equiv r\, dr\, d\theta$$

Hence
$$I^2 = \int_{\theta=0}^{\pi/2} \int_{r=0}^\infty e^{-\alpha r^2} r\, dr\, d\theta$$

$$= \frac{\pi}{4\alpha}$$

$$\therefore\ I = \frac{1}{2}\left(\frac{\pi}{\alpha}\right)^{1/2}$$

and
$$\int_0^\infty v^2 e^{-\alpha v^2}\, dv = \frac{\pi^{1/2}}{4\alpha^{3/2}}$$

We can now proceed to the evaluation of $\bar{v}$ and $\overline{v^2}$.

(1) Mean Velocity

From the above we have

$$\bar{v} = \frac{4\alpha^{3/2}}{\pi^{1/2}} \int_0^\infty v^2 e^{-\alpha v^2} v \, dv$$

$$= -\frac{2}{\pi^{1/2} \alpha^{1/2}} \int_0^\infty (-\alpha v^2) e^{-\alpha v^2} \, d(-\alpha v^2)$$

The value of the definite integral is -1. Hence

$$\bar{v} = \frac{2}{\pi^{1/2}} \left(\frac{2kT}{m} \right)^{1/2}$$

(2) Mean-Square Velocity

In this case we notice the following convenient relation:

$$\int_0^\infty v^2 e^{-\alpha v^2} \, dv = \frac{\pi^{1/2}}{4} \alpha^{-3/2}$$

$$-\frac{\partial}{\partial \alpha} \int_0^\infty v^2 e^{-\alpha v^2} \, dv = \int_0^\infty v^2 \cdot v^2 e^{-\alpha v^2} \, dv$$

$$\therefore \int_0^\infty v^2 \cdot v^2 e^{-\alpha v^2} \, dv = \frac{3\pi^{1/2}}{8\alpha^{5/2}}$$

whence $\quad \overline{v^2} = \dfrac{3\pi^{1/2}}{8\alpha^{5/2}} \cdot \dfrac{4\alpha^{3/2}}{\pi^{1/2}} = \dfrac{3}{2\alpha} = \dfrac{3}{2} \left(\dfrac{2kT}{m} \right)$

Hence $\quad c = (\overline{v^2})^{1/2} = \dfrac{3^{1/2}}{2^{1/2}} \left(\dfrac{2kT}{m} \right)^{1/2}$

We see that $\quad \bar{v} = \left(\dfrac{8}{3\pi} \right)^{1/2} c = 0.921c$

Appendix

THE LARMOR PRECESSION

It is instructive to consider what happens to a charged particle in an orbit as a magnetic field is applied. This is the subject of the classical Zeeman effect (Chapter 3, Section 3). It is not obvious that the field

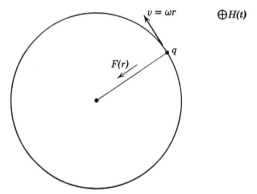

Figure A2.1. To illustrate the Larmor precession theorem.

will simply change the angular velocity ω of the motion without affecting the orbit radius r, but we shall show that this is true to a first approximation.

The field rises from zero to a final value H. We shall suppose the field at some stage to be $H(t)$, and the orbit radius to be r (Fig. A2.1).

The charged particle experiences a tangential electric field E, given by the induction law:

$$2\pi r E = \frac{1}{c}\frac{d}{dt}(\pi r^2 H)$$

The tangential acceleration is thus given by the equation

$$\frac{1}{r} m \frac{d}{dt}(r^2\omega) = qE = \frac{q}{2cr}\frac{d}{dt}[r^2 H(t)]$$

This can be integrated immediately to give

$$r^2\omega = r_0^2\omega_0 + \frac{q}{2mc}r^2 H$$

where r_0, ω_0 are the initial orbit radius and frequency. Thus

$$\omega = \frac{qH}{2mc} + \omega_0 \frac{r_0^2}{r^2}$$

Let us now put $\omega = \omega_0 + \delta\omega$, $r = r_0 + \delta r$.
Then approximately,

$$\omega_0 + \delta\omega = \frac{qH}{2mc} + \omega_0\left(1 - \frac{2\delta r}{r_0}\right)$$

and so

$$\delta\omega = \frac{qH}{2mc} - 2\omega_0 \frac{\delta r}{r_0} \qquad (A2.1)$$

Thus a first-order change of r would cause $\delta\omega$ to depart seriously from the value $qH/2mc$.

Now consider the radial motion. If the central force is $-F(r)$, directed toward the center of the orbit, we have

$$m\ddot{r} - mr\omega^2 = -F(r) - \frac{q}{c}H\omega r$$

In the initial and final states we must have $\ddot{r} = 0$. Hence

$$mr_0\omega_0^2 = F(r_0) \qquad (A2.2)$$

and

$$mr\omega^2 = F(r) + \frac{qH}{c}\omega r$$

Appendix II

In the second of these two equations we replace $F(r)$ by

$$F(r_0) + \frac{dF}{dr} \delta r$$

We also substitute for ω and r as before. Then

$$m(r_0 + \delta r)(\omega_0^2 + 2\omega_0 \delta\omega) \approx F(r_0) + \frac{dF}{dr} \delta r + \frac{qH}{c}(\omega_0 + \delta\omega)(r_0 + \delta r)$$

Multiplying this out, and dropping small terms, we find

$$m\omega_0^2 r_0 + m\omega_0^2 \delta r + 2m\omega_0 r_0 \delta\omega \approx F(r_0) + \frac{dF}{dr} \delta r + \frac{qH}{c} \omega_0 r_0 \quad (A2.3)$$

Subtracting (A2.2) from (A2.3) and dividing through by $m\omega_0 r_0$, we find

$$\omega_0 \frac{\delta r}{r_0} + 2\delta\omega \approx \frac{1}{m\omega_0} \frac{dF}{dr} \cdot \frac{\delta r}{r_0} + \frac{qH}{mc} \quad (A2.4)$$

Substituting for $\omega_0 \delta r/r_0$ from (A2.1), we have

$$\left(\frac{qH}{4mc} - \tfrac{1}{2}\delta\omega\right) + 2\delta\omega \approx \frac{1}{m\omega_0^2} \frac{dF}{dr}\left(\frac{qH}{4mc} - \tfrac{1}{2}\delta\omega\right) + \frac{qH}{mc}$$

Hence

$$\delta\omega\left(1 + \frac{1}{3m\omega_0^2}\frac{dF}{dr}\right) \approx \frac{qH}{2mc}\left(1 + \frac{1}{3m\omega_0^2}\frac{dF}{dr}\right)$$

and so

$$\delta\omega \approx \frac{qH}{2mc} \quad \text{simply} \quad (A2.5)$$

Substituting this value back into (A2.1), we see that

$$\frac{\delta r}{r_0} = 0$$

The angular velocity change thus takes place without any change of orbit radius. This corresponds to the precession theorem of Larmor, according to which the motion after application of a magnetic field H looks the same as the original motion if it is viewed from a system of co-ordinates rotating with the angular velocity $qH/2mc$. But this latter statement of the problem is not really complete unless it is also shown how the transition from the initial state to the final state takes place.

III | *Appendix*

RESONANCE PROCESSES. DISPERSION AND ABSORPTION

In this appendix we collect together some results that have been mentioned or used in connection with anomalous dispersion (Chapter 3, Section 7), absorption by atomic oscillators (Chapter 4, Section 8), and resonance processes in nuclear physics (Chapter 9, Section 18).

We are concerned with systems that have an equation of motion of the type

$$\ddot{z} + 2k\dot{z} + \omega_0^2 z = f_0 e^{j\omega t}$$

under the action of a driving agency f_0 of frequency ω. The solution as outlined in Chapter 4, Section 8, is

$$z = z_0 e^{j(\omega t - \delta)}$$

with $\quad z_0 = \dfrac{f_0}{[(\omega_0^2 - \omega^2)^2 + (2k\omega)^2]^{1/2}}, \qquad \tan \delta = \dfrac{2k\omega}{\omega_0^2 - \omega^2}$

We can put

$$z = z_0 e^{-j\delta} e^{j\omega t} = (z_0 \cos \delta) e^{j\omega t} - j(z_0 \sin \delta) e^{j\omega t}$$

Thus z has an "in-phase" component A_p and a "quadrature" component $-A_q$, such that

$$A_p = z_0 \cos \delta = f_0 \frac{\omega_0^2 - \omega^2}{(\omega_0^2 - \omega^2)^2 + (2k\omega)^2}$$

$$A_q = z_0 \sin \delta = f_0 \frac{2k\omega}{(\omega_0^2 - \omega^2)^2 + (2k\omega)^2}$$

Scattering and dispersion phenomena are governed by A_p, and absorption phenomena are governed by ωA_q. These are shown as functions of driving frequency in Fig. A3.1. It is easy to verify that the turning points of A_p occur at $\omega \approx \omega_0 \pm k$; at these frequencies also the absorbed power (proportional to ωA_q) has half the peak value. The

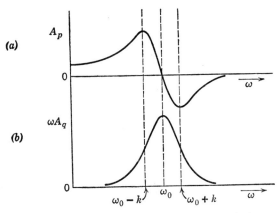

Figure A3.1. (a) A typical dispersion curve. (b) A typical resonance absorption curve.

quantity $2k$ thus characterizes the width of the resonant system for both scattering and absorption.

A typical dispersion curve for a transparent material will show a series of anomalies like Fig. A3.1a if the refractive index is plotted against frequency or wavelength. At each resonant frequency ω_0 there will be a strong absorption.

Similar behavior is found in nuclear scattering and resonance processes. The probability of scattering of neutrons, for example, by a given nucleus will often be observed to vary with neutron energy in a way that can be well described by an anomalous dispersion curve, and the probability of neutron capture will pass through a maximum at the corresponding neutron energy E_0. The equivalence with classical vibrating systems is not exact, but the resonance factor for absorption is given by

$$F(E) = \frac{(\Gamma/2)^2}{(E - E_0)^2 + (\Gamma/2)^2}$$

As written here, $F(E)$ is normalized to unity at $E = E_0$, and the probability of absorption falls to one-half the peak value for $E = E_0 \pm \tfrac{1}{2}\Gamma$. Thus Γ (the energy width of a nuclear resonance) bears essentially the same relation to the resonance energy E_0 that $2k$ (the frequency width) bears to the resonance frequency ω_0.

The effective area presented by a nucleus to an incoming particle is also proportional to $\pi\lambda^2$, where λ is $1/2\pi$ times the de Broglie wavelength of the incident particle (the motion being referred to the center of mass of the system). At the resonance energy, therefore, the cross section may be given essentially by $\pi\lambda_0^2$ [since $F(E_0) = 1$]. It may be noted that this result has its analog in the effective area presented by an atom to light of the resonant frequency in the photoelectric effect (cf. Chapter 5, Section 1). Since λ is inversely proportional to velocity, the effective collision area for a resonance collision at very low energy may be enormously greater than the true nuclear cross section πR^2. This is the basis of the great importance of thermal neutrons in causing nuclear reactions. Such neutrons have wavelengths of the order of 10^{-8} cm (cf. Chapter 7, Section 4), so that we may in principle have a resonance cross section of the order of 10^{-16} cm^2, as against a true nuclear cross section of the order of 10^{-24} cm^2. No example as extreme as this has been found, but there is at least one known case of a neutron capture cross section about one million times the geometrical cross section πR^2.

IV | *Appendix*

THE INTEGRAL IN PLANCK'S RADIATION FORMULA

Planck's formula (Chapter 4, Section 11) requires the evaluation of a certain definite integral. One possible method is given below:

$$I = \int_0^\infty \frac{x^3\, dx}{e^x - 1} = \int_0^\infty x^3 e^{-x}(1 - e^{-x})^{-1}\, dx$$

$$= \int_0^\infty x^3 e^{-x}(1 + e^{-x} + e^{-2x} + \cdots)\, dx$$

$$= \sum_{n=1}^\infty \int_0^\infty x^3 e^{-nx}\, dx$$

Integration by parts shows that

$$\int_0^\infty x^3 e^{-nx}\, dx = \frac{6}{n^4}$$

Hence

$$I = 6 \sum_{n=1}^\infty \frac{1}{n^4} = 6 S_4 \quad \text{say}$$

To carry out the summation of $1/n^4$, we make use of the Fourier analysis of the "sawtooth" function shown in Fig. A4.1. This is

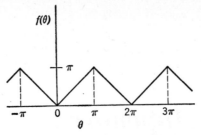

Figure A4.1. A sawtooth function.

defined by the conditions

$f(\theta) = \theta$ $0 < \theta < \pi$ (and similarly in similar intervals)

$f(\theta) = 2\pi - \theta$ $\pi < \theta < 2\pi$ (and similarly in similar intervals)

Since $f(\theta)$ is an even function of θ, it can be expressed in the form

$$f(\theta) = \sum_n b_n \cos n\theta$$

For $n = 0$, we have

$$b_0 = \frac{1}{2\pi} \int_0^{2\pi} f(\theta)\, d\theta = \frac{\pi}{2}$$

For all other n,

$$b_n \int_0^{2\pi} \cos^2 n\theta\, d\theta = \int_0^{\pi} \theta \cos n\theta\, d\theta + \int_{\pi}^{2\pi} (2\pi - \theta) \cos n\theta\, d\theta$$

whence

$$\pi b_n = \frac{1}{n^2} [\cos n\theta]_0^{\pi} - \frac{1}{n^2} [\cos n\theta]_{\pi}^{2\pi}$$

Thus, for n even, $b_n = 0$, and, for n odd, $b_n = -4/\pi n^2$. Hence

$$f(\theta) = \frac{\pi}{2} - \frac{4}{\pi} \sum_{n\text{ odd}} \frac{\cos n\theta}{n^2}$$

In the range $0 < \theta < \pi$ we can thus put

$$\frac{\pi}{2} - \theta = \frac{4}{\pi} \sum_{n\text{ odd}} \frac{\cos n\theta}{n^2}$$

Appendix IV

Let us square both sides of this equation:

$$\left(\frac{\pi}{2} - \theta\right)^2 = \frac{16}{\pi^2} \sum_{m\ \text{odd}} \sum_{n\ \text{odd}} \frac{\cos m\theta \cos n\theta}{m^2 n^2}$$

i.e., $\quad \dfrac{\pi^2}{4} - \pi\theta + \theta^2$

$$= \frac{8}{\pi^2} \sum_{m\ \text{odd}} \sum_{n\ \text{odd}} \frac{1}{m^2 n^2} [\cos(m-n)\theta + \cos(m+n)\theta]$$

If this last equation is now integrated over the range $\theta = 0$ to π, all integrals on the right are of the form $[(\sin px)/p]_0^\pi$, which will be zero unless $p = 0$. Hence we have

$$\frac{\pi^3}{4} - \frac{\pi^3}{2} + \frac{\pi^3}{3} = \frac{8}{\pi^2} \sum_{n\ \text{odd}} \frac{1}{n^4} \int_0^\pi d\theta$$

and so

$$\sum_{n\ \text{odd}} \frac{1}{n^4} = \frac{\pi^4}{96}$$

But $\quad S_4 = \displaystyle\sum_{\text{all}\ n} \frac{1}{n^4} = \left(1 + \frac{1}{3^4} + \frac{1}{5^4} + \cdots\right)$

$$+ \left(\frac{1}{2^4} + \frac{1}{4^4} + \frac{1}{6^4} + \cdots\right)$$

$$= \sum_{n\ \text{odd}} \frac{1}{n^4} + \tfrac{1}{16} S_4$$

$\therefore \quad \tfrac{15}{16} S_4 = \dfrac{\pi^4}{96}, \qquad S_4 = \dfrac{\pi^4}{90}$

and so, finally,

$$\int_0^\infty \frac{x^3\, dx}{e^x - 1} = \frac{\pi^4}{15}$$

V | Appendix

COULOMB SCATTERING: THE RUTHERFORD FORMULA

(1) The Hyperbolic Orbit

We shall consider the scattering of a point particle of positive charge q, mass m, by a scattering center (heavy nucleus) of charge Q and effectively infinite mass. We suppose that the initial path of the light particle is along a line AO (Fig. A5.1) that would miss the heavy nucleus by the perpendicular distance $QN\ (=p)$. Let the velocity

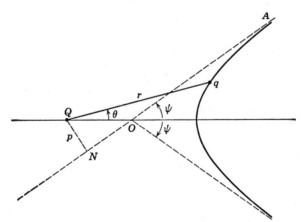

Figure A5.1. A hyperbolic orbit under an inverse square force of repulsion.

Appendix V

of q at large distance from O be v. Then the angular momentum about Q (which must be conserved since the Coulomb force is a central force) is given by

$$L = mvp \qquad (A5.1)$$

Let the position of q be described by polar co-ordinates (r, θ) as shown. Then the angular momentum at any arbitrary point is given by

$$L = mr^2\dot\theta$$

As is usual in central orbit problems we introduce $u = 1/r$. Then

$$\dot\theta = \frac{L}{m}u^2$$

The radial velocity $\dot r$ is then given by

$$\dot r = \frac{d}{dt}\left(\frac{1}{u}\right) = -\frac{1}{u^2}\frac{du}{dt} = -\frac{1}{u^2}\dot\theta\frac{du}{d\theta}$$

Thus

$$\dot r = -\frac{L}{m}\frac{du}{d\theta}$$

The transverse velocity is $r\dot\theta = (L/m)u$. The Coulomb potential energy is $qQ/r = qQu$. Hence, by conservation of energy,

$$\tfrac{1}{2}mv^2 = \frac{L^2}{2mp^2} = \tfrac{1}{2}m\left[\left(\frac{L}{m}\frac{du}{d\theta}\right)^2 + \left(\frac{L}{m}u\right)^2\right] + qQu$$

and so

$$\left(\frac{du}{d\theta}\right)^2 = \frac{1}{p^2} - \frac{2mqQ}{L^2}u - u^2 \qquad (A5.2)$$

The coefficient of u in (A5.2) can be put into a more convenient form by introducing the "distance of closest approach," b. This is defined by a head-on collision such that all the initial kinetic energy E can be stored up as Coulomb potential energy:

$$E = \tfrac{1}{2}mv^2 = \frac{qQ}{b} \qquad (A5.3)$$

From (A5.1), (A5.2), and (A5.3) we then have

$$\frac{2mqQ}{L^2} = \frac{2qQ}{mv^2p^2} = \frac{qQ}{Ep^2} = \frac{b}{p^2}$$

Thus (A5.2) can be written

$$\left(\frac{du}{d\theta}\right)^2 = \frac{1}{p^4}\left[p^2 + \left(\frac{b}{2}\right)^2\right] - \left(u + \frac{b}{2p^2}\right)^2$$

The solution to this equation is a hyperbola defined by

$$u + \frac{b}{2p^2} = \frac{1}{p^2}\left[p^2 + \left(\frac{b}{2}\right)^2\right]^{1/2} \cos\theta \tag{A5.4}$$

Now, for $u \to 0$ ($r \to \infty$), we have $\theta \to \pm\psi$

$$\therefore \quad \cos\psi = \frac{b/2}{[p^2 + (b/2)^2]^{1/2}}$$

whence
$$\tan\psi = \frac{2p}{b} \tag{A5.5}$$

From the geometry of the triangle ONQ in Fig. A5.1 we see that ON is equal to $b/2$.

The total angle ϕ through which q is deflected is the angle between the asymptotes of the hyperbola and is given by

$$\phi = \pi - 2\psi = \pi - 2\arctan(2p/b) \tag{A5.6}$$

(2) The Rutherford Scattering Cross Section

We now consider what happens if a collimated beam of particles q is directed at a "target" of particles Q. We have no control over the value of the "impact parameter" p in any individual collision, and so must consider an averaging process.

Now from (A5.6) we have

$$p = \frac{b}{2}\cot\frac{\phi}{2} \tag{A5.7}$$

If collisions occur for values of impact parameter within a range dp at p, they will result in deflections within the angular range $d\phi$ at ϕ as defined by the above equation. But the probability of such collisions is proportional to the area of an annulus of radius p and width dp, since this is the effective target area presented by Q to the incident beam for producing deflections of this particular type (see Fig. A5.2).

Appendix V

We can therefore introduce a collision area $d\sigma(\phi)$ for scattering into $d\phi$ at ϕ:

$$d\sigma(\phi) = 2\pi p \, dp$$

$$= -\frac{\pi b^2}{4} \cdot \frac{\cos(\phi/2)}{\sin^3(\phi/2)} d\phi \quad (A5.8)$$

(The minus sign simply points to the fact that an increase of p leads to a decrease of ϕ.) With the help of (A5.3) we thus see that the

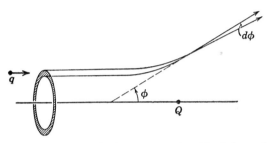

Figure A5.2. Illustrating the effective cross section (shaded annulus) for scattering between ϕ and $\phi + d\phi$.

probability $dw(\phi)$ of scattering into $d\phi$ at ϕ is given by

$$dw(\phi) \sim \frac{q^2 Q^2}{E^2} \cdot \frac{\cos(\phi/2)}{\sin^3(\phi/2)} d\phi \quad (A5.9)$$

Putting $Q = Ze$, $E^2 \sim V^4$, we arrive at the scattering formula as quoted in Chapter 5, Section 3.

In actual experiments it is usually the scattering into a detector subtending a fixed solid angle that is observed. For this purpose we introduce the *differential cross section,* which is the collision area effective for producing scattering into a unit solid angle at ϕ. If we write an element of solid angle as $d\Omega$, we have

$$d\Omega = 2\pi \sin\phi \, d\phi$$
$$= 4\pi \sin(\phi/2) \cos(\phi/2) \, d\phi$$

Hence, from (A5.8):

$$\frac{d\sigma}{d\Omega}(\phi) = \frac{b^2}{16} \cdot \frac{1}{\sin^4(\phi/2)} = \frac{q^2 Q^2}{16 E^2} \cdot \frac{1}{\sin^4(\phi/2)} \quad (A5.10)$$

This is the final form in which the Rutherford scattering formula is usually given.

VI | *Appendix*

USE OF THE CALIBRATION CURVES IN SPECIAL RELATIVITY

(1) Geometry of the Calibration Curves

We have defined

$$\xi = x + ct$$
$$\eta = x - ct$$

i.e.
$$x = \tfrac{1}{2}(\xi + \eta)$$
$$ct = \tfrac{1}{2}(\xi - \eta)$$

Now the calibration curves (Fig. A6.1) are given by

$$\xi\eta = \pm 1$$

$$\therefore \quad \frac{d\xi}{d\eta} = \mp \frac{\xi}{\eta}$$

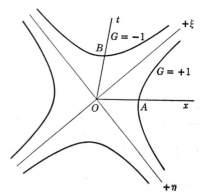

Figure A6.1. The calibration hyperbolas $\xi\eta = \pm 1$.

For the point A, $t = 0$; therefore $\xi = \eta$.

∴ Slope of tangent at $A = -1$

∴ Equation of line through O parallel to tangent is $\xi = -\eta$

i.e. $$\xi + \eta = 0$$

Appendix VI

But this represents the line $x = 0$, and so is identical with the t axis. The converse relation clearly holds also.

The x and t axes make equal angles with the ξ axis, since for OA we have $\eta = +\xi$ and for OB we have $\eta = -\xi$.

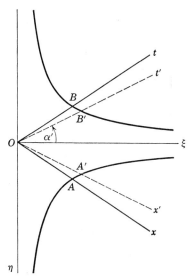

Figure A6.2. Transformation from $S(x, t)$ to $S'(x', t')$.

To transform from system S to system S' we notice that the t' axis is defined by the equation

$$x - vt = 0$$

$$\therefore \quad \xi + \eta = \frac{v}{c}(\xi - \eta)$$

i.e.
$$\frac{\xi}{\eta} = -\frac{1 + \beta}{1 - \beta} = \cot \alpha' \quad \text{say}$$

Thus, for the x' axis in Fig. A6.2,

$$\frac{\xi}{\eta} = -\cot \alpha' = \frac{1 + \beta}{1 - \beta}$$

$$\therefore \quad \frac{x + ct}{x - ct} = \frac{1 + v/c}{1 - v/c}$$

whence
$$t - \frac{vx}{c^2} = 0$$

Since the x' axis is also described by $t' = 0$, we see that

$$t' \propto t - \frac{vx}{c^2}$$

(2) Geometry of the Lorentz Transformations

LENGTH CONTRACTION. The point A (Fig. A6.3) is $\xi_0 = \eta_0 = 1$.

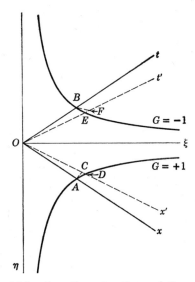

Figure A6.3. Length contraction and time dilation.

The slope of AC is -1. Therefore the equation of AC is

$$\xi - 1 = -(\eta - 1)$$
$$\therefore \quad \xi + \eta = 2$$

The equation of the x' axis is

$$\frac{\xi}{\eta} = \frac{1 + \beta}{1 - \beta} \qquad \text{(previous section)}$$

Therefore, at C,

$$\xi(1 - \beta) = (2 - \xi)(1 + \beta)$$

whence

$$\left. \begin{array}{l} \xi = 1 + \beta \\ \eta = 1 - \beta \end{array} \right\} \quad \therefore \quad OC^2 = 2(1 + \beta^2)$$

Appendix VI

At D,
$$\xi\eta = +1$$
$$\frac{\xi}{\eta} = \frac{1+\beta}{1-\beta}$$

$$\therefore \quad \left.\begin{aligned}\xi &= \left(\frac{1+\beta}{1-\beta}\right)^{1/2} \\ \eta &= \left(\frac{1-\beta}{1+\beta}\right)^{1/2}\end{aligned}\right\} \quad \therefore \quad OD^2 = \frac{2(1+\beta^2)}{1-\beta^2}$$

$$\therefore \quad OC\ (\equiv OA) = (1-\beta^2)^{1/2} \quad \text{when measured in } S'$$

TIME DILATION. The t' axis (Fig. A6.3) has the equation

$$\frac{\xi}{\eta} = -\frac{1+\beta}{1-\beta} \quad \text{(previous section)}$$

The point B is $\xi_0 = -\eta_0 = 1$. The slope of BF (parallel to tangent at E) is $+(1+\beta)/(1-\beta)$. Therefore the equation of BF is

$$\xi - 1 = \frac{1+\beta}{1-\beta}(\eta + 1)$$

Therefore, at F, $\quad \xi - 1 = -\xi + \dfrac{1+\beta}{1-\beta}$

whence
$$\left.\begin{aligned}\xi &= \frac{1}{1-\beta} \\ \eta &= \frac{-1}{1+\beta}\end{aligned}\right\} \quad \therefore \quad OF^2 = \frac{2(1+\beta^2)}{(1-\beta^2)^2}$$

At E, since $\xi\eta = -1$ and $\xi/\eta = -(1+\beta)/(1-\beta)$,

$$\left.\begin{aligned}\xi &= \left(\frac{1+\beta}{1-\beta}\right)^{1/2} \\ \eta &= -\left(\frac{1-\beta}{1+\beta}\right)^{1/2}\end{aligned}\right\} \quad \therefore \quad OE^2 = \frac{2(1+\beta^2)}{1-\beta^2}$$

$$\therefore \quad OF\ (\equiv OB) = \frac{OE}{(1-\beta^2)^{1/2}}$$

VII | *Appendix*

THE GROUND STATE OF THE HYDROGEN ATOM

The radial factor $R(r)$ in the total wave function satisfies the differential equation (Chapter 8, Section 1)

$$\frac{1}{r^2}\frac{d}{dr}\left(r^2\frac{dR}{dr}\right) + \frac{8\pi^2 m}{h^2}\left[E + \frac{Ze^2}{r} - \frac{l(l+1)h^2}{8\pi^2 mr^2}\right]R = 0$$

For the particular case $Z = 1$, $l = 0$, this gives

$$\frac{1}{r^2}\frac{d}{dr}\left(r^2\frac{dR}{dr}\right) + \left(A + \frac{B}{r}\right)R = 0 \qquad (A7.1)$$

where

$$A = \frac{8\pi^2 mE}{h^2} = \frac{2mE}{\hbar^2}, \qquad B = \frac{8\pi^2 me^2}{h^2} = \frac{2me^2}{\hbar^2}$$

With the substitution $R = S/r$, we have the simpler equation

$$\frac{d^2 S}{dr^2} + \left(A + \frac{B}{r}\right)S = 0 \qquad (A7.2)$$

Either (A7.1) or (A7.2) can be used to obtain the radial wave function for an s state (i.e. for $l = 0$).

It may be verified that a solution of the form

$$S(r) = re^{-ar}$$

Appendix VII

satisfies equation (A7.2). Substituting, we obtain

$$(\alpha^2 r - 2\alpha)e^{-\alpha r} + (Ar + B)e^{-\alpha r} = 0$$

which gives the conditions

$$A = -\alpha^2$$
$$B = 2\alpha \qquad (A7.3)$$

We infer that

$$\alpha = \frac{me^2}{\hbar^2}$$

and hence

$$A = \frac{2mE}{\hbar^2} = -\frac{m^2 e^4}{\hbar^4}$$

Thus we find

$$E = -\frac{me^4}{2\hbar^2} = -\frac{2\pi^2 m e^4}{h^2} \qquad (A7.4)$$

This corresponds to the case $n = 1$ in the simple Bohr theory (Chapter 5, Section 4, equation 5.4). Other solutions, more complicated than our assumed form for S, belong to higher n. The quantity $1/\alpha$ has the dimension of a length; it is in fact identical with the radius a_0 of the first Bohr orbit in the old quantum theory:

$$\frac{1}{\alpha} = a_0 = \frac{\hbar^2}{me^2} = \frac{h^2}{4\pi^2 m e^2} \qquad (A7.5)$$

There is no angular variation of wave-function amplitude in this lowest state, and so we may write

$$\psi(r) = Ce^{-\alpha r}$$

where C is a constant. The value of C is defined by the normalization condition:

$$\int \psi^* \psi \, d\tau = 1$$

i.e.,

$$C^2 \int_{r=0}^{\infty} e^{-2\alpha r} 4\pi r^2 \, dr = 1$$

The value of the integral may readily be shown to be π/α^3. Hence

$$\psi(r) = \frac{1}{(\pi a_0^3)^{1/2}} \exp(-r/a_0) \qquad (A7.6)$$

The probability of finding the electron between r and $r + dr$ is given by $p(r)\,dr$, where

$$p(r)\,dr = 4\pi r^2 \psi^* \psi\,dr$$

i.e., $$p(r) = \frac{4}{a_0^3} r^2 \exp(-2r/a_0) \qquad (A7.7)$$

The value of p may be seen to have a maximum at $r = a_0$; its over-all variation with r is shown in Fig. 8.3 (Chapter 8, Section 1).

VIII | *Appendix*

THE BOUND NEUTRON PROBLEM FOR $l = 0$

This problem is analogous to the bound electron problem treated in Appendix VII. We again suppose a radial wave function $R = S/r$, so that S satisfies the equation

$$\frac{d^2S}{dr^2} + \frac{8\pi^2 M}{h^2}[E - V(r)]S = 0$$

We must now distinguish two regions, because the assumed nuclear potential has a discontinuity at $r = r_0$ (Fig. A8.1):

For $0 \leqslant r < r_0$,
$V = -V_0, \qquad E - V(r) = V_0 - \varepsilon$

For $r \geqslant r_0$,
$V = 0, \qquad E - V(r) = -\varepsilon$

The energy $\varepsilon \ (= -E)$ is the binding energy of the neutron.

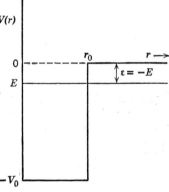

Figure A8.1. The radial variation of nuclear potential, with a weakly bound state of energy E $(= -\varepsilon)$.

The problem is essentially that of the particle in a box (Chapter 7, Section 11). We have the following solutions:

REGION 1 ($r < r_0$):

$$S_1(r) = Ae^{iKr} + Be^{-iKr} \qquad (A8.1)$$

with
$$K = \left[\frac{2M(V_0 - \varepsilon)}{\hbar^2}\right]^{1/2}$$

REGION 2 ($r > r_0$):

$$S_2(r) = Ce^{-\alpha r} \qquad (A8.2)$$

with
$$\alpha = \left(\frac{2M\varepsilon}{\hbar^2}\right)^{1/2}$$

The solutions must give a smooth join for S and dS/dr at $r = r_0$. We also require $S_1(0) = 0$, to keep $R(0)$ and hence $\psi(0)$ finite at $r = 0$. This latter condition requires $B = -A$ in (A8.1). Thus we have

$$S_1(r) = A_1 \sin Kr \qquad (A8.3)$$

For continuity at $r = r_0$, we then have

$$A_1 \sin Kr_0 = Ce^{-\alpha r_0}$$
$$KA_1 \cos Kr_0 = -\alpha Ce^{-\alpha r_0} \qquad (A8.4)$$

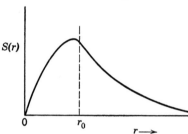

Figure A8.2. The function $S(r)$ corresponding to Figure A8.1. [Note that the radial wave function is $S(r)/r$.]

Equations (A8.4) define the permitted values of the energy, since by division we have

$$\tan Kr_0 = -\frac{K}{\alpha} \qquad (A8.5)$$

Equation (A8.5) can be solved graphically for ε if V_0 and r_0 are known.

The simplest possible type of bound state has a wave function of the form shown in Fig. A8.2, with only one maximum in the internal region. It is this kind of wave function that is characteristic of the deuteron ground state (Chapter 9, Section 11). If V_0 is made appreciably smaller than for this case, the angle Kr_0 remains less than $\pi/2$ at $r = r_0$; the wave function then fails to turn downward in the external region, and a bound state is no longer possible. On the other hand, increase of V_0 or increase of r_0 will encourage the formation of bound states.

Bibliography

GENERAL

*Blackwood, O. H., T. H. Osgood, A. E. Ruark, et al., *An Outline of Atomic Physics*, New York: John Wiley & Sons, 1955.
Bohr, N., *Atomic Theory and the Description of Nature*, Cambridge University Press, 1934.
Born, M., *Atomic Physics*, London: Blackie & Son, 1957.
Harnwell, G. P. and J. J. Livingood, *Experimental Atomic Physics*, New York: McGraw-Hill Book Co., 1933.
Joos, G. and I. M. Freeman, *Theoretical Physics*, London: Blackie & Son, 1951.
Kaplan, I., *Nuclear Physics*, Cambridge, Mass.: Addison-Wesley Press, 1955.
Peaslee, D. C. and H. Mueller, *Elements of Atomic Physics*, New York: Prentice-Hall, 1955.
Richtmeyer, F. K., E. H. Kennard, and T. Lauritsen, *Introduction to Modern Physics*, New York: McGraw-Hill Book Co., 1955.
*Semat, H., *Introduction to Atomic and Nuclear Physics*, New York: Rinehart & Co., 1954.
Shankland, R. S., *Atomic and Nuclear Physics*, New York: Macmillan Co., 1955.
*Stranathan, J. D., *The Particles of Modern Physics*, Philadelphia: Blakiston Co., 1942.
Whittaker, E. T., *A History of the Theories of Aether and Electricity*, London: Thomas Nelson & Sons, Vol. I, 1951; Vol. II, 1954.

CHAPTER 1

Einstein, A., *Investigations on the Theory of the Brownian Movement*, New York: Dover Publications, 1956.
Jeans, J. H., *Introduction to the Kinetic Theory of Gases*, Cambridge University Press, 1940.

* Introductory.

Kennard, E. H., *Kinetic Theory of Gases*, New York: McGraw-Hill Book Co., 1938.
Perrin, J. (trans. F. Soddy), *Brownian Motion and Molecular Reality*, London: Taylor and Francis, 1910.
Sears, F. W., *Thermodynamics*, Cambridge, Mass.: Addison-Wesley Press, 1953.

CHAPTER 2

Bitter, F., *Currents, Fields and Particles*, New York: The Technology Press, Massachusetts Institute of Technology, and John Wiley & Sons, 1956.
Ditchburn, R. W., *Light*, London: Blackie & Son, 1952.
Frank, N. H., *Introduction to Electricity and Optics*, New York: McGraw-Hill Book Co., 1950.
Jenkins, F. A. and H. E. White, *Fundamentals of Optics*, New York: McGraw-Hill Book Co., 1957.
Newton, I., *Opticks*, London: William Innys, 1730; republished by G. Bell & Sons, 1931.
Rossi, B., *Optics*, Cambridge, Mass.: Addison-Wesley Press, 1957.
Whittaker, E. T., *A History of the Theories of Aether and Electricity*, London: Thomas Nelson & Sons, Vol. I, 1951.

CHAPTER 3

Becker, R., *Theorie der Elektrizität, Vol. II (Elecktronentheorie)*, Ann Arbor: J. W. Edwards, Publishers, 1945.
Jenkins, F. A., and H. E. White, *Fundamentals of Optics*, New York: McGraw-Hill Book Co., 1957.
Lorentz, H. A., *The Theory of Electrons*, New York: Dover Publications, 1952.
Millikan, R. A., *Electrons, + and −*, Chicago: University of Chicago Press, 1947.
Stranathan, J. D., *The Particles of Modern Physics*, Philadelphia: Blakiston Co., 1942.
Thomson, J. J., and G. P. Thomson, *The Conduction of Electricity through Gases*, Cambridge University Press, 1928.

CHAPTER 4

Ditchburn, R. W., *Light*, London: Blackie & Son, 1952.
Gurney, R. W., *Introduction to Statistical Mechanics*, New York: McGraw-Hill Book Co., 1949.
Jeans, J. H., *The Dynamical Theory of Gases*, Cambridge University Press, 1925.
Lorentz, H. A., *The Theory of Electrons*, New York: Dover Publications, 1952.
Roberts, J. K. and A. R. Miller, *Heat and Thermodynamics*, London: Blackie & Son, 1951.

CHAPTER 5

Bohr, N., *The Theory of Spectra and Atomic Constitution*, Cambridge University Press, 1922.
Compton, A. H. and S. K. Allison, *X-Rays in Theory and Experiment*, New York: D. Van Nostrand Co., 1935.
Herzberg, G., *Atomic Spectra and Atomic Structure*, New York: Dover Publications, 1944.
Hughes, A. L. and L. A. DuBridge, *Photoelectric Phenomena*, New York: McGraw-Hill Book Co., 1932.

Bibliography 345

Meyer, C. F., *The Diffraction of Light, X-Rays and Material Particles*, Ann Arbor: J. W. Edwards, Publishers, 1949.
Ruark, A. E. and H. C. Urey, *Atoms, Molecules and Quanta*, New York: McGraw-Hill Book Co., 1930.
Rutherford, E., J. Chadwick, and C. D. Ellis, *Radiations from Radioactive Substances*, Cambridge University Press, 1930.
White, H. E., *An Introduction to Atomic Spectra*, New York: McGraw-Hill Book Co., 1934.

CHAPTER 6

Born, M., *Einstein's Theory of Relativity*, London: Methuen & Co., 1924.
Dingle, H., *The Special Theory of Relativity*, London: Methuen & Co., 1950.
Ditchburn, R. W., *Light*, London: Blackie & Son, 1952.
Eddington, A. S., *Space, Time and Gravitation*, Cambridge University Press, 1920.
Einstein, A., *The Meaning of Relativity*, Princeton: Princeton University Press, 1956.
Lorentz, H. A., A. Einstein, H. Minkowski, and H. Weyl, *The Principle of Relativity* (original papers), New York: Dover Publications, 1951.
McCrea, W. H., *Relativity Physics*, London: Methuen & Co., 1947.

CHAPTER 7

Beeching, R., *Electron Diffraction*, London: Methuen & Co., 1936.
D'Abro, A., *The New Physics* (2 vols.), New York: Dover Publications, 1951.
Flint, H. T., *Wave Mechanics*, London: Methuen & Co., 1953.
Frenkel, J., *Wave Mechanics: Elementary Theory*, Oxford: The Clarendon Press, 1936.
Heisenberg, W., *The Physical Principles of the Quantum Theory*, Chicago: University of Chicago Press, 1930.
Heitler, W., *Wave Mechanics*, Oxford: The Clarendon Press, 1956.
Meyer, C. F., *The Diffraction of Light, X-Rays, and Material Particles*, Ann Arbor: J. W. Edwards, Publishers, 1949.
Mott, N. F., *Elements of Wave Mechanics*, Cambridge University Press, 1952.
Pauling, L. and E. B. Wilson, *Introduction to Quantum Mechanics*, New York: McGraw-Hill Book Co., 1935.
Rojansky, V., *Introductory Quantum Mechanics*, New York: Prentice-Hall, 1938.
Thomson, G. P., *Wave Mechanics of Free Electrons*, New York: McGraw-Hill Book Co., 1930.

CHAPTER 8

Born, M., *Atomic Physics*, London: Blackie & Son, 1957.
Gurney, R. W., *Introduction to Statistical Mechanics*, New York: McGraw-Hill Book Co., 1949.
Harnwell, G. P. and W. E. Stephens, *Atomic Physics*, New York: McGraw-Hill Book Co., 1955.
Herzberg, G., *Atomic Spectra and Atomic Structure*, New York: Dover Publications, 1944.
Mayer, J. E. and M. G. Mayer, *Statistical Mechanics*, New York: John Wiley & Sons, 1940.
Roberts, J. K. and A. R. Miller, *Heat and Thermodynamics*, London: Blackie & Son, 1951.

Slater, J. C., *Quantum Theory of Matter*, New York: McGraw-Hill Book Co., 1951.
White, H. E., *Introduction to Atomic Spectra*, New York: McGraw-Hill Book Co., 1934.

CHAPTER 9

Bethe, H. A. and P. Morrison, *Elementary Nuclear Theory*, New York: John Wiley & Sons, 1956.
Evans, R. D., *The Atomic Nucleus*, New York: McGraw-Hill Book Co., 1955.
Glasstone, S., *Source Book on Atomic Energy*, New York: D. Van Nostrand Co., 1958.
Green, A.E.S., *Nuclear Physics*, New York: McGraw-Hill Book Co., 1955.
Halliday, D., *Introductory Nuclear Physics*, New York: John Wiley & Sons, 1955.
Kaplan, I., *Nuclear Physics*, Cambridge, Mass.: Addison-Wesley Press, 1955.

Index

The letter *p* after a page number signifies that the reference is in the problems on that page.

Aberration, 142, 143, 164
Absorptive power, 70
Action, 96
    quantization of, 116
Action integral, 116
Activation energy (in fission), 302
Airy, G. B., 143
Alpha decay, 296–299
Alpha particles, scattering of, 106
Ampère, A. M., 32, 60
Anderson, C. D., 290
Andrade, E. N. da C., 304
Angular momentum, quantization of, 108, 116, 207, 208, 211, 214
    vector addition of, 214, 218–220, 261
Anomalous dispersion, 59, 82, 325
Anomalous scattering, 268
Antiparticle, 291
Antisymmetric wave function, 216
Aston, F. W., 265
Asymmetry energy (nuclear), 275, 288
Atmosphere, isothermal, 10
Atom, definition of, 2
    radius of, 128
    wave properties of, 182
Atomic mass unit, 265

Atomic theory, beginnings of, 1
    summary of, 25
Avogadro, A., 2
Avogadro's number, 7
    determination of, 20–25, 131

Bainbridge, K. T., 266
Balmer, J. J., 103
Balmer series, 103
Barkla, C. G., 122
Bartholinus, E., 38
Bartlett, J. H., 278, 284
Beams, J. W., 102
Becquerel, H., 290, 292
Bell, R. E., 166
Bernoulli, D., 2
Beta decay, 265, 290–294
    and fission, 301
Beta particle, 161, 265
    see also Electron
Beta-ray spectrum, 292–294
Bielz, F., 16
Binding energy, electronic, 108, 225, 266, 267
    nuclear, 266–268, 275, 294, 295, 300
Biot, J. B., 32, 60

347

Black body, 71, 73
Black-body radiation, 71, 73
  spectrum of, 75–80, 85–88, 93–95
  thermodynamics of, 74, 77–79
Bohr, N., 107, 113, 114, 119, 193, 258, 303, 310
Bohr atom, 107–109, 125
Bohr magneton, 212, 235
Boltzmann, L., 7, 10, 74, 85
Boltzmann constant, 7, 95
Boltzmann distribution, 88–91
Boltzmann statistics, 90, 239
Born, M., 189, 248, 250
Bose, S. N., 250
Bose–Einstein statistics, 249–251
Bothe, W., 134
Bound states, atomic, 108, 208, 338–340
  nuclear, 282, 285, 296, 341, 342
Boyle's law, 3, 11
Bradley, J., 142
Bragg, W. L., 129
Bragg equation (Bragg's law), 130, 181
Bratenahl, A., 314
Bremsstrahlung, 124
Broglie, L. de, 175, 179, 182
Brown, R., 3
Brownian motion, 20–25
Bucherer, A. H., 161

Calibration curves in special relativity, 155, 334–336
Cavity radiation, 73
  standing waves of, 86–88
  see also Black-body radiation
Chadwick, J., 264
Charge, accelerated, radiation by, 62, 66, 124
  moving, effective mass of, 63–66
    energy of, 63
    magnetic field of, 60
    magnetic force on, 61
    momentum of, 65
Charge-exchange forces, 278
Charge-exchange scattering, 280
Circuital theorem, 32
Clausius, R., 5
Closed shell, electronic, 224
  nuclear, 284–286
Competing processes, 307–309
Complex conjugate wave function, 190

Compound nucleus, 309–313
Compton, A. H., 131, 132
Compton effect, 131–134, 172p, 304
Compton wavelength, 134, 273
Condon, E. U., 296
Conductivity, electric, of metals, 54
  thermal, of metals, 55
Correspondence principle, 113
Coulomb barrier, 282, 284, 296, 297
Coulomb energy, 275, 277
Coulomb scattering, *see* Rutherford scattering
Coulomb's law, 31
Cowan, C. L., 293
Cowan, R. D., 250
Critical potentials, 120, 136p
Crookes, W., 50
Cross section, differential, 316p, 333
  nuclear, 269–271, 326, 332, 333
  of electron, 67, 127
Crystal lattice, X-ray diffraction by, 128–131
Current in wave mechanics, 191, 203

Davisson, C. J., 179, 180
Day, R. B., 271
de Broglie, L., 175, 179, 182
de Broglie waves, 176, 179–183, 270
Debye, P., 248
Debye temperature, 249
Decay constant, 298
Degeneracy, 118
Degrees of freedom, 85
Delsaux, J., 20
Dempster, A. J., 266
Deuterium, 260
Deuteron, 282
Differential cross section, 316p, 333
Diffuse series, 228
Diffusion, gas kinetic theory of, 18
Dipole moment, for optical transitions, 229–232
  induced, 57, 58
Dirac, P. A. M., 238, 241, 291
Dispersion, anomalous, 59, 82, 325
  optical, 58–60, 325
Displacement current, 34
Displacement law, 76, 78, 95
D lines (sodium), 210, 213, 226
  Zeeman effect of, 237

Index 349

Doppler effect, 141
  relativity theory of, 163
Doublets, in line spectra, 210
  in nuclear levels, 285
Drag coefficient, 144, 163
Drude, P., 54
Drunkard's walk, 23–25
Dulong, P. L., 73, 245

e, measurement of, 52
Eddington, A. S., 102
Eigenfunctions, 195–198
  for harmonic oscillator, 205p
  for hydrogen atom, 206–210, 252p, 338–340
  for particle in a box, 198–201
Eigenstates, 196, 200, 208, 282
Einsporn, E., 136
Einstein, A., 23, 99, 150, 170, 247, 250
Einstein–Lorentz transformations, *see* Lorentz transformations
Einstein–Smoluchowski equation, 23–25
Electrolysis, 44
Electromagnetic mass, 63–66
Electromagnetic theory, 31–36
Electromagnetic waves, 36–41
  energy and momentum of, 39–41
Electron, accelerated, radiation by, 62, 66, 124
  charge of, 52–54
  discovery of, 50
  magnetic moment of, 233–235
  mass of, 66, 109
    velocity dependence, 161, 162
  specific charge of, 51
  spin of, 211, 233
  wave properties of, 179
Electron gas, and Fermi–Dirac statistics, 241–245
  and metallic conductivity, 54–57
Electron radius, 66, 67
Electrons, nuclear scattering of, 271–274
Electrons in metals, 54–57, 241–245
Elliptic orbits, 117–119
Elster, J., 100, 102
e/m, for electron, 51, 52
Emissive power, 70
  of black body, 73
Energy levels, 89
  atomic, 119–121, 226

Energy levels, for Bohr atom, 109, 112, 223
  for harmonic oscillator, 92, 247
  for many-electron atom, 223
  for particle in a box, 200
  in magnetic field, 232, 236, 237
  nuclear, 282–286
Equipartition of energy, 85, 91, 245
  failure of, 86, 88, 245
Equivalent electrons, 219
Estermann, I., 182
Ether, luminiferous, 30
  motion through, 140, 142, 143, 146–148
Ether drag, 145
  see also Drag coefficient
Euler, L., 175
Exchange forces, 277–281
Exchanges, Prévost's theory of, 70, 74
Excited states, atomic, 119–122, 125, 226–228
  nuclear, 303–309, 310–313
Exclusion principle, 215–217, 239
  and nuclear structure, 277, 278, 283
  and periodic table, 220–225
  and two-electron systems, 217–220

Faraday, M., 36, 44, 46
Faraday, definition of, 44
Fermat, P. de, 174
Fermi, E., 241, 284, 293
Fermi–Dirac statistics, 239–241
Fermi gas, of electrons, 242–244
  of nucleons, 283
Field equations, electromagnetic, 35, 36
Fine structure, and electron spin, 210–214
  doublet, 210–213, 222
  relativistic, 118, 221
Fission, 300–303
  induced, 302
  spontaneous, 302
Fitzgerald, G. F., 149, 156
Fizeau, H. L., 144
Forbidden transitions, optical, 237
Foucault, J. B. L., 30
Foucault's pendulum, 138, 171
Franck, J., 119
Franck–Hertz experiment, 119
Fraunhofer lines, 58, 73, 83, 121

Free path, molecular, 13–17
Fresnel, J. A., 30, 144
Friedrich, W., 129

Galilean transformations, 153
Gamma radiation, 69, 303, 307–309
Gamow, G., 296
Gas, pressure of, 2, 5–7
Gauss's theorem, 31
Geiger, H., 105–107, 134, 258
Geitel, H., 100, 102
Gerlach, W., 233, 234
Germer, L. H., 179, 180
Goudsmit, S. A., 211
Gouy, G., 20
Grimaldi, F. M., 29
Group velocity, 177, 178
Gurney, R. W., 296
Gyromagnetic ratio, 235, 238

Haga, H., 122
Half-life, 298
  of neutron, 293
Hallwachs, W., 45
Harmonic oscillator, damped, analysis of, $96p$, $97p$
  eigenfunctions of, $205p$
Harmonic oscillators, absorption by, 81–83
  and thermal radiation, 80–85, 92
  in crystal lattice, 245–248
  in optical dispersion, 58
  quantization of, 92, 114, 247
  radiation by, 80, $96p$
Hartley, W. N., 103
Havens, W. W., 8, 9
Haxel, O., 284
Heisenberg, W., 192, 278
Henkel, R. L., 271
Hertz, G., 119, 121, 122
Hertz, H. R., 39, 45
Heydenburg, N. P., 311
Hincks, E. P., 166
Hoek, M., 146
Hofstadter, R., 272, 273
Hughes, A. L., 100
Hull, G. F., 41
Huygens, C., 29
Hydrogen atom, Bohr theory of, 107–109

Hydrogen atom, ground state of, 338–340
  in wave mechanics, 206–210, 338–340
Hydrogen-like atom, 207, 209, 221
Hyperfine structure, 260–262

Induction law, 36, 322
Inertia, relativity of, 171
Inertial frame, 138
Insulators, optical properties of, 57–60
Invariant, definition of, 163
Ionization potential, 224, 228
Isobar, definition of, 288
Isothermal atmosphere, 10–12
Isotopes, 259, 260
  abundances of, 286–288
Isotopic effect (in line spectra), 260, 261, $315p$
Isotopy, 257

Jeans, J. H., 86
Jensen, J. H. D., 284
jj coupling, 220
Jupiter's moons, 28, 140

Kármán, H. von, 248
K capture, 290
Kelvin, Lord, 72
Kepler, J., 41
Kinetic theory of gases, 2, 5–13
Kirchhoff's law (radiation), 72
K lines (X rays), 123, 126, 127
Knipping, P., 129
Ko, C. C., 7
Kusch, P., 8

Lagrange, J. L., 175
Lagrange multipliers, 90
Lamb, W. E., 238
Landé splitting factor, 236
Laplace, P. S. de, 2, 3
Larmor, J., 323
Larmor precession, 49, 321–323
Latent heat, 4
Laue, M. von, 129
Lawrence, E. O., 102
Least action, principle of, 96, 174, 176
Least time, principle of (Fermat), 174, 176
Lebedew, P. N., 41

Lenard, P., 45, 52, 100, 105
Length, relativistic contraction of, 156, 167, 336
Level width (nuclear), 305–309, 311
Levels, *see* Energy levels
Light, corpuscular theory of, 29
  in gravitational field, 167–170
  linear and circular polarization of, 47
  pressure of, 41, 74
  properties of, 28
  scattering of, 66
  theories of, 28
  velocity of, 38
    independent of source motion, 141
    in moving medium, 145
  wave theory of, 29
Line spectra, 103–105
  of alpha particles, 304
  of gamma rays, 304
  of X rays, 123, 124–127
Liquid drop model, 274
  and fission, 301
Lodge, O. J., 148
Lorentz, H. A., 46, 57, 149, 156
Lorentz–Fitzgerald contraction, 149
Lorentz transformations, 152
  analytic geometry of, 336, 337
Lorenz, L. V., 56, 57
LS coupling, 220
Lummer, O., 76, 79
Lyman, T., 105
Lyman series, 105, 125

Mach, E., 171
Magic numbers, 285, 299
Magnetic field of moving charge, 60
Magnetic moment, intrinsic, 211, 233–235
  nuclear, 261–263, 286
  orbital, 211, 212
  total, 236
Magneton, 212, 235
  nuclear, 263, 265
Majorana, E., 278
Marsden, E., 105–107, 258
Mass, electromagnetic, 63–66
  relativistic variation of, 159–162
Mass–energy equivalence, 159, 175, 266
Mattauch, J., 266
Matter waves, 179–183

Maupertuis, P. L. M. de, 96, 174
Maxwell, J. C., 7, 18, 26, 30, 34–36, 44, 45, 85, 140, 147
Maxwell's field equations, 32, 35, 36
Maxwell velocity distribution, 7, 13, 14, 92, 244, 318
Mayer, M. G., 284
Mean free path, molecular, 13–17
Mean life, of mesons, 165
  of nuclear excited state, 306
Mean-square velocity, molecular, 7, 318, 320
Mean velocity, molecular, 11, $26p$, 318, 320
Mendeléeff, D. I., 105
Mercury (element), excitation of, 121
Mercury (planet), precession of orbit, 170
Mesons, and nuclear forces, 280
  lifetime of, 165
Metals, electric and thermal conductivity of, 54–57
  electron energy distribution in, 241–244
  specific heat of, 243–245
Michel, G., 93
Michelson, A. A., 144, 147, 148
Michelson interferometer, 145
Michelson–Morley experiment, 147, 149
Miller, R. C., 8
Millikan, R. A., 53, 100
Minkowski, H., 156
Molecules, sizes of, 3
  velocity distribution of, 7–13
Morley, E. W., 144
Moseley, H. G. J., 126, 256
Moseley's law, 124–127
Multiplicity, 215, 218
Mu meson (muon), 165

Neutrino, 293
Neutron, beta decay of, 293
  bound states for, 341, 342
  discovery of, 264
  half-life of, 293
  instability of, 265
  magnetic moment of, 265
  mass of, 265
  separation energy of, 282, 284, 295, 302

Neutron, wave properties of, 182, 270, 313, 314, 326
Neutrons, diffraction of, 182, 270, 313, 314
   resonance capture of, 325
   scattering of, 269–271, 313, 314
   thermal, velocity distribution of, 9
Newton, I., 28–30, 38, 171
Newtonian relativity, 137
Newton's laws, 138
   and relativity, 162
Nichols, E. L., 41
Nier, A. O., 266
Normalization, 190, 198, 216
Normal modes for stretched string, 196
Nuclear binding energy, 266–268, 275, 294, 295
Nuclear charge formula, 276
Nuclear cross section, 269–271, 326, 332, 333
Nuclear energy levels, 282–286, 304, 306, 311, 312
   width of, 305–309, 311, 325
Nuclear fission, 300–303
Nuclear forces, 276–281, 312
   exchange character of, 277–281
   range of, 276
   saturation of, 277
   strength of, 276
Nuclear magneton, 263
Nuclear mass formula, 274–276
Nuclear potential well, 279, 281–284
   bound states for, 341, 342
Nuclear radii, 255, 268–274
Nuclear shell model, 284–286, 316p
Nuclear stability, 287–291, 294–296, 300
Nucleon, definition of, 266
Nucleus, charge distribution of, 271–274
   density of, 255
   discovery of, 105–107
   excited states of, 303–309, 310–313
   instability of, 294–303
   liquid drop model of, 274, 301
   magnetic moment of, 261–263
   motion of, and spectra, 111
   optical model of, 313, 314
   shell model of, 284, 286
   size of, 255
   spin of, 261–264
Nuclide, definition of, 286

Optical model (nuclear), 313, 314
Orbital capture, 290
Orbits, quantized, 107–110, 116–119
Orthogonality, 197, 198
Oscillators, *see* Harmonic oscillators

Pairing energy, 288
Partial width, 306–309
   for neutron emission, 308
Particle in a box, 198, 205p, 341
Paschen, F., 105
Paschen series, 105
Pauli, W., 217, 261, 293
Pauli's exclusion principle, *see* Exclusion principle
Penetrability of potential barrier, 204
Periodic table, 224, 225
Perrin, J. B., 22–25, 50
Petit, A. T., 73, 245
Phase integral, 116
Phase space, 114–117
Phonon, 250
Photoelectric effect, 45, 99–103, 123, 304
   time scale of, 101, 102
Photoelectric equation, 99–101
Photon, 99, 250
   see also Quantum
Pi meson (pion), 281
Pitkanen, P. H., 205
Planck, M., 92–96, 107, 114
Planck's constant, 93, 95, 108
Planck's formula, 93
   definite integral evaluated, 327–329
Plücker, J., 50
Point charges in motion, 60–63
Polarization, linear and circular, 47
Positive rays, 258
Positron, 291
Potential barrier, 203, 284, 296, 297
Potential step, 202
   nuclear, 307, 308
Potential well, 199
   nuclear, 279, 281–284, 341, 342
Poynting's vector, 40, 65, 74
Pressure of light, 41
Prévost, P., 70
Principal quantum number, 108, 118, 208, 221
Principal series, 228

Index 353

Pringsheim, E., 76, 79
Probability, and wave mechanics, 189–191, 195, 200
  in kinetic theory, 14
  in statistical mechanics, 89
Probability current, 191, 203
Proton, magnetic moment of, 265
  mass of, 265
Pryce, M. H. L., 316

Quantization, of angular momentum, 108, 117, 184, 207–209, 214, 285
  of energy, 92, 109, 118, 125, 208, 221
  of spin, 211
Quantum, definition of, 96
  in gravitational field, 167–169
Quantum numbers, for Bohr atom, 108, 118
  for nucleon states, 285
  for two-electron systems, 217–220
  from wave mechanics, 200, 207, 208
Quantum statistics, 239–241, 249–251
Quantum theory, and black-body radiation, 92–95
  and hydrogen atom, 107–109

Radiation, by accelerated charge, 62, 66
  pressure of, 41, 74, 77
  thermal, 69
  see also Black-body radiation
Radiative capture, 309
Radiative transitions, 107–109, 228–233, 303, 307–309
Radioactivity, 265, 290, 292, 296–300, 304
Radium, 300
Rainwater, J., 8, 9
Random walk, 23–25
Rayleigh, Lord, 86, 101, 257, 314
Rayleigh–Jeans law, 84–88, 93, 248
Rearrangement collisions, 309
Red shift, gravitational, 168
Reines, F., 293
Relativity, general, 170
  special, 149–167
Relativity of inertia, 171
Resonance, in nuclear reactions, 310–312, 325
  in optical dispersion, 59, 325
Resonance absorption, 81–83, 325

Resonance processes, 324–326
Retherford, R. C., 238
Ritchie, W., 71
Robson, J. M., 293
Roentgen, W. C., 122
Römer, O., 140
Rotary polarization, 47
Rowland, H. A., 61
Rubens, H., 93
Russell–Saunders coupling, 220
Rutherford, E., 53, 106, 255, 264, 304, 309
Rutherford scattering, 106, 268, 330–333
Rydberg, J. R., 103, 227
Rydberg constant, 104, 109
  and nuclear mass, 112, 113, 260

Saturation (of nuclear forces), 277
Scattering, of alpha particles, 106, 268
  Rutherford theory, 330–333
  of electrons by nuclei, 271–274
  of fast neutrons, 269–271, 313, 314, 316p
  with charge exchange, 280
  of light, 66
  of X rays, 127, 131–134
Scattering cross section (Rutherford), 332, 333
Schrödinger, E., 185
Schrödinger's equation, 185–188, 313
  three-dimensional, 206
  time-dependent, 188
Screening by orbital electrons, 125, 222
Secular equilibrium, 299
Sedimentation equilibrium, 20–22
Selection rules, 227, 231, 232
  nuclear, 305
Semiempirical mass formula, 274–276, 295, 301
Separation energy (in nuclei), 283, 284, 295, 302, 307, 312
Sharp series, 228
Shell model (nuclear), 284–286, 316p
Shells of electrons, 109, 110
Shielding, electronic, 125, 222, 223
Simultaneity, and special relativity, 151
  definition of, 150
Singlet state, 218
Smoluchowski, M. von, 23

Soddy, F., 257
Solids, specific heat of, 245–249
Sommerfeld, A., 118, 221
Specific heat, for conduction electrons, 243–245
  of solids, 245–249
Spectra, 103
  see also Spectrum, Line spectra
Spectral series, 103–105, 227
Spectroscopic notation, 110, 214
Spectrum, definition of, 70
  of alpha particles, 304
  of atomic hydrogen, 104
  of beta particles, 292–294
  of black-body radiation, 75, 76
  of excited mercury atoms, 122
Spin, of electron, 211
  of neutron and proton, 263
  of nucleus, 261–264
Spin-orbit interaction, atomic, 213, 222
  nuclear, 284–286
Stability rules (nuclear), 288
Stationary states, and radiative transitions, 229–231
  in atom, 107, 184
Statistical mechanics, 84, 239
  and Boltzmann's distribution, 88–92
  see also Statistics
Statistical weight of energy level, 91
Statistics, Boltzmann, 90, 239
  Bose–Einstein, 249–251
  Fermi–Dirac, 239–241
Stefan, J., 74
Stefan–Boltzmann law, 74–76
Stefan's constant, 74, 75
  and Planck's formula, 94
Stellar aberration, 142, 143, 164
Stern, O., 182, 233, 234
Stern–Gerlach experiment, 233
Stirling's formula, 90
Stokes's law, application of, 22, 23, 52
Stoletow, A., 45
Stoney, C. J., 45
Suess, H. E., 284
Surface energy (nuclear), 274–276, 301
Surface tension, 4
Symmetric wave function, 216

Temmer, G. M., 311
Term diagram, 226–228

Theory of exchanges, 70, 74
Thomas, L. H., 238
Thomson, G. P., 179, 180
Thomson, J. J., 51, 53, 66, 68, 105, 258, 259
Thomson cross section, 66, 127
Time, relativistic dilation of, 157, 165, 337
Tin, isotopes of, 287
Total angular momentum, 214, 218–220
  of nucleus + electrons, 261
Townsend, J. S., 52
Transitions, radiative, in atom, 109, 125, 227–233, 236, 237
Transmission coefficient (for potential step), 307, 308
Transport phenomena, for electrons in metals, 54–57
  in gases, 17–20, 26p
Triplet state, 218
Tunnel effect, in fission, 302
  in optics, 189
  in wave mechanics, 203, 296–298
  nuclear, 296–298, 302
Two-electron systems, 215–220
Tyndall, J., 73

Uhlenbeck, G. E., 211
Ultraviolet catastrophe, 85, 86, 88
Uncertainty principle, 191–195
  and nuclear level width, 305, 306
Uniform-temperature enclosure, 71
Uranium, instability of, 294, 295, 299–303
Urey, H. C., 261

Velocities, molecular, 3
Velocity addition in special relativity, 158
Velocity distribution, molecular, 7–10
Viscosity, gas kinetic theory of, 17
Volume energy (nuclear), 274–276

Water, range of molecular force in, 5
  refractive index of, 60
Wave function, definition, 185
  for hydrogen-like atom, 209
  properties of, 188–191
  symmetric or antisymmetric, 216, 239
  see also Eigenfunctions

Index 355

Wave number, 86, 104, 109, 126, 242
Wave packet, 192, 204p
Wave vector, 86, 248
Wave velocity, 177
Wheeler, J. A., 303
White dwarf stars, 169
Width (of nuclear levels), 305–309, 311, 325
Wiedemann–Franz law, 56
Wien, W., 77, 94
Wien's displacement law, 76, 78, 95
Wigner forces, 277
Wilson, H. A., 53
Wind, C. H., 122
Work function, 99, 243, 283

X rays, 122–131
   continuous spectrum, 122–124

X rays, diffraction of, 128
   discovery of, 122
   line spectrum, 123, 124–127
   scattering of, 127, 131–134
   wavelengths of, 122, 123, 127, 128
X-ray spectrometer, 130

Yennie, D. R., 273
Young, T., 2, 3, 30
Young's experiment, 194

Zahn, C. T., 162
Zartman, I. F., 7
Zeeman, P., 46, 50
Zeeman effect, anomalous, 235–237
   classical, 46–50, 321–323
   by wave mechanics, 232, 233
Zero-point energy, 247